Excel

SCIENCE STUDY GUIDE

Year 7

Get the Results You Want!

Geoffrey Thickett & Jim Stamell

Reprinted 2015, 2017, 2019, 2021, 2023, 2025

ISBN 978 1 74125 391 7

Pascal Press
PO Box 250
Glebe NSW 2037
www.pascalpress.com.au

Publisher: Vivienne Joannou
Project editor: Mark Dixon
Edited by Leanne Poll
Cover and page design by DiZign Pty Ltd
Typset by Precision Typesetting (Barbara Nilsson)
Printed by Vivar Printing/Green Giant Press
Photos by Dreamstime and iStockphoto except Figure 1.19, p. 14, © Office of Environment and Heritage NSW; Figure 1.20, p. 15, © M Fagg, Australian National Botanic Gardens; Figure 3.23, p. 69, © Martin Shields/Science Photo Library

Contents

CHAPTER 4

Forces

CHAPTER 5

Investigations and problem solving

How to use this book

The Australian Curriculum

This study guide covers the complete course of the Year 7 Science Australian Curriculum including:

- all core knowledge
- all required problem-solving skills
- all aspects of the scientific method.

The book is divided into sections based on the Australian Curriculum's three learning strands and substrands. The three learning strands in the Australian Curriculum are:

- Science Understandings (Biological Sciences; Chemical Sciences, Earth and Space Sciences; Physical Sciences)
- Science as a Human Endeavour
- Science Inquiry Skills.

The Science as a Human Endeavour strand is covered in Chapters 1 to 4, plus each substrand of the Science Understandings strand is treated as follows:

- Chapter 1—Biological Sciences
- Chapter 2—Chemical Sciences
- Chapter 3—Earth and Space Sciences
- Chapter 4—Physical Sciences.

Chapter 5 covers the Science Inquiry Skills strand.

Tips for tests and examinations

Preparation

In order to prepare for your school exams, you will need to examine the Contents section at the front of the book and then identify the relevant sections of this book before you begin to revise and practise exam-style questions.

Each topic in your school programs for Year 7 may have integrated content from more than one of these different areas of study. It is unlikely that you will have studied the content in this order.

You need to allow sufficient time for this thorough revision—at least one week for a class test and at least three weeks for a major examination. Revise the content of the chapter or section and attempt all questions. Use the supplied answers to determine which areas you need further work in.

Test/exam questions

Multiple-choice questions

In order to answer these sorts of questions, make sure that you do the following.

- Read the stem of the question thoroughly.
- Look carefully at any diagrams, flow charts or tables, and interpret them thoroughly.
- Choose the letter of the best response and not just a correct independent statement; if you don't know the answer, make the most logical choice you can.

Free-response questions

In order to answer this type of question effectively, follow the guidelines below.

- Highlight the verbs and key words in the question and respond accordingly.
- Don't waste time by restating the question; keep your answers concise.
- Label any diagrams that you draw with a pencil and ruler.
- Line graphs should occupy more than 80% of the available grid space.
- All graphs must have a title and the axes should have linear scales and be appropriately labelled with titles and units.
- Any experimental methods must be written as a series of numbered sentences in the present or past tense.
- Repeating the experiment at least five times or more can improve the reliability of experimental results.
- For questions involving the scientific method, ensure that you state the dependent and independent variables, the variables you controlled (kept the same) and the experiment that you used as a control.
- When drawing a table of data, ensure that the table is fully bounded by lines to create columns and rows. The column headings and units should occupy the first row.

Features of this book

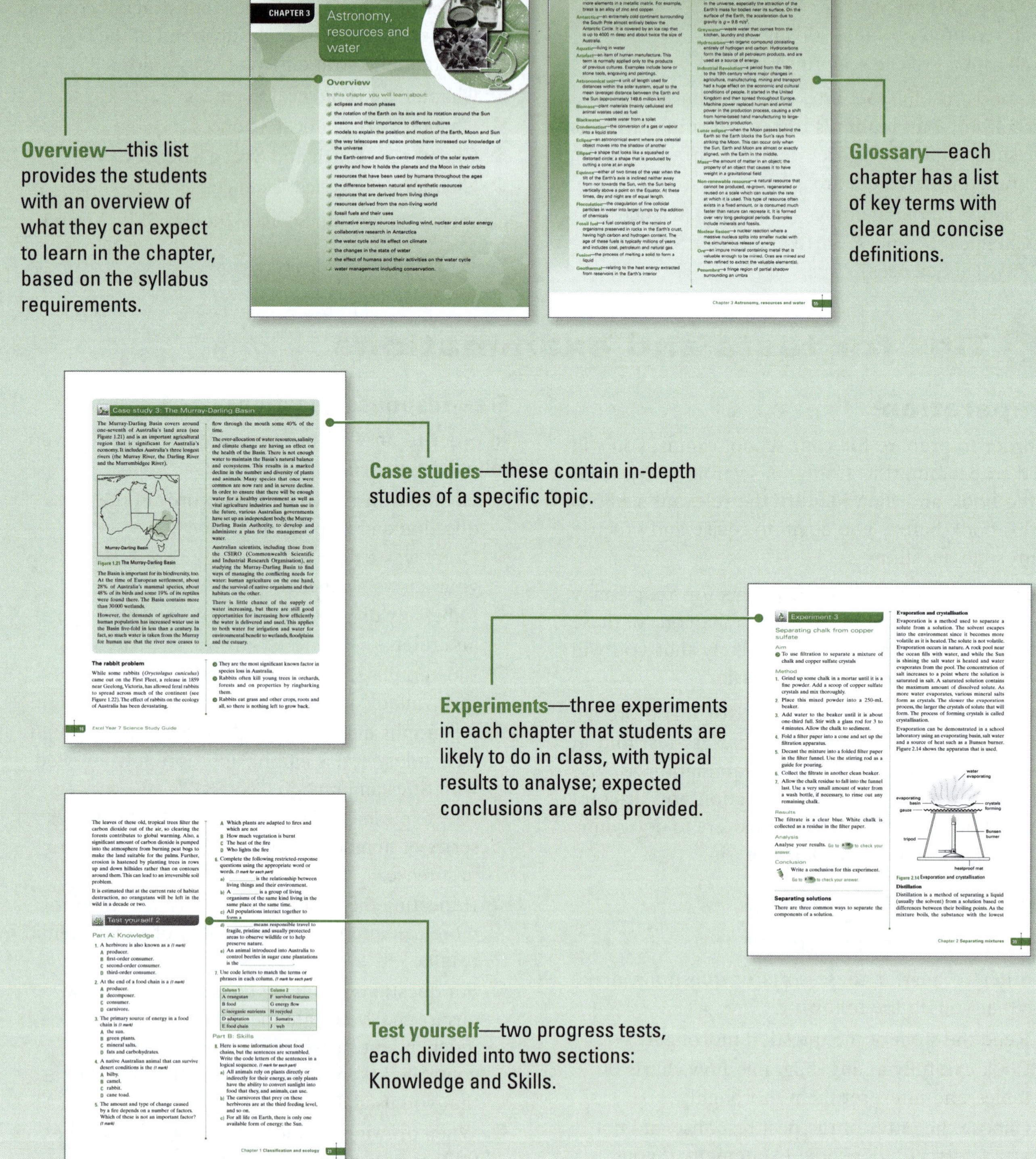

You will be thoroughly prepared for examinations and tests when you use this study guide. This guide is an effective revision and study program for exams and class tests in Year 7.

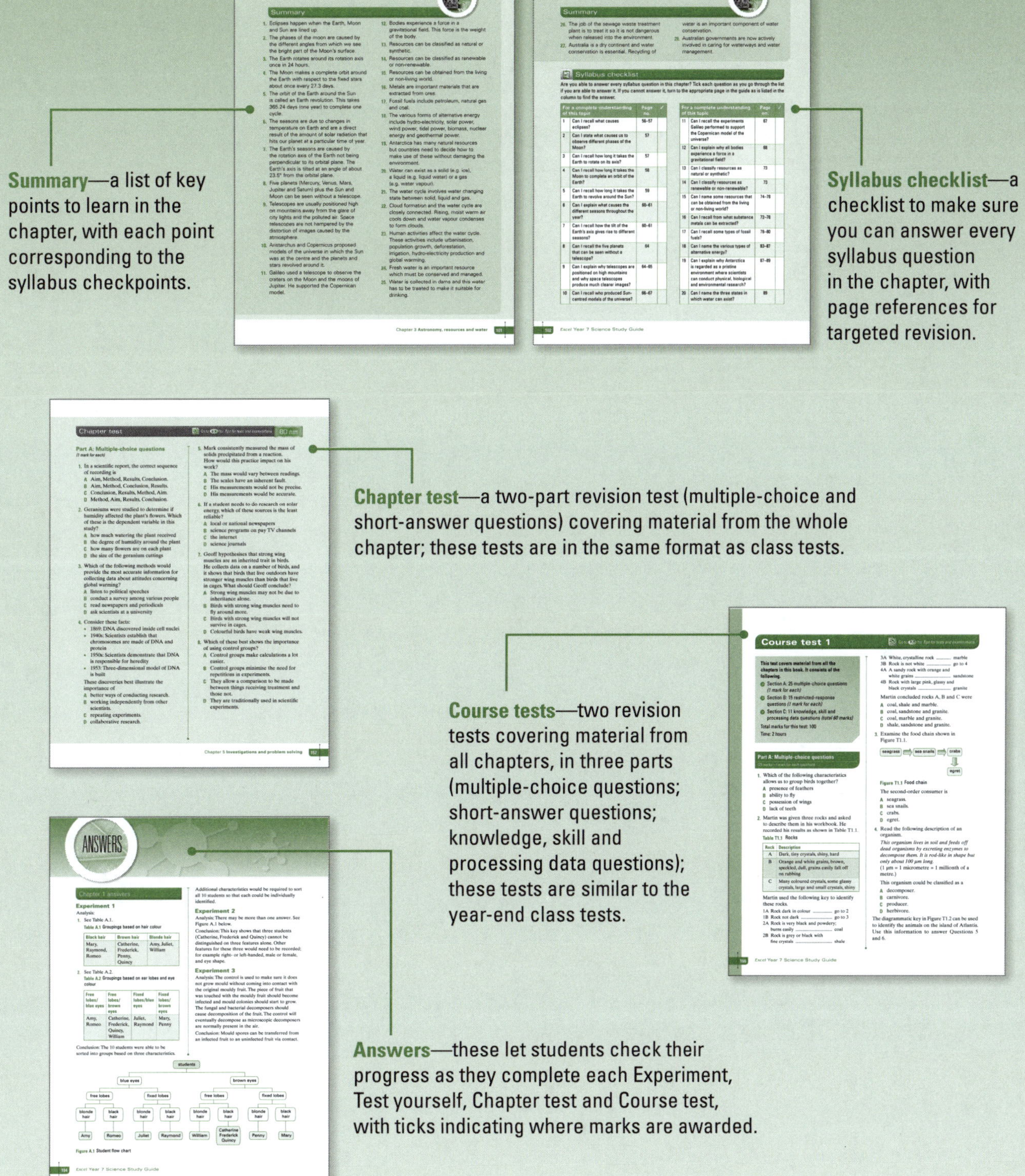

Summary—a list of key points to learn in the chapter, with each point corresponding to the syllabus checkpoints.

Syllabus checklist—a checklist to make sure you can answer every syllabus question in the chapter, with page references for targeted revision.

Chapter test—a two-part revision test (multiple-choice and short-answer questions) covering material from the whole chapter; these tests are in the same format as class tests.

Course tests—two revision tests covering material from all chapters, in three parts (multiple-choice questions; short-answer questions; knowledge, skill and processing data questions); these tests are similar to the year-end class tests.

Answers—these let students check their progress as they complete each Experiment, Test yourself, Chapter test and Course test, with ticks indicating where marks are awarded.

CHAPTER 1

Classification and ecology

Overview

In this chapter you will learn about:

- the principles of classification
- the classification of vertebrates and mammals
- the five kingdom classification system
- classification levels and binomial nomenclature
- the use of keys in classification
- food chains and feeding relationships in natural communities
- the classification of consumers
- the importance of microorganisms as decomposers
- the effect of human activity on natural habitats
- the effect of agriculture on natural communities in the Murray-Darling Basin
- how introduced species such as rabbits and cane toads affect natural communities
- the land can be sustainably managed
- how Aborigines used fire
- how rainforests have been destroyed in Borneo to grow economic crops.

Glossary

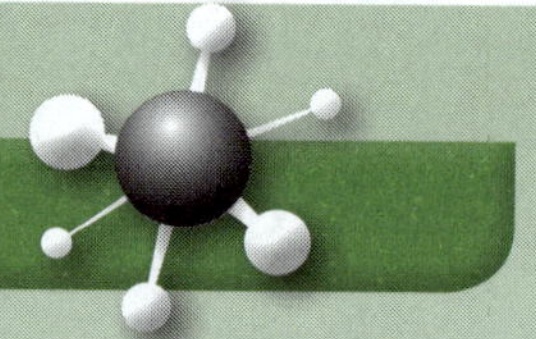

Carnivore—an animal that derives its energy and nutrient requirements from a diet consisting mainly or exclusively of animal tissue

Classification—the process of sorting into groups with similar characteristics

Consumer—an organism in a food chain that relies upon feeding on other organisms for survival

Decomposer—an organism that breaks down dead or decaying organisms and, in so doing, carries out the natural process of decomposition. An organism that obtains energy from dead or waste organic matter, such as bacteria and fungi.

Deforestation—clearing naturally occurring forests by logging and burning. This creates non-forest land for uses such as crops, pasture, or urban development.

Dichotomous keys—keys for classification that provide two choices at each step

Feral animal—an animal that has escaped from domestication and returned, partly or wholly, to a wild state

Food chain—a representation of the predator–prey relationships between species within an ecosystem or habitat

Fungus—non-photosynthetic organisms with a cell wall and specialised tissues that feed by decomposing the remains of dead organisms

Herbivore—an animal that is adapted to eat plants

Invertebrate—an animal without a bony spinal column

Kingdom—highest level of classification. Living things are often classified into one of the five kingdoms

Land management—the process of managing the use and development of land resources in a sustainable way

Monera—very simple cellular organisms in which the nucleus is not surrounded by a membrane

Omnivore—an animal that eats both plants and animals as its primary food source

Plants—organisms that make their own food by photosynthesis

Producer—an organism, mainly a green plant, which produces complex organic compounds (such as carbohydrates, fats and proteins) from simple inorganic molecules using the energy from light (by photosynthesis). It is able to make its own food and can fix carbon dioxide.

Protista—unicellular (and some multicellular) organisms that do not have specialised tissues

Vertebrate—an animal with a bony column enclosing the spinal cord

1.1 Reasons for classification

Throughout history, people have been grouping things in ways that made sense to them. The groups they chose were often different because they chose different features by which to group them. If other people liked the groupings, they used them as well.

Why do we group things? The following are some possible answers.

- Items can be found easily and quickly. In a library, books are classified as fiction or non-fiction. In the non-fiction section, books are further classified into subject areas such as Natural Sciences and Mathematics (500–599), History and Geography (900–999). In the Natural Sciences, further subdivisions are used to classify different branches of the sciences; for example Astronomy (520–529), Chemistry (540–549) and Botany (580–589).
- Similar items can be compared easily. In a supermarket all the different brands of cereals are placed in one section and the shopper can compare them according to price, size and contents.

- Material is presented in a neat and organised form. In a newspaper the contents are organised into groups such as state news, national news, world news and sporting news.
- Some features of the item can be predicted because you know which group it belongs to. If you are told that a new animal has been discovered, and it belongs to the cat family, then you can make some predictions about its appearance, eating habits and behaviour from information you already know about cats.

1.2 Similarities and differences

Things can be classified into groups according to the similarities and differences in their observable characteristics.

Example

Examine the drawings of four leaves shown in Figure 1.1.

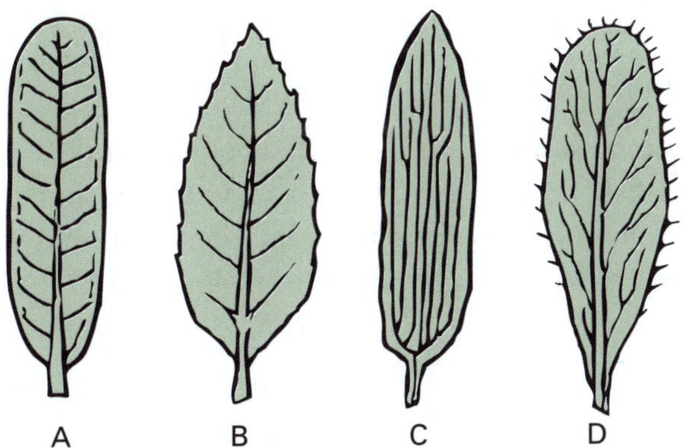

Figure 1.1 Some types of leaves

The similarities and differences between the leaves are as follows.

- Leaves A, B and D have one central vein with veins branching off the central vein. Leaf C has many parallel veins.
- Leaves B and C have pointed tips and leaves A and D have rounded tips.

Experiment 1

Similarities and differences between people

Aim

- To sort 10 people in your class into groups using identified characteristics

Method

1. Divide into small groups and decide on the set of characteristics that would best help you sort the class into groups.
2. As a class, decide what characteristics to survey.
3. Draw up a table with the selected characteristics as headings.
4. Choose 10 students to sort. Complete your table.
5. Examine the table carefully and group the students with similarities together. Which characteristic produces the smallest number of groups?

Typical results

Table 1.1 shows an example of typical results for this experiment.

Table 1.1 Typical results for sorting classes by characteristics

Student name	Hair colour	Eye colour	Ear lobes (free or fixed)
Amy	blonde	blue	free
Catherine	brown	brown	free
Frederick	brown	brown	free
Juliet	blonde	blue	fixed
Mary	black	brown	fixed
Penny	brown	brown	fixed
Quincy	brown	brown	free
Raymond	black	blue	fixed
Romeo	black	blue	free
William	blonde	brown	free

Analysis

You can now group the students with similarities. Draw up two tables. In the first table group the students on the basis of different hair colour. In the second table, group the students on the basis of ear lobe shape and the colour of their eyes. This should be a four-column table.

Go to p. 184 to check your answer.

Conclusion

Now write your own conclusion.

Go to p. 184 to check your answer.

Case study 1: Vertebrates

Vertebrates are animals that have a backbone. Vertebrates can be classified into five groups based on similarities and differences in their observable characteristics.

The five groups of vertebrates are fish, amphibians, reptiles, birds and mammals.

The vertebrates have a number of features in common.

- A vertebral column made out of bone or cartilage runs down the back of the animal.
- A brain and spinal cord is protected by the skull and the hollow vertebral column.
- Blood circulates around the organism through blood vessels.
- Oxygen is obtained through gills or lungs (see Figure 1.2).
- There are two pairs of limbs. All vertebrates have two forelimbs (arms) and two hind limbs (legs). In some vertebrates, these limbs are almost non-existent (e.g. legless lizards).
- Two eyes form images.
- Two separate sexes exist in each species.

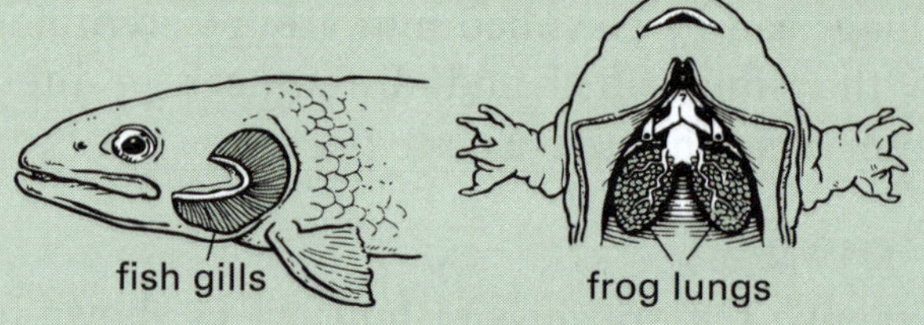

Figure 1.2 Fish gills and frog lungs

Table 1.2 summarises the major characteristics of these five groups of vertebrates.

Note: research has identified some snakes and fish that give birth to live young rather than laying eggs. Examples include the green anaconda snake and the lemon shark.

Table 1.2 Vertebrate groups

	Fish	Amphibians	Reptiles	Birds	Mammals
Limbs	fins	legs	legs (no legs in snakes)	wings, legs	legs, arms
Body covering	wet, scaly skin	slimy skin	dry, scaly skin	feathers	fur/hair
Breathing structure	gills	gills/lungs	lungs	lungs	lungs
Method of reproduction	soft, jelly-covered eggs (see note in text)	jelly-covered eggs	mostly soft-shelled egg (see note in text)	hard-shelled egg	live young
Examples	barramundi salmon	frog salamander	snake crocodile	emu swan	dingo cow

Case study 2: Mammals

The most important features of mammals are:

- the presence of hair or fur
- glands that can secrete milk.

There are three different groups of mammals: monotremes, marsupials and placentals.

Monotremes lay eggs. They are the most primitive of the mammals. The only living monotremes are the platypus and echidna (see Figure 1.3). The word *monotreme* means 'single opening'. This refers to the single opening of the body used for elimination of wastes, as well as for reproduction. Female monotremes do not have teats. Milk emerges from mammary glands onto the surface of special areas of the skin. Platypuses and echidnas do not have teeth.

Figure 1.3 Platypus and echidna

Marsupials are pouched mammals. There are many examples in Australia, such as the wallaby, kangaroo, wombat and koala. The young are born in a very immature state and make their way to the mother's pouch (marsupium) where they attach themselves to a teat to feed and develop until they are mature enough to leave the pouch.

Placentals are the most advanced and numerous of the mammals. In the mother's womb, the foetus is joined to the mother by a placenta. The young are born alive in a more advanced state of development than the young of marsupials. Placental young feed from the milk obtained from the teat of mammary glands. Of all the mammals, the placentals are the most numerous. They include lions, bears, camels, porpoises, whales, sheep, bats and, of course, humans. Of all the mammals, placentals have the greatest capacity to learn.

1.3 Changing methods of biological classification

About 350 BC, Aristotle developed a classification system of plants and animals.

- Animals: those organisms that move around and eat other organisms to obtain food.
- Plants: those organisms that do not move and make their own food by using the sun's energy which is absorbed by the green chlorophyll in their leaves.

Aristotle classified animals into two groups: animals with blood and backbones and those without blood. These distinctions correspond closely to the modern distinction between vertebrates and invertebrates. Animals with blood were further divided into live bearing (humans and mammals) and egg bearing (birds and fish). Invertebrates ('animals without blood') included insects, crustaceans and molluscs. With one of his pupils, Theophrastus, he classified plants according to size as trees, shrubs and herbs. Since only about 1000 plants and animals were known at the time, this grouping system was very useful for their purposes.

For many years, mushrooms, toadstools and moulds were thought to be special types of plants. Eventually in 1969 it was finally decided to classify them as a separate kingdom from plants.

- Fungi: unicellular or multicellular organisms that have no chlorophyll and which feed off other living things; their cell walls differ from plants in that they are composed of chitin instead of cellulose.

There were two reasons for creating this classification. First, unlike green plants fungi do not make their own food; they do not contain chlorophyll. Second, fungi feed by decomposing other organisms (usually dead ones) and absorbing their nutrients.

The invention of microscopes during the 17th century helped biologists to look more closely at the cells of living things. Kingdom protista was finally established in the late 19th century as biologists confirmed that these organisms should not be classified as plants or animals. The invention of the powerful electron microscope allowed biologists to recognise a new kingdom of organisms in 1938. This group included bacteria and were known as kingdom monera. These organisms had no membrane surrounding the genetic material inside the cell.

This discovery led to the introduction of two new classification groups.

- Protista: single-celled (unicellular) organisms and a few multicellular organisms (e.g. some algae) that had very simple cellular structure.
- Monera: very simple cellular organisms (e.g. bacteria) in which the nucleus (information centre) of the cell was not surrounded by a membrane.

Today the five kingdoms model of living things (animal, plant, fungus, protista and monera; see Figure 1.4) is gradually being modified as new genetic information is being obtained. Many biologists have proposed changes to the way protists and monerans are classified.

1.4 Classification levels and biological nomenclature

Biologists use the word *organism* to describe any living thing. They group organisms together according to the characteristics they share. In order to place a living thing in a particular category, biologists compare features such as its physical structure, the way its body functions, how it behaves and how it relates to its surroundings.

Millions of different kinds of living things have been identified. Biologists call these different kinds species. The general rule is that if two animals cannot successfully breed with each other to produce fertile offspring, they belong to different species. A species is often defined as a group of organisms capable of interbreeding.

Each species of living thing is classified into various categories or levels. These classification levels are shown in Figure 1.5.

Figure 1.5 Levels of classification

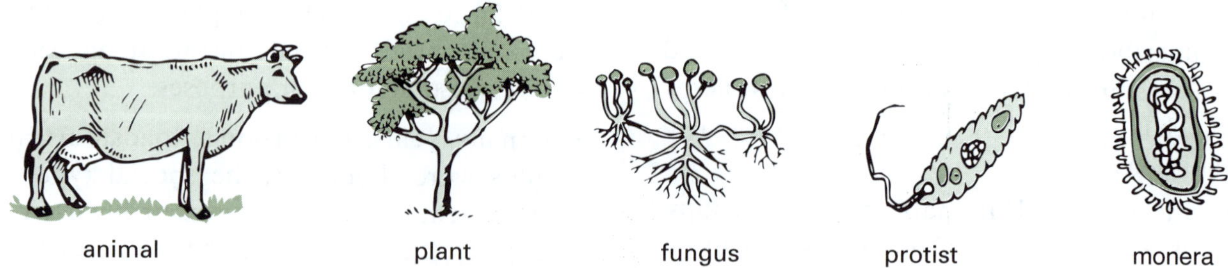

Figure 1.4 Living things in the five kingdoms (not drawn to scale)

Kingdom is the most general level. Within each kingdom there are a number of sublevels called phyla (plural of phylum). Within a phylum there may be subphyla. Phyla or subphyla are then divided into a number of classes and so on.

Two examples of classification are shown in Tables 1.3 and 1.4.

Example 1: humans

Table 1.3 Classification of humans

Classification level	Name	Further information
Kingdom	Animals	eats food and moves about
Phylum	Chordates	a hollow nerve cord runs down the body from the head
Subphylum	Vertebrates	the nerve cord is inside a bony vertebral column
Class	Mammals	hair and mammary glands are present
Order	Primates	humans, apes and monkeys are the most advanced of all animals
Family	Hominids	human-like characteristics
Genus	*Homo*	human (or man)
Species	*sapiens*	thinking or wise

Thus humans are classified by biologists as *Homo sapiens*. The genus and species classification level is used to name all organisms. It is an example of binomial nomenclature, which means 'two-word naming system'.

We use this two-name system in recording names on the school roll. For example, if there are two students at a school with the same surname (e.g. Aaron Presley and Minnie Presley) then the surname *Presley* is written first as it is a more general grouping and the first names *Aaron* and *Minnie* are written last as they are their specific names. The computerised school roll classifies students on their surname, not their first name.

Example 2: lemon-scented gum tree

Table 1.4 Classification of the lemon-scented gum tree

Classification level	Name	Further information
Kingdom	Plants	makes its own food by photosynthesis
Phylum	Tracheophytes	transports water and nutrients throughout its body through narrow tubes
Class	Angiosperms	produces flowers for reproduction
Order	Myrtales	a groups of evergreen plants called 'myrtles' that have flower parts in multiples of four or five
Family	Myrtaceae	woody plants with aromatic leaves
Genus	*Eucalyptus*	tree produces eucalyptus oil
Species	*citriodora*	oil has the smell of citrus fruit

The lemon-scented gum tree is called *Eucalyptus citriodora* using the binomial system of nomenclature.

1.5 Using keys in classification

Scientists use a variety of methods to classify living things. One method is to use a key which allows you to classify things by making choices grouped in pairs or couplets. Each step of the key offers two choices (1a and 1b, 2a and 2b, and so on). These choices are based on important characteristics of the object. By making the correct choice and following the directions, you should be able to identify the object or living thing. Keys that provide two choices at each step are called dichotomous keys.

Example 1: written key—classifying objects according to shape

In this example, each of the figures in Figure 1.6 has a body and head of a particular geometric shape. The key can be used to identify each figure.

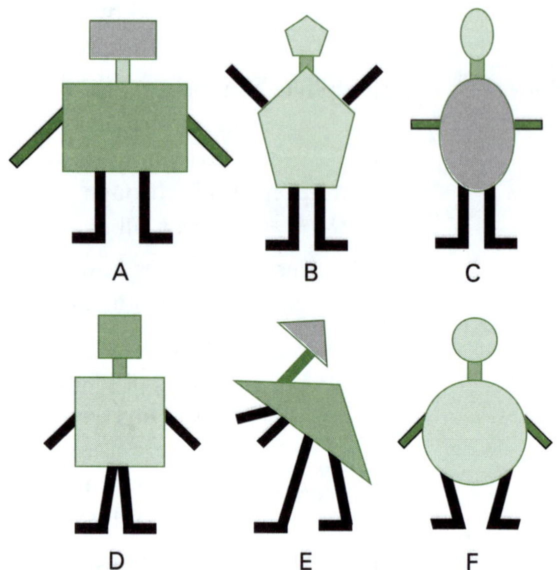

Figure 1.6 Geometric shapes to classify

1A Head and body have straight edges .... go to 3
1B Head and body do not have straight edges .... go to 2
2A The head and body are round in shape .... figure F
2B The head and body are elliptical in shape .... figure C
3A The head and body have four sides .... go to 4
3B The head and body do not have four sides .... go to 5
4A The head and body are square .... figure D
4B The head and body are rectangular .... figure A
5A The head and body have five sides .... figure B
5B The head and body have three sides .... figure E

Keys can also be presented in diagrammatic rather than written form. Diagrammatic keys have the structure of a flow chart.

Example 2: diagrammatic key—vertebrate classification

Figure 1.7 shows a simple diagrammatic key to distinguish the five families of vertebrates. At each step, you have to make a decision between two alternatives. Thus, if you classify a frog then the first choice is that it has no feathers. The next choice is that it has no scales and the last choice is that it has no hair or fur.

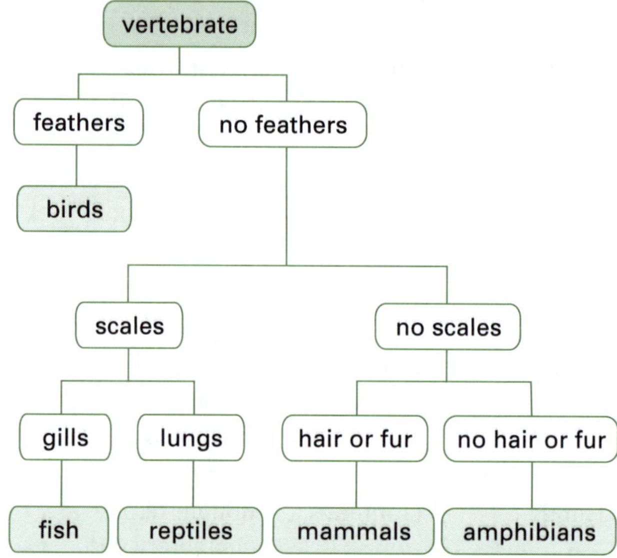

Figure 1.7 Classification of the vertebrates

Experiment 2

Developing a key to identify students in a class

Aim

- To use the data collected about the characteristics of 10 people in the class in order to develop a key to identify them

Method

Use the results from Table 1.5 to develop a diagrammatic key.

Table 1.5 Characteristics of 10 people in the class

Student name	Hair colour	Eye colour	Ear lobes (free or fixed)
Amy	blonde	blue	free
Catherine	brown	brown	free
Frederick	brown	brown	free
Juliet	blonde	blue	fixed
Mary	black	brown	fixed
Penny	brown	brown	fixed
Quincy	brown	brown	free
Raymond	black	blue	fixed
Romeo	black	blue	free
William	blonde	brown	free

Analysis

This is a dichotomous key and so you need to organise the information collected into opposing groups, such as students who have blue eyes and students who have brown eyes. Or you could form two groups using free or fixed ear lobes. Draw your own diagrammatic key based on this data. Go to p. 184 to check your answer.

Conclusion

Now write your own conclusion for this experiment.

Go to p. 184 to check your answer.

Test yourself 1

Part A: Knowledge

1. Which kingdom contains organisms that are multicellular, have no chlorophyll and absorb nutrients from decaying tissue? *(1 mark)*
   A plant
   B fungi
   C protista
   D monera

2. To which kingdom do truffles, yeast and tinea belong? *(1 mark)*
   A monera
   B protista
   C fungi
   D plant

3. To which vertebrate class do whales and bats belong? *(1 mark)*
   A mammals
   B birds
   C amphibians
   D reptiles

4. A tortoise and a goanna are classified into which vertebrate group? *(1 mark)*
   A amphibians
   B fish
   C reptiles
   D dinosaurs

5. A wombat is classified into which mammalian group? *(1 mark)*
   A placental
   B marsupial
   C monotreme
   D vertebrate

6. Complete the following restricted-response questions using the appropriate word. *(1 mark for each part)*
   a) Diagrammatic keys have the structure of a .... chart.
   b) Keys that provide two choices at each step are called .... keys.
   c) Vertebrates obtain their oxygen via .... or lungs.
   d) In all vertebrates the .... is protected by the skull.
   e) The eggs of reptiles are mostly .... shelled.

7. Use the code letters to match the terms or phrases in each column. *(1 mark for each part)*

Column 1	Column 2
A salmon	F kangaroo
B reptile	G botany
C mammal	H egg laying
D leaves	I fish
E echidna	J crocodile

Part B: Skills

8. Use the key provided to identify the five dinosaurs shown in Figure 1.8. *(5 marks)*

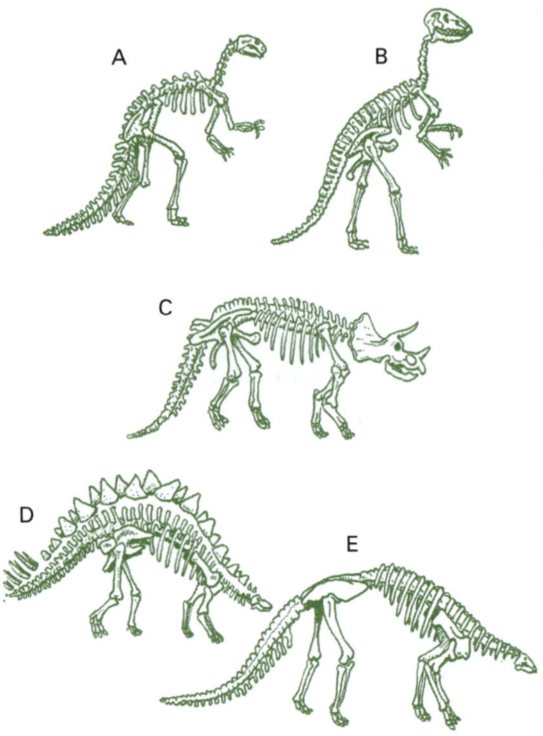

Figure 1.8 Skeletons of some dinosaurs

1A Walks on all fours .... go to 2
1B Walks on two hind legs .... go to 4
2A Has bony plates down back........ Stegosaurus
2B No bony plates down back .... go to 3
3A Has a horn .... Triceratops
3B No horn .... Nodosaurus
4A Bone at hip facing upwards .... Camptosaurus
4B Bone at hip almost horizontal.... Iguanodon

9. Table 1.6 shows three species, A, B and C.

Table 1.6 Classification of three different species

	A	B	C
Kingdom	Animalia	Animalia	Animalia
Phylum	Arthropoda	Chordata	Arthropoda
Class	Insecta	Mammalia	Insecta
Order	Lepidoptera	Carnivora	Coleoptera
Family	Danaidae	Canidae	Chrysomelidae
Genus	*Danaus*	*Canis*	*Leptinotarsa*
Species	*plexippus*	*familiaris*	*decemlineata*

a) Which two species are the more closely related? *(1 mark)*

b) From the information given, can you tell what kind of organisms they are? *(3 marks)*

10. Use the key provided to replace the letters K, L, M and N with the names of the insects shown in the Figure 1.9. *(4 marks)*

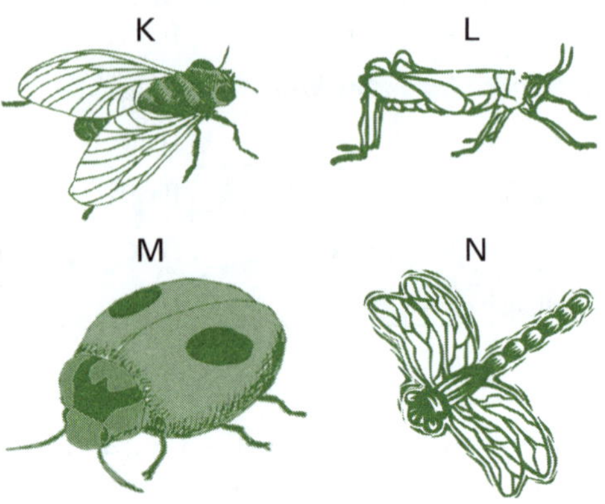

Figure 1.9 Some types of insects

1A Wings with hard outer covering.. go to 2
1B Wings without hard outer covering.. go to 3
2A Body has elongated shape .... grasshopper
2B Body has round shape.................... lady beetle
3A Wings point towards rear of body .... fly
3B Wings point out from side of body .... dragonfly

11. The scheme in Figure 1.10 shows some differences between butterflies and moths.

butterflies		moths
thin, hairless body	has six legs	wide, furry body
antennae have swellings on end	two antennae	antennae thick and feathery
wings upright when at rest	three body parts	wings horizontal when at rest
wings usually colourful	two pairs of wings	wings usually dull
most active during day	hatches from egg	most active at night
	compound eyes	

Figure 1.10 Similarities and differences between butterflies and moths

a) List two features: *(2 marks for each part)*
   i) common to both butterflies and moths
   ii) common to only butterflies
   iii) common to only moths.

b) A boy found what he thought was a butterfly but the wings were not very colourful. Could this still be a butterfly? Why? *(2 marks)*

c) What is another feature about wings that can distinguish butterflies from moths? *(1 mark)*

d) How can you tell that Figure 1.11 shows a butterfly and not a moth? *(1 mark)*

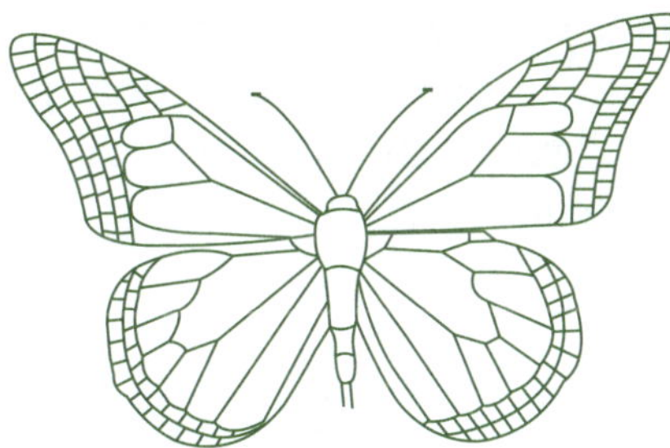

Figure 1.11 A butterfly

12. Examine the leaves in Figure 1.12. Identify the leaves that have the following features:
    a) leaves with four points on each edge *(1 mark)*
    b) leaves with a network of branching veins *(1 mark)*
    c) leaves with round shape *(1 mark)*
    d) leaves with a rounded tip *(1 mark)*
    e) leaf divided into leaflets. *(1 mark)*

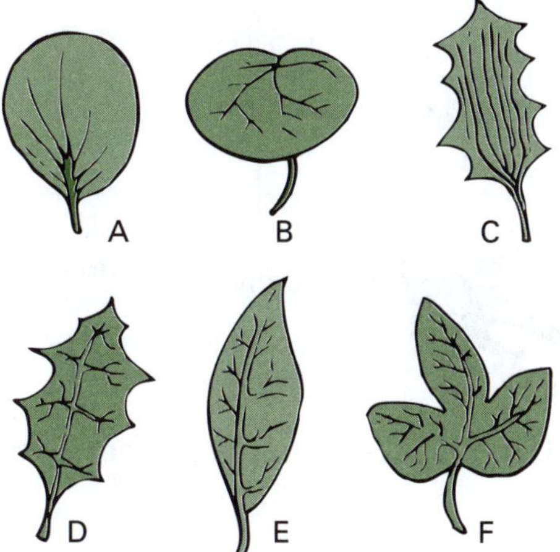

Figure 1.12 Different types of leaves

13. Read the following description of a vertebrate and classify it into its correct group. *(1 mark)*

    Constant body temperature; has a beak, but no teeth; breathes using lungs; eggs fertilised inside the female's body; eggs covered in a hard, protective shell.

14. Create a diagrammatic key of the different types of screws in Figure 1.13 to identify them by code letter. *(4 marks)*

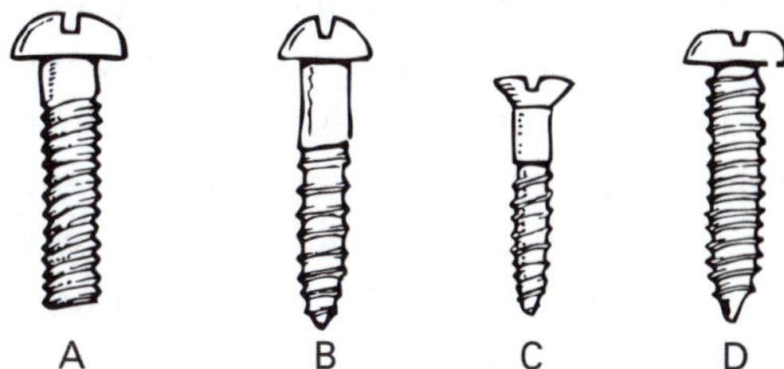

Figure 1.13 Different types of screws

Go to p. 185 to check your answers.

1.6 Food chains and feeding relationships

All living things need energy to survive. It is this cycle of energy that determines which organisms survive and which die. The ultimate source of energy is the Sun. Plants are able to trap the energy in sunlight and turn it into energy-rich substances that are eaten by animals. Little by little the energy is converted to heat, and the matter taken in by the succession of organisms is returned to the environment to be recycled.

A food chain shows how each living thing gets its food. For example, Figure 1.14 shows a simple food chain that links grass, the cows (that eat the grass) and humans (that eat the cows). Each link in this chain is food for the next link. A food chain always starts with a green plant and ends with an animal.

The arrows in a food chain indicate the direction of flow of food (energy).

Figure 1.14 A simple food chain

Feeding relationships

Organisms can be described by where in the food chain they are found. Producers are green plants, which are able to use light energy from the sun to manufacture (produce) their own food (simple sugars) using carbon dioxide and water. Since animals cannot make their own food, they must eat plants and/or other animals. They are known as consumers.

There are three kinds of consumers.

- Herbivores: animals that only eat plants (producers). They are also known as first-order consumers; for example sheep and cattle.
- Carnivores: animals that eat other animals. If those animals eat herbivores, they are called second-order consumers; for example dingos and lions.
- Omnivores: animals (including humans) that eat both other animals and plants; for example, chickens eat plant material such as seeds and leaves but they also eat snails and worms.

Figure 1.15 shows how producers and consumers are linked.

Food chains and food webs

A food chain shows the pathway along which energy and nutrients are passed. Food chains are not usually very long as the energy passed from one level to the next drops dramatically. It is estimated that only about 10% of the energy from one level passes on to the next. This is because organisms use much of the energy they acquire to grow, move and carry on all their daily functions and produce heat.

There cannot be too many links in any single food chain because the animals at the end of the chain would not get enough food (and hence energy) to stay alive.

While a food chain links one organism to another, the real world is far more complicated. Most animals eat from a variety of sources. A food web is more realistic in showing interactions in a community (see Figure 1.16). It is made up of interlocking food chains.

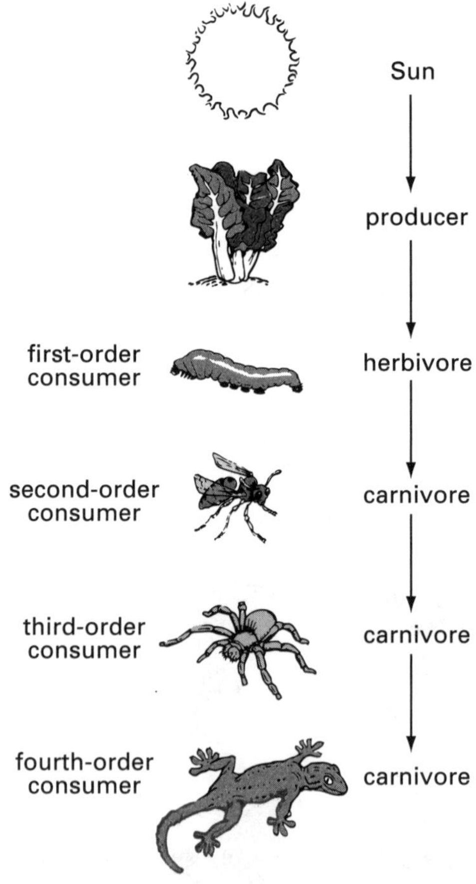

Figure 1.15 How producers and consumers are linked

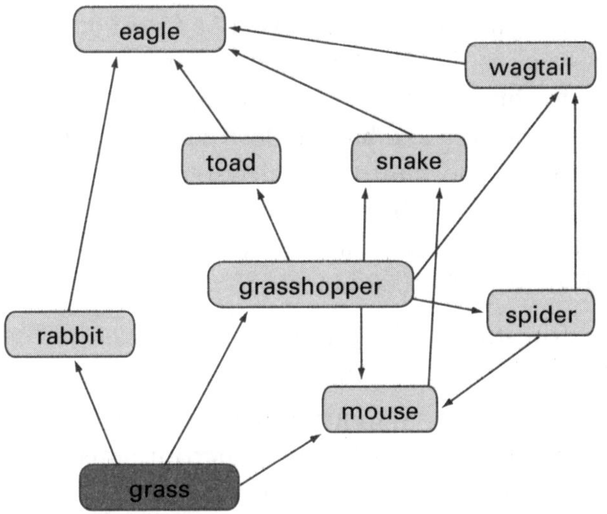

Figure 1.16 A food web

1.7 Microorganisms and food chains

Decomposers are the organisms of decay. They are usually microorganisms such as bacteria and fungi that release simple substances back into the environment to be used by producers.

Bacteria are among the smallest living organisms and are the most numerous of decomposers. There are billions of them in each gram of soil (see Figure 1.17). Bacteria can thrive in any environment, such as the cold Antarctic, the intense heat of a steamy geyser or in the acid of your stomach. They will break down a whole host of things, from a dead animal carcass to an ocean surface oil slick.

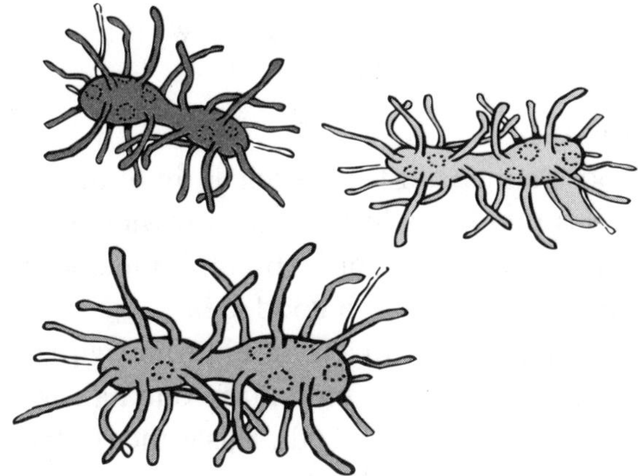

Figure 1.17 Typical bacteria found in the soil

Earlier we said that a food chain begins with a plant and ends with an animal. Actually, it begins with a producer and ends with a decomposer (see Figure 1.18). The producer brings energy and materials into the chain. The decomposers release the last amount of matter back into the environment to be recycled in another food chain. The decomposer uses the last remaining energy.

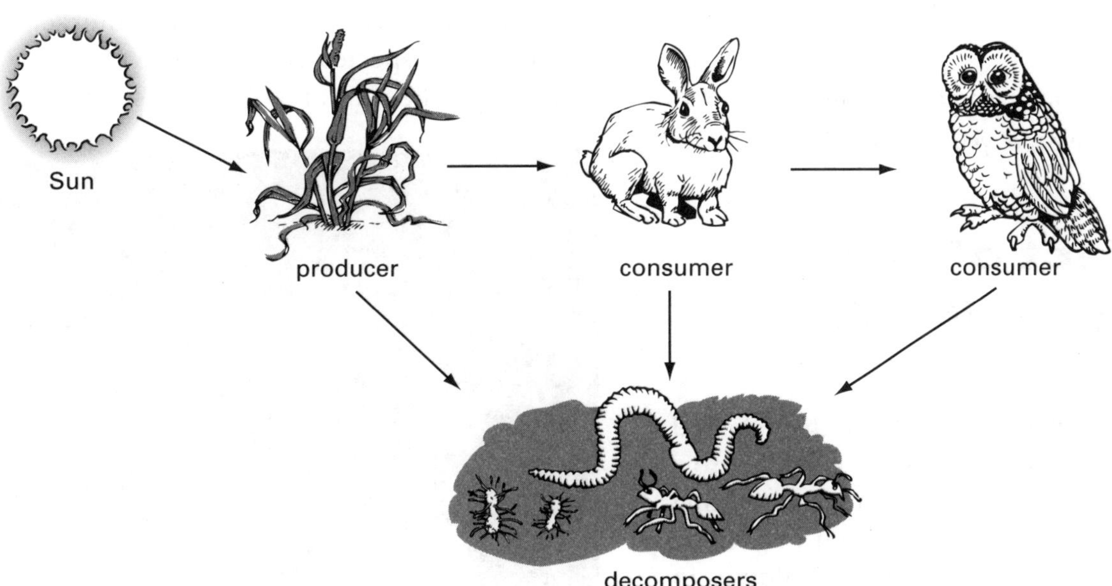

Figure 1.18 A food chain including decomposers

Experiment 3

Observing decomposers break down fruit

Aim

- To culture and observe how microbes decompose fruit

Method

1. Choose a piece of fruit (such as an orange or lemon) that has a little mould growing on its surface. Take care not to breathe the spores. Using a face mask is a good idea.
2. Take a fresh piece of a different fruit (with no mould) and cut it in half. Place one half in a plastic bag and tie the ends. This is the control. Lightly touch the other half on the surface of the mouldy fruit and seal it in the other bag. Label the bags.
3. Observe the bags for several days and see if the mould grows on this new piece of fruit. Compare it to the control.

Results

The control fruit either did not grow mould or did so at a much lower rate than the others that were infected.

Analysis

Analyse the results. Go to p. 184 to check your answer.

Conclusion

Now write your own conclusion.

Go to p. 184 to check your answer.

1.8 Effect of human activity on local habitats

When settlers arrived in Australia, they cleared some land to build their towns and cities, to grow plantations and for agriculture. Logging (cutting down trees for timber) became an important activity as many houses and other buildings in the early days were made from wood. Wood is still an important building material.

Deforestation is the removal of a forest or stand of trees so that the land is afterwards converted to a non-forest use. For example, worldwide deforestation and habitat destruction is severely crippling the rainforest ecosystem. Rainforests are currently being destroyed at a rate of 310 000 km^2 each year. (Compare this to the area of Victoria, 228 000 km^2.)

Deforestation has a number of effects on the natural environment.

- Plant life is degraded due to a loss of nutrient-rich litter and microorganisms to decompose organic matter.
- The loss of shade causes leaching, soil erosion (see Figure 1.19) and drying at a faster rate.
- The hard surface caused by baking prevents water from soaking in, causing excess runoff and flooding.
- Deforestation affects the composition of the atmosphere, as well as rainfall. Because there are fewer plants to transpire less water, there is less rainfall and increased droughts. So what plants do remain grow slower and are more degraded.
- Cutting down rainforest trees causes a loss of species. The canopies of tropical rainforests have rich animal and plant diversities. They create new sources of food, new shelters, new hiding places and new areas for interaction with other species. In fact, some 70 to 90% of life in the rainforest is found in the trees.

Figure 1.19 Soil erosion

The Australian environment

There are five most frequently recorded processes considered to be pressures on the biodiversity in Australian regions.

1. Effect of grazing animals. About half of all clearing and land development for agriculture, and for cities, has happened in the last 40 years (much of it in NSW and Queensland). This has affected the habitats for living things and the availability of water.
2. Weeds. Once land is cleared, weeds take the opportunity to gain hold. Many of these weeds are introduced species (Figure 1.20 for example). Fertilisers on agricultural land feed weeds, or fertiliser is washed into streams where pond weeds can take hold. Native plants have adapted to poor soils and won't tolerate high fertiliser levels.
3. Feral animals. When animals such as cats, dogs, pigs, goats and rabbits are returned to an untamed state from domestication, they can damage the unique and fragile Australian environment. For example, feral cats have reduced the numbers of native wildlife on the mainland by killing them. With its 300 million years of geographic isolation and separate evolution, native plants and animals are being challenged by introduced species.
4. Effects of fire. Australia's unique plants are adapted to alternating patterns of fire frequency and intensity, and water availability. Some plants, for example, require fire to set seed. These requirements are being altered by human activities.
5. Habitat fragmentation. For both plants and animals, breaking up large areas of habitat into smaller pieces means that fewer species can be accommodated in the new 'islands'. In addition, fragmentation can slow or prevent the mixing of genes among individuals in populations. This can result in inbreeding that reduces the reproductive fitness and long-term health of populations.

Figure 1.20 Bitou bush is a noxious introduced weed in the Australian bush. It is among the worst pest plants in the Australian coastal environment, restricting access to beaches and destroying native bushland.

1.9 Introduced species

Being isolated from the rest of the world for many millions of years, Australian plants and animals have adapted to living in a mainly dry environment with poor soils. Between 20 000 to 18 000 years ago, some 80% of the continent became acutely arid. There has been little recovery since then.

Humans, especially European migrants in the last two centuries, have made dramatic changes to the environment.

- Grassland loss in Australia began some 300 million years ago but this loss has dramatically increased with European settlement.
- Clearing forests for crops or pastures for introduced animals like cows and sheep, and overgrazing by these animals, has seen the total amount of plant material decrease.
- Introduced pests like the rabbit have further damaged the environment, making large tracts of land no longer suitable for farming.

Case study 3: The Murray-Darling Basin

The Murray-Darling Basin covers around one-seventh of Australia's land area (see Figure 1.21) and is an important agricultural region that is significant for Australia's economy. It includes Australia's three longest rivers (the Murray River, the Darling River and the Murrumbidgee River).

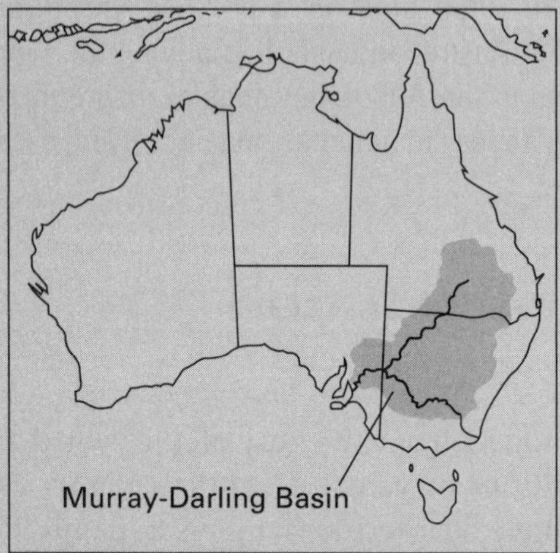

Figure 1.21 The Murray-Darling Basin

The Basin is important for its biodiversity, too. At the time of European settlement, about 28% of Australia's mammal species, about 48% of its birds and some 19% of its reptiles were found there. The Basin contains more than 30 000 wetlands.

However, the demands of agriculture and human population has increased water use in the Basin five-fold in less than a century. In fact, so much water is taken from the Murray for human use that the river now ceases to flow through the mouth some 40% of the time.

The over-allocation of water resources, salinity and climate change are having an effect on the health of the Basin. There is not enough water to maintain the Basin's natural balance and ecosystems. This results in a marked decline in the number and diversity of plants and animals. Many species that once were common are now rare and in severe decline. In order to ensure that there will be enough water for a healthy environment as well as vital agriculture industries and human use in the future, various Australian governments have set up an independent body, the Murray-Darling Basin Authority, to develop and administer a plan for the management of water.

Australian scientists, including those from the CSIRO (Commonwealth Scientific and Industrial Research Organisation), are studying the Murray-Darling Basin to find ways of managing the conflicting needs for water: human agriculture on the one hand, and the survival of native organisms and their habitats on the other.

There is little chance of the supply of water increasing, but there are still good opportunities for increasing how efficiently the water is delivered and used. This applies to both water for irrigation and water for environmental benefit to wetlands, floodplains and the estuary.

The rabbit problem

While some rabbits (*Oryctolagus cuniculus*) came out on the First Fleet, a release in 1859 near Geelong, Victoria, has allowed feral rabbits to spread across much of the continent (see Figure 1.22). The effect of rabbits on the ecology of Australia has been devastating.

- They are the most significant known factor in species loss in Australia.
- Rabbits often kill young trees in orchards, forests and on properties by ringbarking them.
- Rabbits eat grass and other crops, roots and all, so there is nothing left to grow back.

- Rabbits are responsible for serious erosion problems as they eat native plants, leaving the topsoil exposed and vulnerable to erosion. Removing this topsoil devastates the land since it takes many hundreds of years to regenerate.
- Rabbits also affect populations of native animals. For example, the rabbit-eared bandicoot or bilby needs a constant supply of seeds and roots. Feral animals such as rabbits graze or degrade vegetation that provides food and shelter for them and for other native animals. If vegetation is destroyed or eaten by feral animals, the bilby and other native species are placed under great pressure and may die out or significantly decrease in numbers in that area.
- Rabbits can produce many young and spread rapidly across much of the country. Australia has ideal conditions for a rabbit population explosion. With mild winters, rabbits are able to breed the entire year. And with widespread farming practices, areas that may have been desert, scrub or woodlands were instead turned into vast areas with low vegetations, and this creates ideal habitats for rabbits.

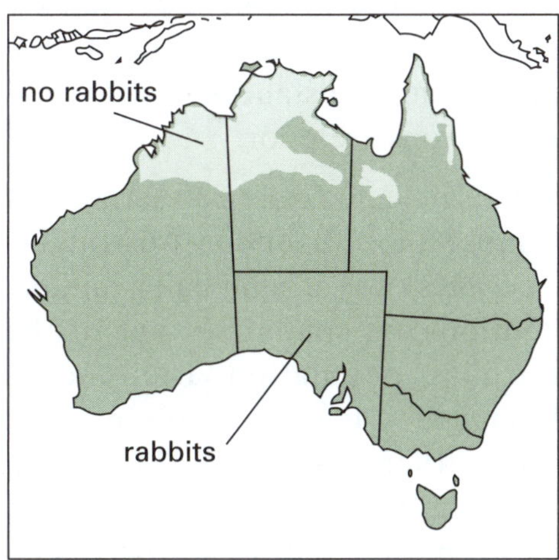

Figure 1.22 The spread of feral rabbits in Australia

Control measures

A number of control measures (such as shooting, trapping, poisoning or digging up warrens) were tried in the early part of the 20th century but with limited success. Even a rabbit-proof fence (almost 2000 km long) was built from north to south in Western Australia in an effort to keep them out.

It wasn't until 1950 when myxomatosis (a disease caused by the myxoma virus) was deliberately released into the rabbit population that numbers dropped from around 600 million to about 100 million. Soon after, genetic resistance in the remaining rabbits allowed the population to recover to 200 to 300 million by 1991.

So back to the drawing board! The CSIRO developed and released calicivirus (also known as rabbit haemorrhagic disease virus or RHDV) in 1996. This virus is more successful in controlling rabbits at higher temperatures.

But don't worry! If you have a cute, cuddly pet rabbit, there is a legal vaccine in Australia to protect for RHDV.

The cane toad problem

The cane toad (*Bufo marinus*) was introduced from South America into the Queensland cane fields in 1935 (see Figure 1.23). It was mistakenly thought that the toad would control several kinds of beetles whose larvae were damaging sugar cane plants. Instead, this poisonous pest is thought to have decimated some species of native Australian animals. In humans the poison can cause intense pain, temporary blindness and inflammation, and it may kill small animals.

Figure 1.23 The cane toad. Average-sized adults are 10 to 15 cm long, and can weigh over 1 kg. Male cane toads are smaller and wartier than females.

Since then, cane toads have spread rapidly into New South Wales, west into the Northern Territory and have even reached the wetlands of heritage-listed Kakadu National Park (see Figure 1.24). Most cane toads are found in urban areas, and in areas with grassland or woodland. While they are basically land animals, they require access to water for re-hydration, and breeding.

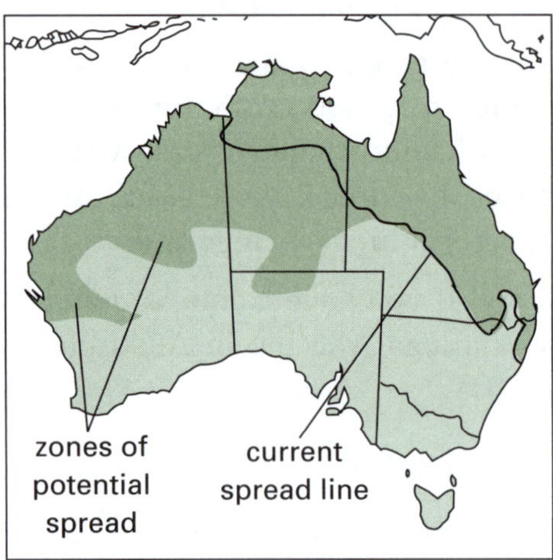

Figure 1.24 The line represents the current limit of the cane toad in northern and eastern Australia. The pale green coloured area shows where it could soon spread.

Problems with cane toads include the following:

- Female toads lay up to 35 000 eggs at a time and can breed twice a year. Cane toads need between 6 and 18 months to reach sexual maturity and live for about 5 years.
- Cane toads have large swellings on each shoulder, which can squirt poison when threatened. Cane toads have no known predators in Australia.
- Cane toads affect the environment. For example, it has been observed that there is a decline in quoll numbers and native frogs in areas where large numbers of cane toads are found. While cane toads are voracious feeders, eating mainly insects and small animals, they can also compete for food with native animals. They are also known to spread diseases such as salmonella.

There is no effective control method that can be applied to the vast area where cane toads have spread. In some areas, communities have applied bounty systems. But the problem with bounties is that sometimes native frogs are caught by mistake.

CSIRO scientists are working to find a biological control method using toad genes. Other university scientists are studying the toad's impact on native fauna and discovering the ways in which native species are adapting to their presence, or trying to find a sex pheromone (a chemical secreted by an animal that influences the behaviour or development of others of the same species) in cane toads that may be used to disrupt their breeding cycle.

1.10 Sustainable land management

Before Europeans arrived in Australia, environmental conditions and Aboriginal hunting largely controlled the sizes of animal populations. Many animals are well adapted to surviving in these harsh and dry environments. For example, during dry periods some species of kangaroo stop reproducing but quickly resume when the rains come. Other animals are adapted to travel large distances to find food when local conditions are poor.

At the time of permanent European settlement the total number of Indigenous Australians (Aboriginal peoples) was around half a million. Their distribution was similar to that of the current Australian population. Their needs were simple and they lived in harmony with the land. Today there are some 23 million people living on this continent, with many of us demanding a lot more from the environment. These increased expectations place a lot more stress on the land with problems occurring if we are too short-sighted.

Bushfires

Bushfires are a natural feature of the dry Australian landscape. Lightning commonly sets the countryside ablaze and has done so for millions of years. Fires are a natural way of clearing old growth and returning some of the sparse minerals back into the soil. Since Aboriginal settlement, over 40 000 years ago, periodic use of fire has been used to drive out animals from grassy and wooded areas to be caught for food.

The 700 or so species of eucalypts have leaves containing flammable oils. These vaporise quickly when heated and then burn. The bark of these plants can ignite easily, but then drops off leaving the core intact. The thick bark insulates the tissues below from the heat of the fire. New shoots soon appear. You only need to drive through an area a few weeks after a bushfire to see the natural regeneration of some plant species. By comparison, many forests elsewhere in the world are dead once a fire passes through.

After fire, some plant species' seedlings prosper. For some plants, fire liberates the seeds but in other plants it destroys them. The mountain ash needs the site to be thoroughly burnt and exposed to full sunlight before it can regenerate.

Fires can change the composition of a community. The amount and type of change depends on a number of factors, some of which include:

- which plants are fire-adapted and which are not
- what level of vegetation is burnt
- how hot the fire is
- how often the fire occurs.

European settlement has led to ecological consequences. As fires are generally suppressed, undergrowth can sometimes build up leading to more intense and far more damaging fires. Forestry and fire departments have learned that periodic controlled burning of areas can sometimes reduce the amount of flammable undergrowth, and allow some plants to complete their life cycles. Fires also destroy fauna and their food and homes.

Aboriginal use of fire

What European settlers learned about the importance of fire in the Australian landscape, Aboriginal peoples knew for thousands of years. Aborigines created fire by rotating one wooden stick while its end was pressed into a depression in a stick lying on the ground. Friction produced sufficient heat to create a fire when dried grass was added (see Figure 1.25).

Figure 1.25 Traditional aboriginal method for making fire. The drill stick is rotated rapidly while being pressed into a small socket on the hearth stick. This stick rests on a small pad of suitable tinder such as dried kangaroo dung or other easily flammable fibrous substance. Heat from the friction ignites a fine charcoal-like dust that piles up. When the tinder is burning slowly it is wrapped in teased grass fibres and gently blown until it bursts into flame. Some Aboriginal groups may also have used flint and pyrites to strike a spark.

Some Aboriginal people still use fire to manage their land and forests in some parts of Australia. Many see managing fire as part of their responsibility as custodians of the land. Besides cooking and keeping warm on cold nights, fire is used for a variety of purposes:

- to clear rubbish plants from the area around food plants
- to remove scrub near camp sites so that soft young grasses grow that attract animals like kangaroos and wallabies, which can be hunted for food
- to remove older plant parts and encourage new growth; the nutrients in the ash of the fire can be recycled locally
- to frighten game out of scrub and thickets so hunting parties can trap them
- to allow people to move easily around the land
- to communicate with other groups of Aboriginal people
- for spiritual purposes associated with the land.

Palm oil harvesting

The oil palms are used in commercial agriculture in the production of palm oil. Oil is extracted from both the pulp of the fruit (palm oil, an edible oil) and the kernel (palm kernel oil, used in foods and for making soap). Palm oil is also used in some biofuels (oil is mixed in with petrol and other liquids). The high oil yield of oil palm trees (over 7000 L/ha each year) has made it a common cooking ingredient in South-East Asia and the tropical belt of Africa. And, being cheap, it is increasingly used in the commercial food industry in many parts of the world.

Borneo and Sumatra are two large islands in South-East Asia that produce a lot of palm oil (see Figure 1.26). Sumatra (area 473 000 km^2) is owned by Indonesia, while Borneo (area 743 000 km^2) is owned by several countries, the largest part being Indonesian.

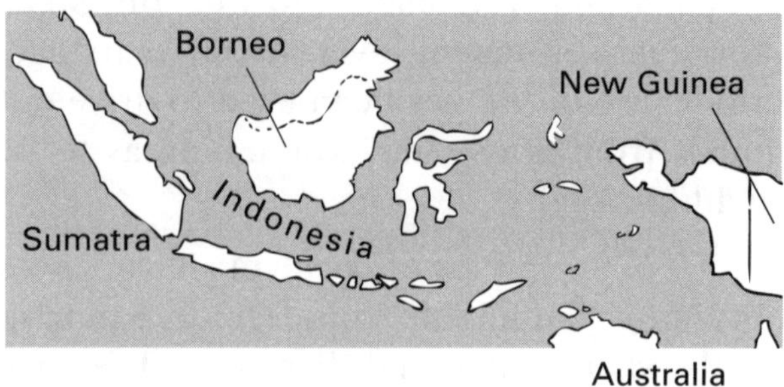

Figure 1.26 Location of Sumatra and Borneo in South-East Asia

There is a lot of pressure on the governments of these islands to clear land and plant more oil palms. Over the last few decades, oil palm plantations have rapidly spread across South-East Asia and are a source of important economic benefits as they bring in a lot of money and provide employment in Indonesia. This has led to the:

- rapid destruction of rainforests, which are the homes of orangutans and many other animals
- tropical forest cover disappearing at a rate of more than 2.8 million hectares per year! (See Figure 1.27.)

These rainforests are biodiversity 'hotspots', home to tigers, sun bears, elephants, rhinoceroses, clouded leopards and proboscis monkeys and orangutans. On Sumatra there is now more than four times as much land cultivated with oil palms as there is orangutan habitat remaining.

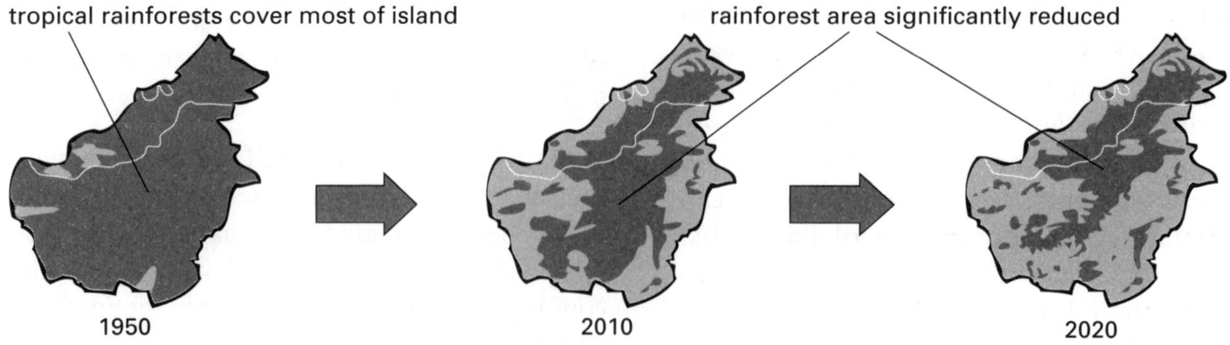

Figure 1.27 Rainforest areas left in Borneo

The leaves of these old, tropical trees filter the carbon dioxide out of the air, so clearing the forests contributes to global warming. Also, a significant amount of carbon dioxide is pumped into the atmosphere from burning peat bogs to make the land suitable for the palms. Further, erosion is hastened by planting trees in rows up and down hillsides rather than on contours around them. This can lead to an irreversible soil problem.

It is estimated that at the current rate of habitat destruction, no orangutans will be left in the wild in a decade or two.

Test yourself 2

Part A: Knowledge

1. A herbivore is also known as a *(1 mark)*
- **A** producer.
- **B** first-order consumer.
- **C** second-order consumer.
- **D** third-order consumer.

2. At the end of a food chain is a *(1 mark)*
- **A** producer.
- **B** decomposer.
- **C** consumer.
- **D** carnivore.

3. The primary source of energy in a food chain is *(1 mark)*
- **A** the sun.
- **B** green plants.
- **C** mineral salts.
- **D** fats and carbohydrates.

4. A native Australian animal that can survive desert conditions is the *(1 mark)*
- **A** bilby.
- **B** camel.
- **C** rabbit.
- **D** cane toad.

5. The amount and type of change caused by a fire depends on a number of factors. Which of these is not an important factor? *(1 mark)*
- **A** Which plants are adapted to fires and which are not
- **B** How much vegetation is burnt
- **C** The heat of the fire
- **D** Who lights the fire

6. Complete the following restricted-response questions using the appropriate word or words. *(1 mark for each part)*
- **a)** .... is the relationship between living things and their environment.
- **b)** A .... is a group of living organisms of the same kind living in the same place at the same time.
- **c)** All populations interact together to form a ....
- **d)** .... means responsible travel to fragile, pristine and usually protected areas to observe wildlife or to help preserve nature.
- **e)** An animal introduced into Australia to control beetles in sugar cane plantations is the ....

7. Use code letters to match the terms or phrases in each column. *(1 mark for each part)*

Column 1	Column 2
A orangutan	F survival features
B food	G energy flow
C inorganic nutrients	H recycled
D adaptation	I Sumatra
E food chain	J web

Part B: Skills

8. Here is some information about food chains, but the sentences are scrambled. Write the code letters of the sentences in a logical sequence. *(1 mark for each part)*
- **a)** All animals rely on plants directly or indirectly for their energy, as only plants have the ability to convert sunlight into food that they, and animals, can use.
- **b)** The carnivores that prey on these herbivores are at the third feeding level, and so on.
- **c)** For all life on Earth, there is only one available form of energy: the Sun.

d) Plants then, are the first feeding level and plant eaters, or herbivores, feed on these plants (second feeding level).
e) The key idea you need to learn from food chains and food webs is that energy is transferred.
f) Only plants can change this energy into a form that can be used by animals; that is, they 'feed' off sunlight.

9. Here is a simple food chain:
   grass seeds ➡ mouse ➡ owl
   a) What does such a food chain show? *(1 mark)*
   b) Where does the energy in the grass seeds come from? *(1 mark)*
   c) What is indicated by the direction of the arrows? *(1 mark)*
   d) What is the difference between a food chain and a food web? *(1 mark)*

10. Here is a simple food chain:
   bulrush ➡ muskrat ➡ weasel ➡ great-horned owl
   a) Name the:
      i) producer *(1 mark)*
      ii) herbivore *(1 mark)*
      iii) first-order consumer *(1 mark)*
      iv) third-order consumer. *(1 mark)*
   b) Does all the energy originally in the bulrush plant make its way to the owl? Explain. *(4 marks)*
   c) If the energy originally comes from the sun, why is it not included in the food chain? *(2 marks)*

11. Many eucalypts are well adapted to areas of poor soil and little water. Their leaves hang vertically during the hottest part of the day and are light and silver coloured. How does this help in preventing moisture loss? *(3 marks)*

12. For each of the following, complete the food chain using the list provided.
   a) List: lion; whale; grasshopper; eagle; kookaburra
      grass ➡ .... ➡ frog ➡ snake *(1 mark)*
   b) List: whale; seal; shark; crab; worm
      seaweed ➡ shellfish ➡ .... ➡ seagull *(1 mark)*

13. Here is some information about four organisms living in Australia. Name an environment each would be best suited to.
   a) Mulga tree: a hardy tree or shrub that can grow in relatively infertile clay soils without much rain. It is also long living (up to 200 years) so it can stick around while the rains come for it to set seed. *(1 mark)*
   b) Red kangaroo: feeds during the cool of the night and rests in the shade during the hottest part of the day. It has been known to travel great distances in search of food. *(1 mark)*
   c) Bilby: can burrow underground to a depth of 1 to 2 m where the temperature is about 10 °C cooler. As the air in the burrow is more moist, the bilby loses less water. It gets most of its moisture from food such as insects, roots, seeds and bulbs. *(1 mark)*
   d) Spinifex: leaves are hard and folded into tight cylinders. This reduces the surface area and slows down the rate of heating and water loss. Spinifex roots spread out deep and wide in search of water. *(1 mark)*

14. Figure 1.28 shows a food, or biomass, pyramid.

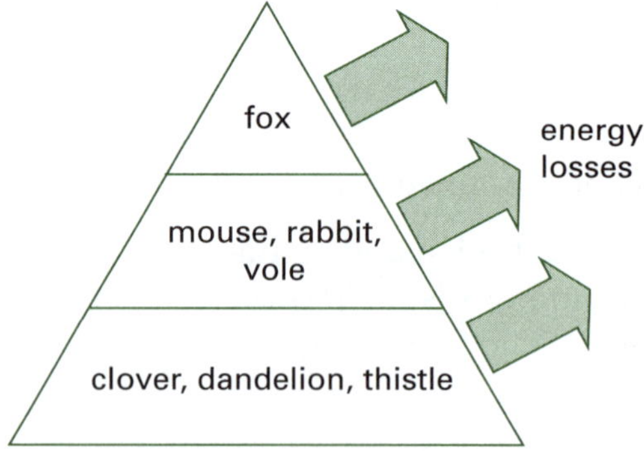

Figure 1.28 Food pyramid

   a) What does this diagram show? *(1 mark)*
   b) Explain the shape of the pyramid. *(3 marks)*

Go to p. 185–186 to check your answers.

Summary

1. Classification is used to: (a) help find things easily; (b) compare things; and (c) organise things.
2. Classification is based on similarities and differences.
3. Fish, amphibians, reptiles, birds and mammals are all classified as vertebrates as they all have a vertebral column.
4. The five kingdom system is commonly used to classify all living things.
5. The five kingdoms are animals, plants, fungi, protista and monera.
6. Binomial nomenclature is used to name living things. The first name is its genus and the second name is its species.
7. All living things need energy to survive. The ultimate source of energy is the Sun.
8. A food chain shows how each living thing gets its food.
9. Organisms can be described by where in the food chain they are found. These are producers, consumers (herbivores, carnivores, omnivores) and decomposers.
10. Decomposers break down dead plants and animals.
11. Deforestation causes a number of effects on the natural environment including soil erosion and drying.
12. The five pressures on Australian biodiversity are grazing animals, weeds, feral animals, fire and habitat fragmentation.
13. Many introduced species including rabbits and cane toads have had a large negative effect on the Australian environment.
14. Fires can change the composition of a community.
15. The destruction of rainforests in Borneo has led to changes in biodiversity and climate.

Syllabus checklist

Are you able to answer every syllabus question in this chapter? Tick each question as you go through the list if you are able to answer it. If you cannot answer it, turn to the appropriate page in the guide as is listed in the column to find the answer.

	For a complete understanding of this topic	Page no.	✓
1	Can I give three reasons for classifying things?	2–3	
2	Can I explain why classification systems are based on similarities and differences?	3	
3	Can I describe the similarities and differences between vertebrates?	4	
4	Can I recall the names of the five kingdoms?	5–6	
5	Can I recall what the five kingdom system is commonly used for?	6–7	

	For a complete understanding of this topic	Page no.	✓
6	Can I recall what the binomial system of nomenclature is used for?	7	
7	Can I state the ultimate source of energy that is used to power all living things?	11	
8	Can I write a food chain from a list of organisms?	11–12	
9	Can I name and state the role of producers, consumers and decomposers in a food chain?	12	
10	Can I recall what the job of decomposers is?	13	

	For a complete understanding of this topic	Page no.	✓
11	Can I explain the effects of deforestation?	14	
12	Can I recall the five pressures on Australian biodiversity?	15	
13	Can I describe the effect on the Australian environment of the introduction of rabbits and cane toads?	16–18	

	For a complete understanding of this topic	Page no.	✓
14	Can I recall how fires can change the composition of a community?	19–20	
15	Can I explain how the destruction of rainforests in Borneo has led to changes in biodiversity and climate?	20–21	

Chapter test

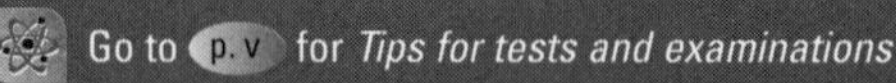
Go to p. v for *Tips for tests and examinations*

80 MIN

Part A: Multiple-choice questions

(1 mark for each)

1. Consider the following group of animals: elephant; giraffe; cow; horse. The similarity between these animals is that they all

A have large molar teeth for grinding grass.
B are marsupials.
C are cold blooded.
D are carnivorous.

Use the information in Table 1.7 to answer the next three questions.

Year 7 girls were looking for similarities and differences among their friends. The girls in one group came up with the following table.

Table 1.7 Characteristics of six friends

Student	Blonde hair	Brown eyes	Fair skin	Curly hair	Lobed ears
Julia	yes	no	yes	no	yes
Zoe	no	yes	no	no	yes
Linh	no	yes	no	no	no
Emily	yes	yes	yes	no	no
Stella	yes	no	yes	no	no
Lauren	no	no	yes	yes	yes

2. Which girl has most features like Julia?

A Zoe
B Stella
C Linh
D Lauren

3. The girls were then divided into two groups.

Group 1	Group 2
Julia	Zoe
Stella	Linh
Lauren	Emily

The characteristic used to divide them was

A hair colour.
B eye colour.
C fair skin.
D lobed ears.

4. Which of the following characteristics would be the most difficult to change permanently?

A hair colour
B eye colour
C curly hair
D lobed ears

5. A living thing has the following characteristics:

- dry, scaly skin
- lays shell-covered eggs
- lungs
- four legs.

This living thing could be a

A frog.
B snake.
C goanna.
D shark.

6. Consider the snail and octopus shown in Figure 1.29.

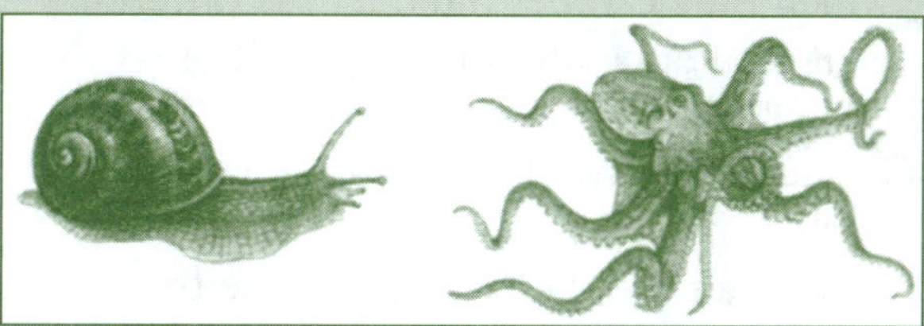

Figure 1.29 Snail and octopus

The feature that the animals have in common is that they both

A have shells.
B are vertebrates.
C are carnivores.
D have muscular feet for movement.

Questions 7 to 10 refer to the food web in Figure 1.30.

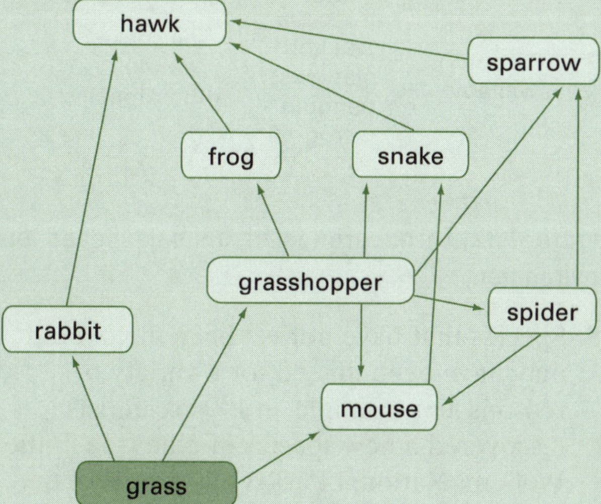

Figure 1.30 Food web

7. The producer in this food web is the
A grass.
B mouse.
C rabbit.
D hawk.

8. The first-order consumers are
A rabbits, frogs and sparrows.
B rabbits, grasshoppers and mice.
C rabbits , frogs and snakes.
D rabbits, grasshoppers and frogs.

9. The mouse is classified as
A a carnivore.
B an omnivore.
C a herbivore.
D a decomposer.

10. Why is the rainforest in Sumatra and Borneo being cut down?
A To grow profitable oil crops like palm oil
B To make way for large cities
C To reduce the mosquito population
D To search for crude oil and natural gas

Part B: Short-answer questions

11. Complete the following restricted-response questions using the appropriate word. *(1 mark for each part)*
a) Wallabies are classified as .... because they have fur and mammary glands.
b) The biological classification level between class and family is called ....
c) Bacteria are classified in the .... called Monera.
d) Humans are classified in the .... called Chordate as they have a hollow nerve cord.
e) Grass is classified in the .... kingdom because it makes its own food.

12. Use the code letters to match the terms or phrases in each column. *(1 mark for each part)*

Column 1	Column 2
A Earthworm	F Garfish
B Gills	G Decomposer
C Protista	H Invertebrate
D Whale	I Single-celled
E Fungus	J Vertebrate

Use the following key to answer Questions 13 and 14.

1A Thread all the way up to the head.......... go to 2
1B Thread covers only part of body............. go to 7
2A Thread tapered to point at end............... go to 3
2B Body not tapered to point at end........... go to 4
3A Thread all the way to point tip............... go to 6
3B Thread stops before flatenned point........thread-cutting screw
4A Head larger than body in diameter........ go to 5
4B Head same thickness as body...................... set screw
5A Head thick with socket opening.......... socket cap-head screw
5B Head designed with screwdriver slots........machine screw
6A Phillips slot (+) in head........................Phillips sheet-metal screw
6B Single slot on head for flat screwdriver................ slotted sheet-metal screw
7A Head flattened on top............................. go to 8
7B Head rounded on top.............................. go to 9
8A Head circular.............flat-head wood screw
8B Head hexagonal (six-sided) in shape..................... cap screw
9A Body tapers to a point....................... round-head wood screw
9B Body not tapered to a point.................wing bolt, round head screw

13. Use the key to identify the screw illustrated in Figure 1.31. *(1 mark)*

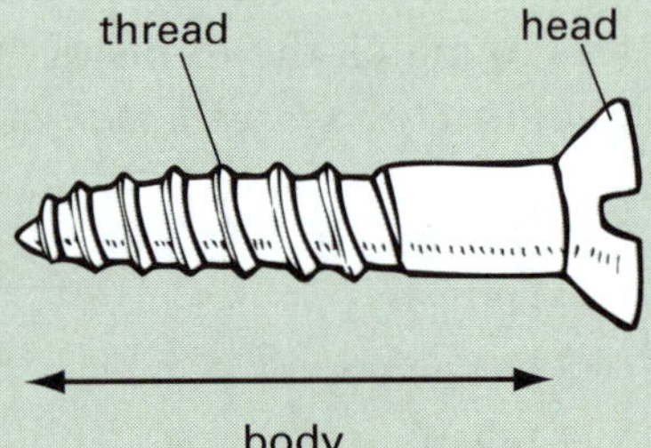

Figure 1.31 Screw

14. A screw has the following features. Identify the screw. *(1 mark)*

The thread of the screw extends from the tip to just under the head of the screw. The thread has a uniform diameter along its length and the threaded body of the screw has the same diameter as the head of the screw.

15. Various animals have been classified into three groups based on the environment in which they can be found. These different environments are labelled with the code letters (A, B and C) in Figure 1.32. Some animals can live in more than one environment. This is shown where the circles overlap. Identify these three environments. *(3 marks)*

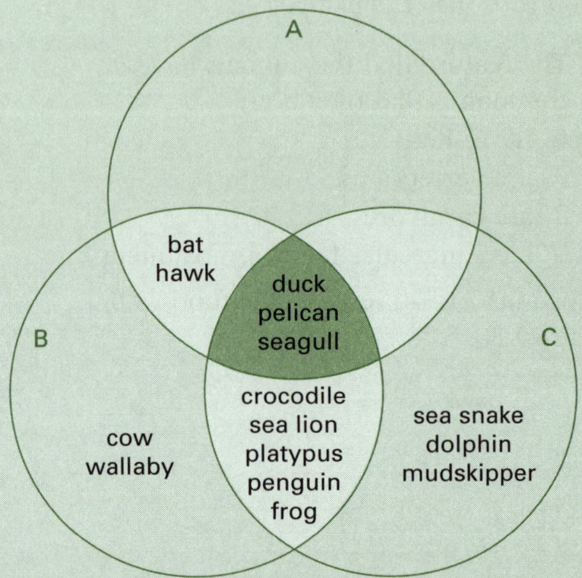

Figure 1.32 Three groups of animals based on environment

16. Species that have not yet been discovered may have been missed for a variety of reasons. For example, in 1994 scientists discovered a new species of pine tree in the Wollemi National Park. Only fossils of the tree had been found previously. Suggest four reasons why the Wollemi pine had not been discovered until recent times. *(4 marks)*

17. Convert the following written dichotomous key for four animals into a diagrammatic key or flow chart. *(3 marks)*

1A Four legs present .... go to 2
1B More than four legs .... go to 3
2A Body covered in hair or fur ... kangaroo
2B Body not covered in hair or fur... frog
3A Six legs present .... bee
3B More than six legs present ..... spider

18. In 1798, when a specimen of a platypus was sent from Australia to London, scientists were on their guard. What a strange creature, they thought. It had the body of an otter, the beak of a duck and the tail of a beaver. They quickly went to work to try to prove it a fake, but soon realised it was genuine. It surprised them even more when it was discovered a few years later that this mammal actually laid eggs.

Construct a two-column table of examples from the random statements below of ways in which the platypus is similar to placental mammals and ways in which the platypus is different from placental mammals. *(2 marks)*

- A platypus lays eggs.
- The body of a platypus is like an otter's.
- The tail of a platypus is like a beaver's.
- Their mammary glands have no teat.
- The beak of a platypus is like a duck's.
- The body of a platypus is covered with fur.
- A platypus has no teeth.
- Mammary glands are present.
- Milk trickles onto a skin patch.

19. Consider the classification key for girls in a music class shown in Figure 1.33.

a) Describe Katya's appearance. *(1 mark)*

b) Describe Lily's appearance. *(1 mark)*

c) In what ways are Ivy and Lauren different? *(1 mark)*

d) Some characteristics used in this classification would not be used by a scientist, either because they are not easily observed or because they do not stay the same as time passes. What characteristics are they? Explain why they would not be used. *(1 mark)*

20. Tess has made a teddy bear in class. It has been completed except for the eyes.

The pattern says that the bear requires 6-mm eyes. She wants to give the bear black, shoe-button leather eyes.

a) Work out how much the eyes will cost using the following key. *(1 mark)*

b) Use the key to work out the most expensive type of teddy bear eyes. *(1 mark)*

1A Plastic eyes .... go to 2
1B Non-plastic eyes (glass or leather) .... go to 3
2A Safety eyes (black) .... go to 4
2B Safety eyes (coloured)............ go to 5
3A Glass .... go to 6
3B Leather .... go to 7
4A Size (2–12 mm) ...... cost $3.00 per pair
4B Size (13–24 mm) .... cost $4.00 per pair
5A Size (2–12 mm) ...... cost $3.30 per pair
5B Size (13–24 mm) ..... cost $4.60 per pair

(cont.)

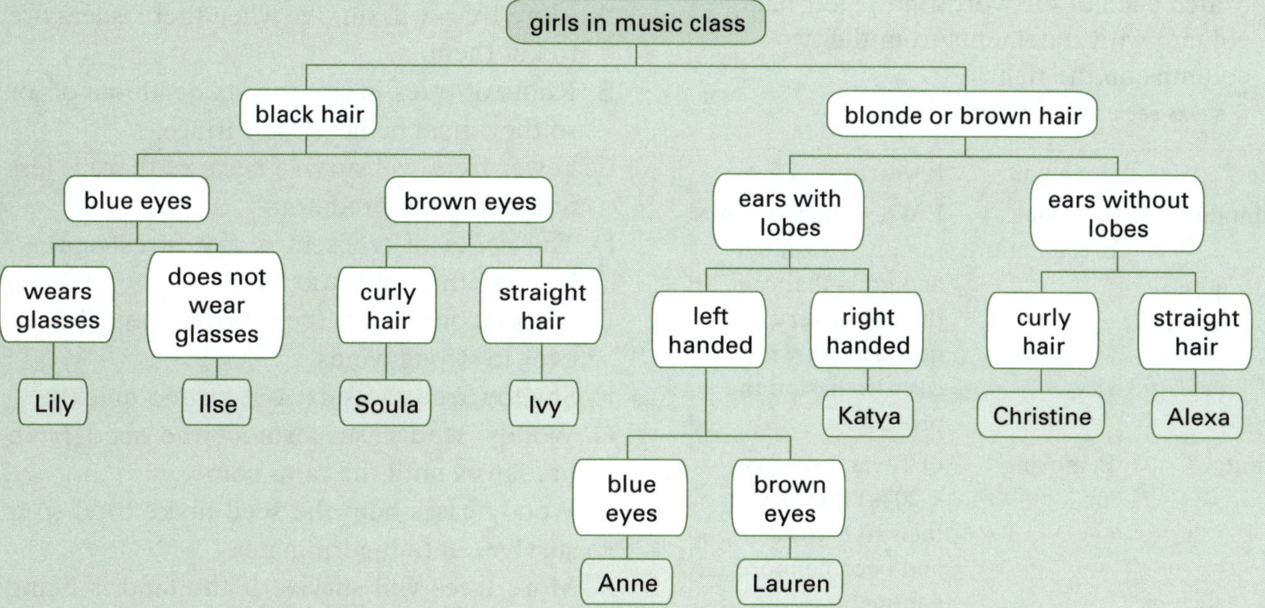

Figure 1.33 Classification key for girls in a music class

6A Safety eyes (black) .... go to 8
6B Safety eyes (coloured) .... go to 9
7A Shoe button (black) .... go to 10
7B Boot button (black) .... go to 11
8A Size (2–14 mm) .... cost $3.70 per pair
8B Size (15–26 mm) ..... cost $4.20 per pair
9A Size (5–10 mm) ...... cost $2.00 per pair
9B Size (11–20 mm) ..... cost $3.00 per pair
10A Size (1–6 mm) .... cost $4.40 per pair
10B Size (7–12 mm) ..... cost $6.20 per pair
11A Size (2–6 mm) .... cost $3.40 per pair
11B Size (7–10 mm) ..... cost $4.70 per pair

21. Why is vertebral column a better word to use than backbone when describing vertebrates? *(1 mark)*

22. State two essential features of plants that places them in a separate kingdom to bacteria, protists and fungi. *(2 marks)*

23. Identify the class and order to which humans belong. Identify two other vertebrates that belong to this order. *(4 marks)*

24. The lemon-scented gum is called *Eucalyptus citriodora*.

a) Identify the classification levels demonstrated by this name. *(2 marks)*

b) Two other plants are called *Backhousia citriodora* and *Eucalyptus alba*. Which of these plants is more closely related to *Eucalyptus citriodora*? Explain. *(2 marks)*

25. Match each of the words in the left-hand column with statements from the two columns on the right.
(2 marks each; total 8 marks)

Word	Definition	How it affects fires
humidity	A. liquid precipitation	P. When high, it can rapidly dry timber and grass that can then burn very quickly. Hot winds can also hasten the process.
wind	B. moving air	Q. If very low (say < 20%) it causes fuels to dry out and become more flammable.
rainfall	C. degree of hotness or coldness of a body or environment	R. Moisture in twigs, leaves and even logs can make burning more difficult.
temperature	D. a measure of the moisture in the air	S. It provides oxygen needed for a fire to burn. Greater speed increases oxygen and the flame intensity. It also carries burning embers downwind, spotting (starting) new fires.

26. In North and South America the large cat *Felis concolor* is known as a panther, cougar, mountain lion, puma or león. Why is the scientific name better for communication between scientists? *(1 mark)*

Read the following text. Then select from the list of statements which one can be used to answer Questions 27 to 30. In each case provide a simple explanation of your choice.

Many trees in dry forests have roots that are shallow and spread widely. These trees tend to produce lots of seeds with woody cases. Many of these seeds also need fire to break them open and germinate.

Statement list

A. More trees will survive when there is less fire to kill them.
B. Roots of trees in dry forests need lots of air so they must be near the surface.
C. Fewer trees will survive because there is less fire to kill any predators.
D. The roots of trees in a dry forest spread because the soil is too hard.
E. Without deep roots the trees are easily blown over in strong winds.
F. Shallow tree roots dry out far too quickly.
G. Woody seed cases protect the seed from predators until the rains come.
H. Woody cases help the seed make food so it survives until it germinates.
I. More trees will survive if the land is being correctly managed.

J. Woody cases help to protect the seed from fire.

K. Shallow roots are easily eaten by burrowing animals.

L. Shallow roots allow the trees to photosynthesise faster as they can absorb more sunlight.

M. Trees will survive as the hard seed casings will not open when the rains come.

N. In a dry forest there is more moisture in the surface soils for the roots to absorb.

O. Woody cases around the seed assist the absorption of water into the seed.

P. Shallow tree roots snap easily if disturbed.

27. What is an advantage of the type of root system described in the text? *(2 marks)*

28. What is a disadvantage of this type of root system? *(2 marks)*

29. What is an advantage of the woody case around the seed? *(2 marks)*

30. With land management, the threat of fire has been reduced. How has this affected the survival of these trees? *(2 marks)*

31. Organise the following into a simple food chain:

a) mouse, grass, hawk, snake *(1 mark)*

b) tiger snake, long-necked tortoise, algae, tadpole *(1 mark)*

c) snail, cabbage leaf, cat, blue-tongue lizard. *(1 mark)*

32. Consider this food chain:

phytoplankton ➡ small fish ➡ seal ➡ killer whale

Name a:

a) carnivore *(1 mark)*

b) herbivore *(1 mark)*

c) first-order consumer *(1 mark)*

d) producer. *(1 mark)*

33. Why are there more herbivores than carnivores? *(2 marks)*

34. What do decomposers do? Why are they important? *(2 marks)*

35. a) What do you think is meant by the term biodiversity? *(1 mark)*

b) Account for why a forest has a large biodiversity compared to, say, a field of wheat or corn. *(2 marks)*

36. A number of non-government organisations such as Landcare and Greening Australia are carrying out revegetation programs. Local, state and Australian governments support them. Explain how these might be useful in reducing pressures on biodiversity. *(4 marks)*

37. How is fragmentation of forests a threat to biodiversity? *(4 marks)*

38. Identify and discuss some of the issues involved in the health and sustainability of the Murray-Darling Basin. *(4 marks)*

39. a) What are feral animals? *(1 mark)*

b) Why was the rabbit so successful in colonising much of Australia? *(1 mark)*

c) Which parts of Australia has the rabbit not colonised? *(1 mark)*

d) Why do you think conventional control measures before the 1950s was not very successful? *(1 mark)*

e) The myxoma virus was tested to be species specific before it was released into wild rabbit populations. What does species specific mean, and why was this important? *(2 marks)*

40. a) Was the cane toad species specific when introduced into Queensland? *(1 mark)*

b) How is the cane toad affecting native and other wildlife? *(2 marks)*

c) What are some precautions you need to take when catching cane toads? Will the bounty technique work in eliminating cane toads? *(2 marks)*

d) Should you destroy cane toads, or have an expert identify them first? Explain. *(1 mark)*

e) What do you think is meant by biological control? *(1 mark)*

Go to pp. 186–189 to check your answers.

CHAPTER 2

Separating mixtures

Overview

In this chapter you will learn about:

- the differences between mixtures and pure substances
- homogeneous and heterogeneous mixtures
- solutes, solvents and solutions
- concentrated and dilute solutions
- physical methods used to separate the components of suspensions and solutions
- the application of physical separation methods in the home and in industry.

Glossary

Atom—the smallest unit of an element

Centrifugation—fast separation of the components of a suspension by spinning them rapidly in a circle

Chemical reaction—substances rearrange their atoms and form new substances

Chromatography—separating the components of a solution due to different absorbencies on a solid material and differing solubilities of the solutes in various solvents

Distillation—a physical separation process in which a solution of different liquids is separated based on differences in their boiling points

Evaporation—a physical separation process used to separate a soluble solute from its solution by heating and vapourising the solvent; crystals of the solute form once all the solvent has evaporated

Filtration—a physical separation technique used to separate the solid components of a suspension from the fluid in which they are suspended; the particle size of the solid components are too great to fit through the pores of the filter

Froth flotation—separating minerals from unwanted rock using detergent froth to which the minerals stick

Heterogeneous mixtures—mixtures that are not uniform throughout

Homogeneous mixtures—mixtures that are uniform throughout

Mixture—a material composed of two or more different substances that are mixed together but not chemically combined

Pure substance—a material that cannot be separated into simpler materials by physical means

Solute—a substance which dissolves in a solvent (e.g. a liquid) to form a solution; solute particles are less in number than solvent particles

Solution—a mixture which forms when a solute dissolves in a solvent

Solvent—a substance (often a liquid) that dissolves the solute; solvent particles are in greater amounts than solute particles

Suspension—particles of a material which remained suspended in a fluid and do not quickly sediment under gravity

Transparent—having the property of allowing light rays through without the light rays being scattered

Volatile—evaporating very rapidly; readily forms a vapour

2.1 Pure substances and mixtures

We live in a world of mixtures. Everywhere you look you see examples of objects or substances mixed together. For example:

- a bowl of fruit may contain a mixture of apples, oranges and bananas
- methylated spirits is a mixture of ethanol and a little methanol
- the air we breathe is a mixture of nitrogen, oxygen, argon, water vapour, carbon dioxide, dust particles and many other materials.

A mixture is a material composed of two or more different substances that are mixed together but not chemically combined. Mixtures have the following characteristics.

- Each component of the mixture retains its own unique properties. For example, if powdered iron and powdered sulfur are ground together, they form a fine powder. The iron component still retains its magnetic properties and the sulfur its non-magnetic properties. This can be shown by holding a bar magnet on top of the mixture. The iron particles leave the mixture and adhere to the magnet, as shown in Figure 2.1.

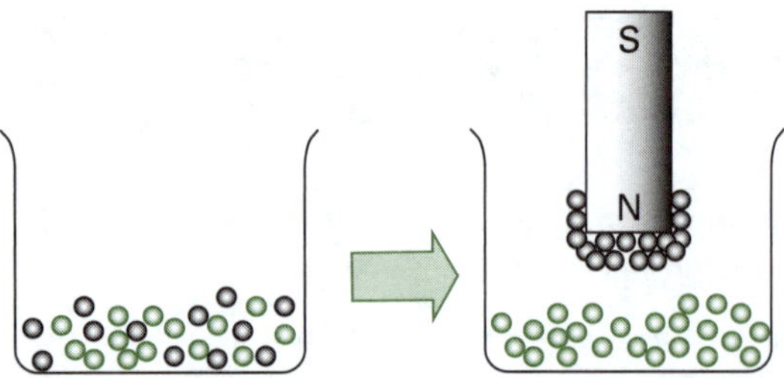

Figure 2.1 Mixture of iron powder and sulfur powder

- Mixtures can have variable compositions. Brass is a mixture of zinc and copper. The percentages of zinc and copper vary somewhat and produce brasses with different properties. Thus a mixture of 15% zinc and 85% copper is called red brass, and is used in cheap jewellery that is to be gold plated. A mixture of 40% zinc and 60% copper called Muntz metal is much harder and used for nuts and bolts.
- The components of a mixture can be separated by various physical methods. For example, cooked spaghetti in a pot of hot water is a mixture. The spaghetti can be easily separated from the water by using a colander.

Physical methods of separation do not alter the chemical nature of a material or its components. Physical separation methods include filtration, evaporation, distillation and magnetic separation. We will examine these methods in detail in a later section of this chapter.

Not all matter can be separated by these physical methods. These forms of matter are called pure substances. Pure substances have the following characteristics.

- Pure substances have a constant composition throughout the sample. A sample of pure gold has identical gold atoms arranged throughout the crystalline sample.
- The properties of the pure substance are constant throughout the sample. The melting point of one section of a pure gold sample is the same as the melting point of any other region of the gold sample.
- Pure substances cannot be separated by physical separation techniques. A sample of pure gold cannot be separated into anything simpler by any physical process such as filtration, evaporation or distillation.

Figure 2.2 illustrates the concept of mixtures and pure substances using particle models. Mixtures consist of a variety of different particles that are not bound together and which have a variable composition throughout the sample. There are two types of pure substances. In the first type the particles are very simple and cannot be broken down into anything simpler. In the second type the particles are composed of different types of simple particles but in a fixed proportion. The simple particles are joined strongly to each other and cannot be easily separated.

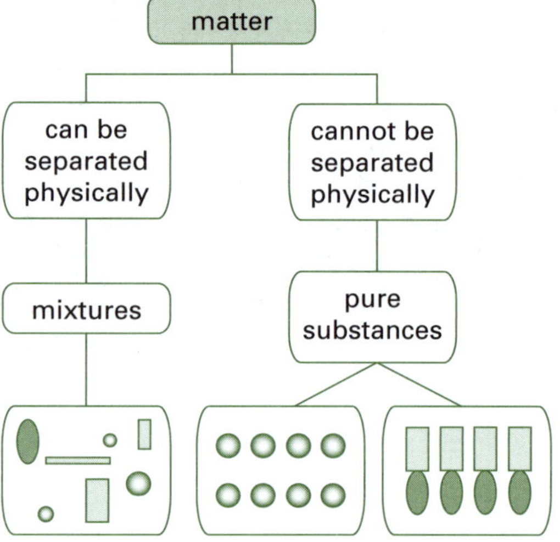

Figure 2.2 Mixtures and pure substances

Example: mixtures and pure substances

Mixtures can be further divided into two groups.

- Homogeneous mixtures are mixtures that are uniform throughout.
- Heterogeneous mixtures are mixtures that are not uniform throughout.

As shown in Figure 2.3, in a homogeneous mixture the particles of each substance are uniformly mixed. There is only one state or phase of matter present. Thus, salt water is a homogeneous mixture which has a constant composition throughout. In a heterogenous mixture there is no uniformity and two or three different states of matter may be present. Thus chalk powder suspended in water is a heterogeneous mixture.

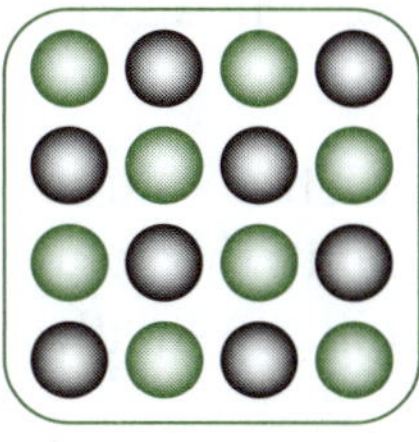

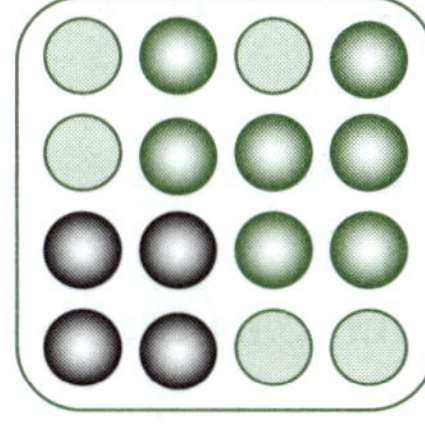

Figure 2.3 Homogeneous and heterogeneous mixtures

Table 2.1 lists some common examples of pure substances as well as homogeneous and heterogeneous mixtures.

Table 2.1 Pure substances and mixtures

Pure substances	Homogeneous mixtures	Heterogeneous mixtures
gold	salt water	milk
silver	soft drink	smoke
mercury	petrol	foam
oxygen	brass	fog
nitrogen	bronze	paint
water	white gold	emulsions
carbon dioxide	wine	blood

Silver and water are pure substances as each has a constant composition (see Figure 2.4). In the case of silver, all the particles are identical silver atoms. In the case of water, all the particles are water molecules.

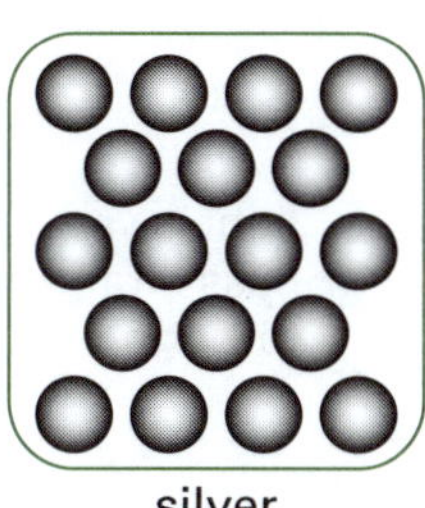

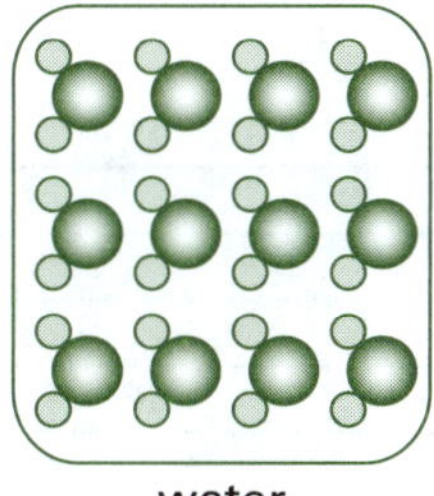

Figure 2.4 Simplified particle models of silver and water

Experiment 1

Homogeneous and heterogeneous mixtures

Aim

- To investigate different types of mixtures

Method

Part A: Mixing in a mortar

1. Select two pieces of chalk of contrasting colours (such as dark blue and white). Break small pieces of each colour into a mortar and start to grind the chalk into a powder using the pestle.
2. After 30 seconds, remove a small sample using a spatula and spread evenly on a glass slide or in a glass Petri dish. Label this as sample 1.
3. Continue grinding and mixing. Remove samples each minute for 5 minutes and place them in separate dishes as before. Label each one as sample 2, sample 3 and so on up to sample 6.
4. Use a magnifying glass or stereomicroscope to examine each sample removed from the mortar and see if you can distinguish any blue and white grains.

Part B: Examining rocks under a stereomicroscope

1. Collect samples of granite and shale.
2. Set up a stereomicroscope and lamp so that each specimen of rock is well lit when placed on the microscope stage.
3. Examine under high magnification. Adjust the focus as required.
4. Examine each rock in turn and look for different grains or crystals.

Part C: Solutions

1. Place a little copper sulfate solution in a Petri dish. Place the dish on the stage of a stereomicroscope.
2. Observe the solution. Is it heterogeneous or homogeneous?

Results

Part A: As the grinding continues, it becomes harder and harder to see individual pieces of the two chalk colours.

Part B: The granite shows distinct crystals of different colours using the naked eye as well as under the microscope. The shale has very fine particles to the naked eye. Tiny grains can be seen under high power.

Part C: The solution is a uniform colour under the microscope.

Analysis

Analyse these results. Go to p. 190 to check your answer.

Conclusion

Write a suitable conclusion.

Go to p. 190 to check your answer.

2.2 Solvents, solutes and solutions

When a teaspoon of sugar is added to a glass of water and the mixture is stirred, the sugar dissolves to produce a sweet-tasting, colourless, clear liquid. The solid grains of sugar have disappeared and become part of the liquid. This is confirmed by examining the liquid under very high magnification. This homogeneous mixture is called a solution. In this example, the solution is made by mixing the solute (the sugar crystals) with the solvent (the liquid water). The sugar is said to be soluble in the water. Figure 2.5 shows a simple particle model of a solution.

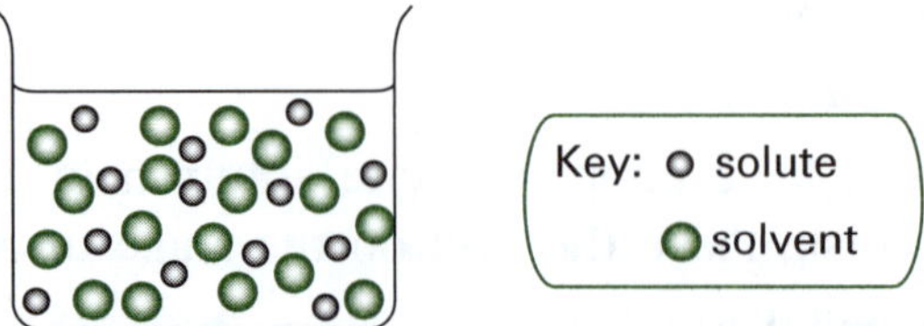

Figure 2.5 Particle model of a solution

Water is such a common liquid that it acts as a solvent for many solids. Sea water contains many different solutes, with salt being the most common.

Waxy and greasy solids do not dissolve in water. They can be dissolved to produce a solution using other liquid solvents such as kerosene, petrol or turpentine. Oils commonly dissolve in other similar oils. Thus safflower oil will dissolve in canola oil.

The solvent is always the part of the solution that is present in the largest amount; the solute is always that part that is present in the smallest amount.

Solutions can be classified into various types:

- solid–liquid solutions (e.g. salt dissolved in water)
- liquid–liquid solutions (e.g. lavender oil dissolved in alcohol)
- gas–liquid solutions (e.g. carbon dioxide dissolved in water forms soda water)
- gas–gas solution (e.g. oxygen dissolved in nitrogen)
- solid–solid solution (e.g. copper dissolved in silver forms sterling silver—to prepare the solution the silver must first be melted, the copper added and then the mixture stirred until it dissolves; it is then cooled until it solidifies).

Solutions of gases in gases, liquids in liquids and solids in liquids are transparent. This means that rays of light will pass through the solution without being deflected or scattered, as shown in Figure 2.6. This is evidence that the solute particles in the solution are extremely small. Although these solutions are transparent or clear, they may be coloured.

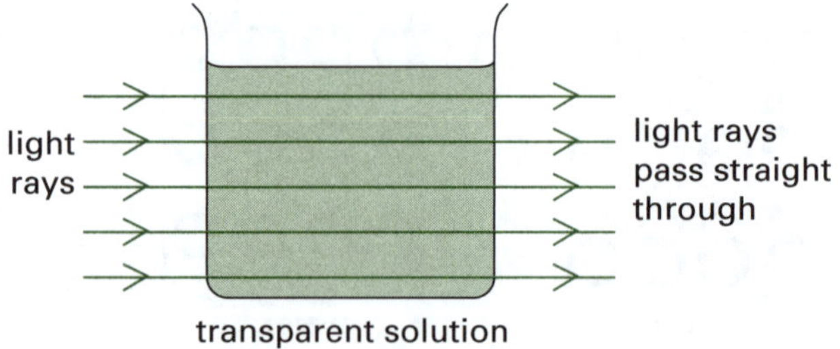

Figure 2.6 Solutions formed by solids dissolving in liquids or liquids dissolving in liquids are transparent.

Concentrated and dilute solutions

The strength or concentration of a solution refers to the amount of solute present in a fixed amount of the solution, as shown in Figure 2.7. When you make 100 millilitres of salty water, you can dissolve only a few crystals of salt or a scoop of salt. The solution that contains very little solute and a great amount of solvent is said to be dilute.

When the amount of solute increases, the solution becomes concentrated. Adding more solvent can dilute concentrated solutions. If a solute is brightly coloured, it is easy to distinguish dilute solutions from concentrated solutions, as the concentrated solution is a deeper colour.

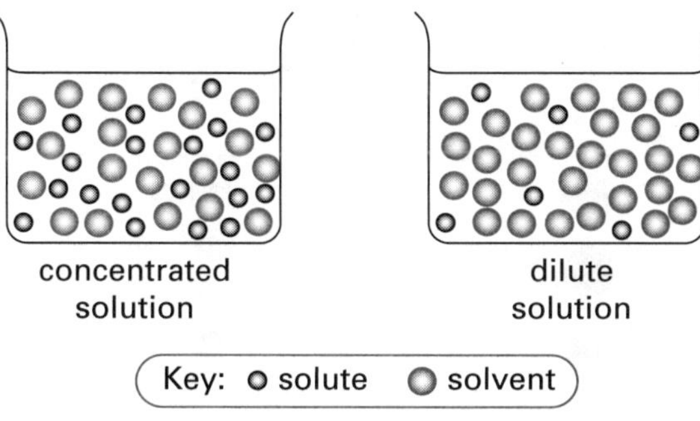

Figure 2.7 Concentrated and dilute solutions

Experiment 2

Testing different solvents

Aim

- To determine which substances are soluble in different solvents

Method

- Solvents to test: water, kerosene
- Substances to test: salt, olive oil, wax shavings, baking soda (sodium hydrogen carbonate), paraffin oil

1. Use water as your first solvent. Measure 5 mL of water into five labelled test tubes.
2. Add a small amount of each test substance (about the size of a rice grain or five drops if a liquid) to each tube. Stopper the tube and shake for 1 minute. Check to see if the substance has dissolved.
3. Repeat steps 1 and 2 for kerosene. The test tubes must not have any water in them for the remaining experiments.
4. Examine your collected data.

Results

Salt and baking soda dissolve in water but olive oil, paraffin oil and wax do not.

Olive oil, paraffin oil and wax dissolve in kerosene but they do not dissolve in water.

Conclusion

Use the results to write a suitable conclusion for this experiment.

Use the results to write a suitable conclusion for this experiment.

Go to p. 190 to check your answer.

Test yourself 1

Part A: Knowledge

1. What is methylated spirits an example of? *(1 mark)*
   A pure substance
   B heterogeneous mixture
   C solute
   D solution
2. What is carbon dioxide an example of? *(1 mark)*
   A pure substance
   B homogeneous mixture
   C heterogeneous mixture
   D solute
3. What do dilute solutions contain? *(1 mark)*
   A a large amount of solute dissolved in a small amount of solvent
   B a small amount of solute dissolved in a large amount of solvent
   C solute and solvent mixed in equal proportions
   D solvent and solution mixed in equal proportions

4. What is milk classified as? *(1 mark)*
   A pure substance
   B homogeneous mixture
   C heterogeneous mixture
   D solution

5. Mixtures *(1 mark)*
   A can be separated physically.
   B cannot be separated physically.
   C always contain at least three different components.
   D always allow light to pass through them without scattering.

6. Complete the following restricted-response questions using the appropriate word. *(1 mark for each part)*
   a) Pure substances have a .... composition throughout the sample.
   b) The melting point of one section of a pure platinum sample is the .... as the melting point of any other region of the platinum sample.
   c) Mixtures consist of a variety of different particles that are not bound together and which have a .... composition throughout the sample.
   d) Water is an example of a .... substance.
   e) Iron can be separated from sulfur using a ....

7. Use code letters to match the terms or phrases in each column. *(1 mark for each part)*

Column 1	Column 2
A air	F pure substance
B oxygen	G sulfur
C zinc and copper	H filtration
D physical process	I mixture
E non-magnetic	J brass

Part B: Skills

8. Can tap water be classified as a mixture? Explain. *(2 marks)*

9. Classify the substances A to F shown in Figure 2.8 as mixtures or pure substances. *(2 marks)*

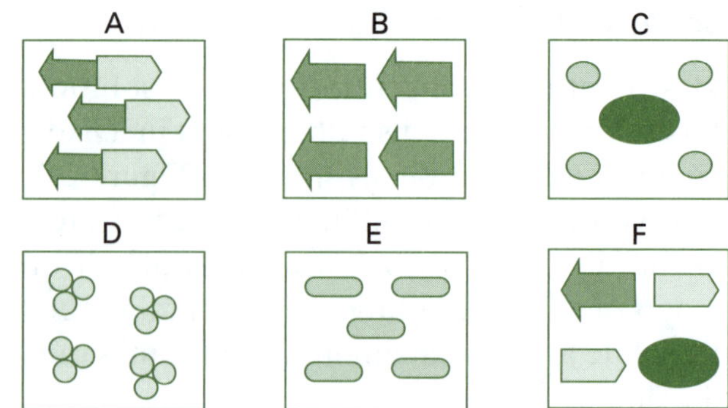

Figure 2.8 Mixtures and pure substances

Questions 10 and 11 refer to the following figure.

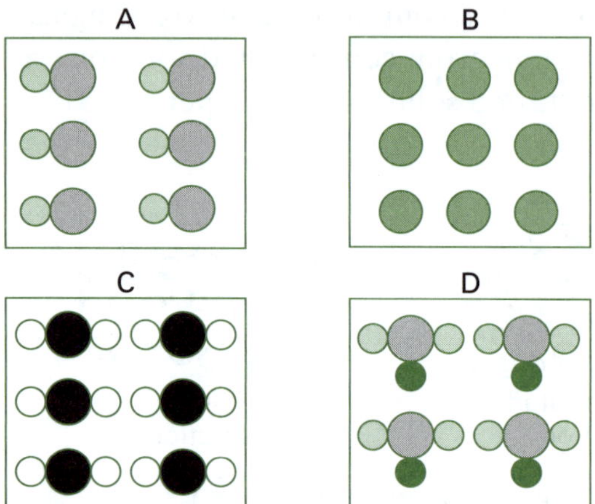

Figure 2.9 Particle diagrams

10. Carbon dioxide is a pure substance that is made of particles containing one carbon atom and two oxygen atoms. Identify which particle model is carbon dioxide in Figure 2.9. Explain your reasoning. *(2 marks)*

11. Examine the particle diagrams in Figure 2.9. Which ones are pure substances? Explain your answer. *(2 marks)*

12. Two salt solutions were prepared. Solution X was made by dissolving 5 g of salt in 100 mL of water at 20 °C. Solution Y was made by dissolving 2 g of salt in 20 mL of water at the same temperature. Determine which salt solution is more concentrated. *(2 marks)*

13. A solid–liquid solution such as sugar crystals dissolved in water is quite common. The sugar is the solid component and the water is the liquid component. At room

temperature, mercury can be dissolved in silver and copper to form a mixture called an amalgam. This substance can be used as a tooth filling.

a) What state is the solute in the amalgam solution? *(1 mark)*

b) What state is the solvent in the amalgam solution? *(1 mark)*

c) Classify the type of solution formed in this example. *(1 mark)*

14. In a two-stroke mower, oil is mixed with the unleaded petrol to create a suitable two-stroke fuel. Name the following.

a) solute *(1 mark)*

b) solvent *(1 mark)*

c) solution *(1 mark)*

Go to p. 190 to check your answers.

2.3 Physical separation techniques

Mixtures are made up of combinations of different substances not joined together by any chemical means. This means they can be separated using a variety of physical separation techniques.

We will examine two types of mixtures that can be separated:

- suspensions
- solutions.

Separating suspensions

A suspension is a heterogeneous mixture in which one substance does not dissolve in another but produces a cloudy mixture in which the different substances slowly separate. The cloudiness of the suspension shows that light rays are being scattered by the large solute particles, as shown in Figure 2.10.

If an insoluble solid is added to a liquid, and the mixture is left to stand, the particles of the solid settle to the bottom of the vessel. Large, heavy

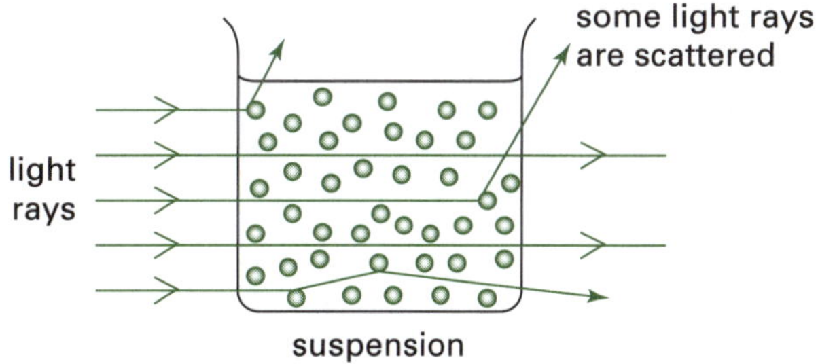

Figure 2.10 Suspensions scatter light rays and look cloudy.

particles will settle faster than smaller, lighter particles. The solid that collects on the bottom of the vessel is called the sediment.

There are two common and simple ways to separate the components of a suspension.

Decanting

Decanters have been used for centuries to serve wines that contain sediments in the original bottle or wine barrel (see Figure 2.11). These sediments could be due to age, or a wine that was not filtered or clarified during the winemaking process. This is rarely a problem for most modern wines since many wines no longer produce significant amounts of sediment as they age.

Figure 2.11 Transferring wine into a decanter

If the particles in a suspension are heavy, they will settle rapidly under gravity to form a sediment on the bottom of a container such as a beaker. Decanting involves slowly tipping off most of the liquid into another container while

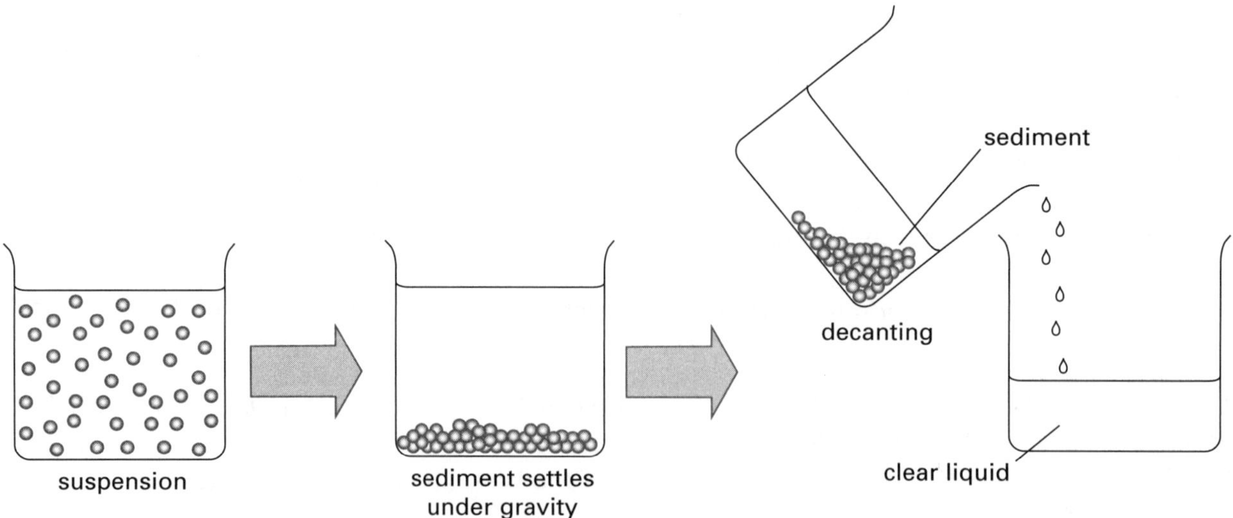

Figure 2.12 Decanting

ensuring that none of the sediment is carried over, as shown in Figure 2.12. This procedure is often done before filtration to remove most of the liquid so the filtration step is faster.

Filtration

In the kitchen, a colander is used to separate spaghetti from the water it was boiled in. Similarly, a strainer can be used to remove tea leaves when pouring tea from a kettle. Modern tea bags allow liquid to pass through while holding back solid tea leaves. In a car, oil filters clean the oil which circulates around the engine. Petrol filters remove solid impurities from fuel. Air filters remove dust and other solid suspensions in air before it goes into the engine. Home vacuum cleaners remove dust from the carpet using filters.

In the laboratory, filter paper is used for filtering suspensions. There are various grades of filter paper depending on the size of the particles to be filtered. The filter paper is folded to fit the filter funnel, making sure it is not torn in the process. The suspension is first allowed to settle and then most of the clear liquid is decanted. Then the sediment and remaining liquid are carefully poured into the filter. The liquid that passes through the filter paper is called the filtrate. The solid material that remains behind in the filter paper is the residue. This process is shown in Figure 2.13.

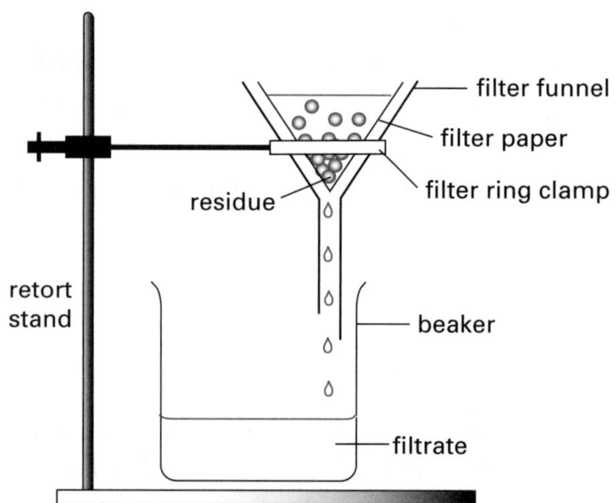

Figure 2.13 Filtration

Filtration will separate the insoluble solids from the liquid as long as the particle size of the solids is greater than the size of the pores in the filter paper.

Experiment 3

Separating chalk from copper sulfate

Aim

- To use filtration to separate a mixture of chalk and copper sulfate crystals

Method

1. Grind up some chalk in a mortar until it is a fine powder. Add a scoop of copper sulfate crystals and mix thoroughly.
2. Place this mixed powder into a 250-mL beaker.
3. Add water to the beaker until it is about one-third full. Stir with a glass rod for 3 to 4 minutes. Allow the chalk to sediment.
4. Fold a filter paper into a cone and set up the filtration apparatus.
5. Decant the mixture into a folded filter paper in the filter funnel. Use the stirring rod as a guide for pouring.
6. Collect the filtrate in another clean beaker.
7. Allow the chalk residue to fall into the funnel last. Use a very small amount of water from a wash bottle, if necessary, to rinse out any remaining chalk.

Results

The filtrate is a clear blue. White chalk is collected as a residue in the filter paper.

Analysis

Analyse your results. Go to p. 190 to check your answer.

Conclusion

Write a conclusion for this experiment.

Go to p. 190 to check your answer.

Separating solutions

There are three common ways to separate the components of a solution.

Evaporation and crystallisation

Evaporation is a method used to separate a solute from a solution. The solvent escapes into the environment since it becomes more volatile as it is heated. The solute is not volatile. Evaporation occurs in nature. A rock pool near the ocean fills with water, and while the Sun is shining the salt water is heated and water evaporates from the pool. The concentration of salt increases to a point where the solution is saturated in salt. A saturated solution contains the maximum amount of dissolved solute. As more water evaporates, various mineral salts form as crystals. The slower the evaporation process, the larger the crystals of solute that will form. The process of forming crystals is called crystallisation.

Evaporation can be demonstrated in a school laboratory using an evaporating basin, salt water and a source of heat such as a Bunsen burner. Figure 2.14 shows the apparatus that is used.

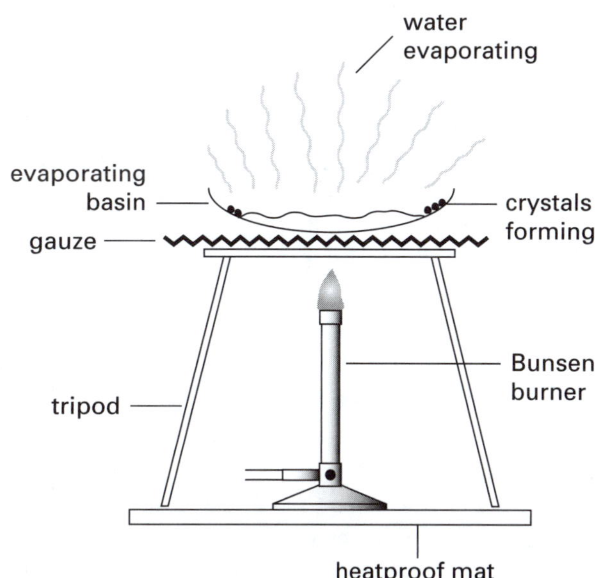

Figure 2.14 Evaporation and crystallisation

Distillation

Distillation is a method of separating a liquid (usually the solvent) from a solution based on differences between their boiling points. As the mixture boils, the substance with the lowest boiling point boils off first. A thermometer is used to check the temperature.

The vapour passes down the condenser where it is cooled. The condenser has an outside jacket where cold water can circulate. Near the end of the condenser, the vapour has been turned back into a liquid where it can be collected into a beaker or flask. The substance that collects in the beaker or flask is called the distillate.

Figure 2.15 shows the apparatus used in the school laboratory to carry out a simple distillation.

For simple distillation, the substance to be separated has to have a widely different boiling point. If salt water is distilled, then pure distilled water is collected as the distillate. The salt remains behind in the boiling flask as the boiling point of the salt is so high that it does not boil.

Simple distillation may also be used to partly separate alcohol from water. Alcohol (ethanol) boils at 78.5 °C, while water boils at 100 °C. If the solution is heated to 79 °C, and the temperature kept there, most of the alcohol will boil off and be collected. Of course, some water will evaporate and be collected too, but the amount of alcohol in the distillate will be greatly increased.

Distillation is the main method for purifying liquids. It is used in a large variety of industries such as producing perfume, alcoholic drinks and cooking oils. For example, distillation is used in making brandy and whisky; in brandy the alcohol concentration is increased to about 70%.

Chromatography

Chromatography is a method of separating solutes in a solution when the solutes have different absorbencies and different solubilities in various solvents.

In column chromatography, the solution to be separated is poured onto the top of a column of absorbent powdered solid, followed by more solvent. The different components in the sample mixture move through the column at different

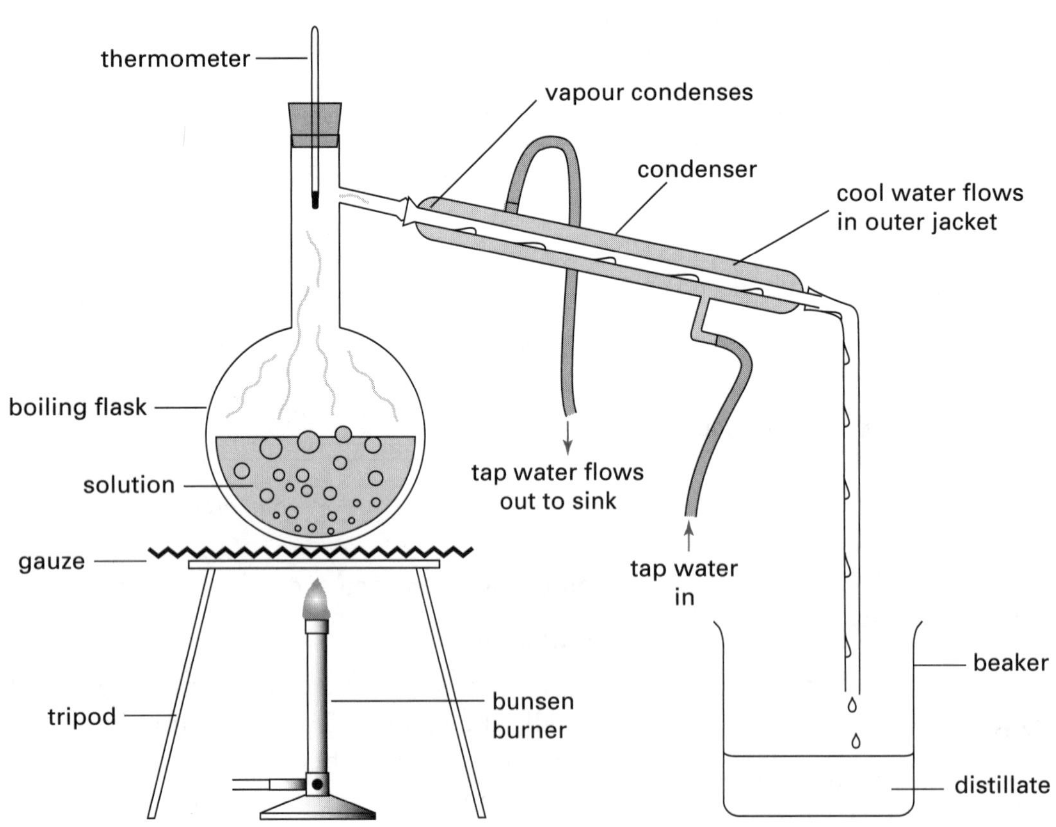

Figure 2.15 Distillation

speeds as they are absorbed to different extents by the material of the column. If the components are coloured they appear as bands of different colours. As the components come out the other end they can be collected separately and then analysed. Figure 2.16 shows an example of column chromatography. The more soluble and less absorbent the component of the mixture, the faster it moves through the column.

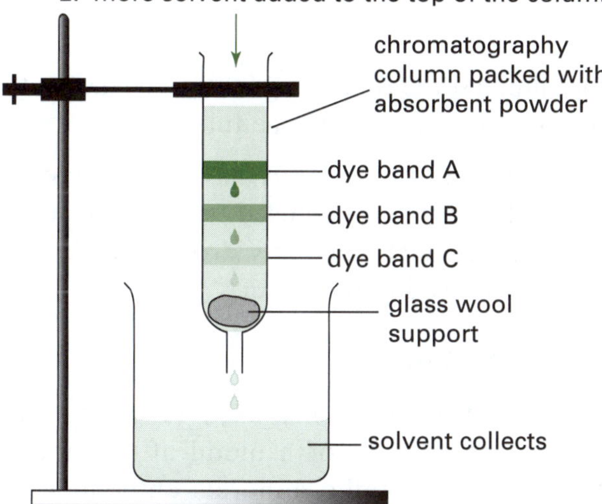

Figure 2.16 Column chromatography

Paper chromatography can be used to separate the colours in ink, or to see which different coloured food dyes are present in various lollies. In the case of lollies with dyed sugar coatings, a wet paint brush can be used to remove some food dye from each coloured lolly. The dye can then be applied from each of these extracts to different positions along a base line on the chromatography paper. This method is very useful when only small quantities of the solution are available. Figure 2.17 shows an example of paper chromatography. In this case, the dyes are mixed together in a solvent. The mixture of dyes is applied along a line near the base of the paper and allowed to dry. The paper is then placed in a beaker with a small amount of solvent which is absorbed by the paper and moves up the paper. The different solubilities of the dyes in the solvent and their different absorbencies on the paper cause separation into dye bands.

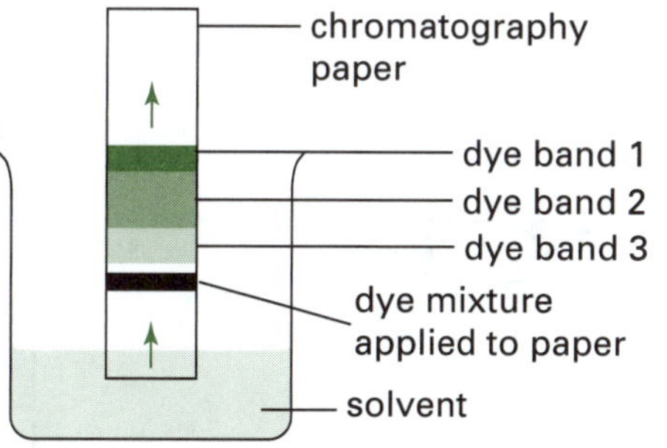

Figure 2.17 Paper chromatography

2.4 Practical applications of separation techniques

The separation methods discussed in the previous section are frequently used in industry and the home. In this section, we will examine some of these practical applications.

Centrifugation

Have you ever been on a fast spinning merry-go-round or other ride at a funfair and felt like you were being thrown to one side? This idea is used in a centrifuge. A centrifuge is a machine that uses a fast spinning motion to separate suspensions. The rapid spinning motion increases the rate at which solid particles undergo sedimentation (see Figure 2.18) or emulsions separate into separate layers.

Separating blood components

In order to separate blood components such as red blood cells, a sample is placed in a centrifuge and spun. The heavier red blood cells settle to the bottom of the tube, leaving a straw-coloured liquid (called plasma) above. The white blood cells and platelets are in a layer just above the red blood cells, as shown in Figure 2.19. Platelets are important in blood clotting and are used by hospitals for patients who suffer haemophilia (a genetic disease in which people bleed internally after even small knocks to the body). The Australian Red Cross collects blood from donors and prepares blood components for distribution to hospitals. Plasma is important in rehydrating

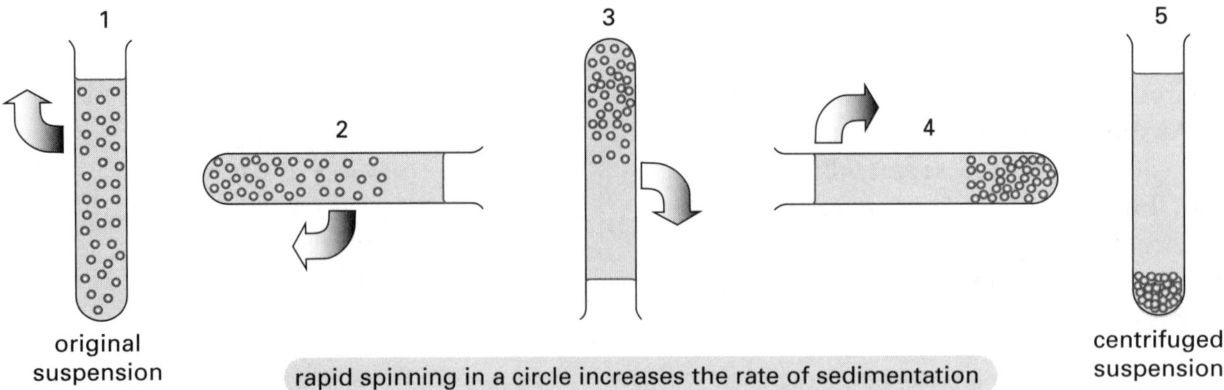

Figure 2.18 Centrifugation

burn victims. It also contains antibodies which help to prevent infection.

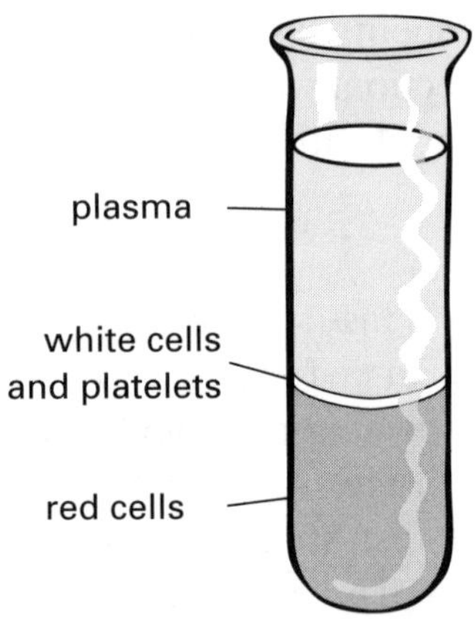

Figure 2.19 Centrifuged blood

Separating honey

Centrifugation is used to separate honey from the honeycomb. First the caps on the cells in the honeycomb are removed and the honeycomb frames are put inside a barrel and spun. The honey flies out to the sides of the barrel and gradually settles to the bottom.

Separating cream from milk

Milk is a dispersion or emulsion of butter fat droplets in water. Fresh milk will separate if left to stand. The fatty component called cream floats to the top as it is less dense than water.

The cream can be separated more rapidly from whole milk by centrifuging. When whole milk is spun in a centrifuge, the less dense cream moves to the top and the heavier watery layer (containing some remaining butter fat) moves to the bottom of the centrifuge tube.

The milk you normally buy has had the cream removed and is homogenised to create a stable emulsion in which very tiny droplets of butter fat are dispersed in the water.

Froth flotation

Froth flotation is mainly used to purify and concentrate the ore of a metal after mining. Ores contain metallic minerals embedded in unwanted rocky material.

The finely crushed ore is added to water to create a suspension. Frothing agents are added to the water. These frothing agents are different types of oils that wet the ore better than the other impurities. The other impurities (called gangue) may include clay, sand and rock.

When air is blown through this mixture, the mineral grains stick to the air bubbles. The froth carries the valuable minerals to the surface where it is suspended and can be skimmed off. This process is shown in Figure 2.20. The collected minerals are then sent to a smelter to extract the metal.

While versions of froth flotation were tried previously, modern froth flotation techniques were invented in the early 1900s in Australia to recover zinc minerals and later for separating zinc from lead minerals. It is now a valuable method of physical separation used widely around the world.

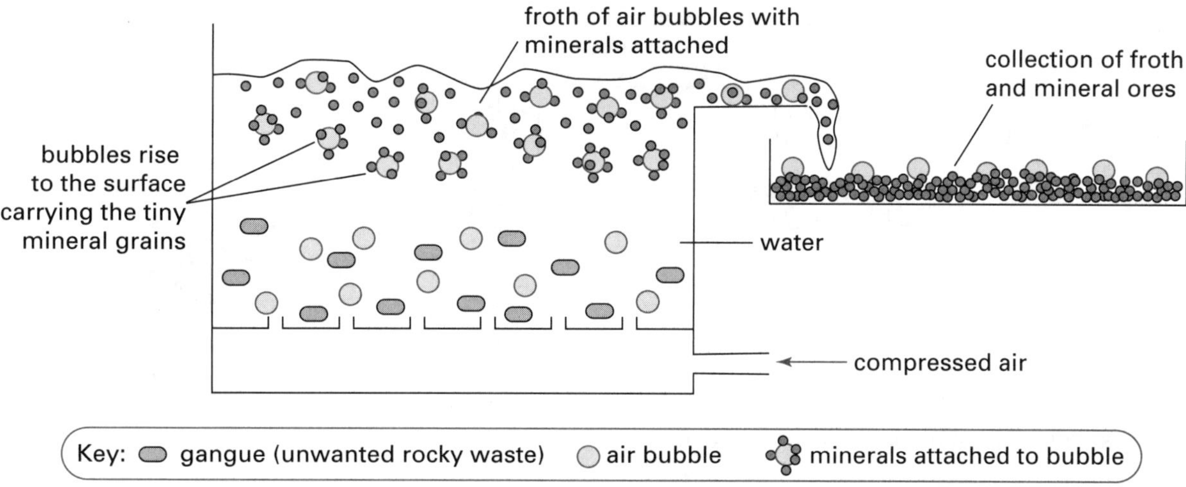

Figure 2.20 Froth flotation

Separation techniques in winemaking

The main steps in bulk winemaking are listed below but there are other variations of this method that can be used.

1. The grapes are crushed and pressed to release the juice which escapes from the press leaving the skins, stems and seeds behind. The juice is pumped to a holding tank.
2. The juice is chilled to promote sedimentation of small particles. These are then filtered off.
3. The juice is transferred to large vats. Yeast is added and air is excluded to allow the fermentation of the sugars in the grape juice to occur. This fermentation process converts the sugars to alcohol. This step lasts about 3 to 4 weeks.
4. The temperature of the fermented mixture is dropped to near freezing to crystallise out unwanted tartaric acid crystals and then gelatine is added to flocculate fine, colloidal tannin particles. In the process of flocculation, the very fine tannin particles stick onto the gelatine making much larger and heavier particles. These heavier particles then undergo sedimentation.
5. The sediment and yeast are filtered from the fermented mixture to produce a clear white wine.

Extracting natural products from plants

Plants contain many important chemicals that can be extracted and used as medications, perfumes and food products and as components of body creams and shampoos. Chemists have developed various methods to extract these natural products from the plant tissue. Some examples are discussed below.

Steam distillation of eucalyptus oil

Various plant oils such as eucalyptus oil can be extracted from their leaves by steam distillation. The leaves are shredded and steam is passed through the leaves in the still. A mixture of the volatile oil and steam then passes through the condenser where the steam is cooled to liquid water and the oil vapour turns into oil droplets. This mixture passes into the collection chamber where the insoluble oil separates from the water as it is less dense. The oil floats to the top and is drawn off. This process is shown in Figure 2.21.

Extraction of sugar from sugar cane

Figure 2.22 shows the separation steps involved in extracting raw cane sugar crystals from sugar cane. This separation process involves crushing, evaporation, crystallisation and centrifugation. Additional steps are required to obtain white table sugar.

Sorting waste and recycling

Most local councils in Australia provide householders with a garbage bin and a recycling bin. The recycling bin is used for paper, cardboard, metal, glass and certain plastic waste. The flow chart in Figure 2.23 shows the steps in sorting this recyclable waste at the collection

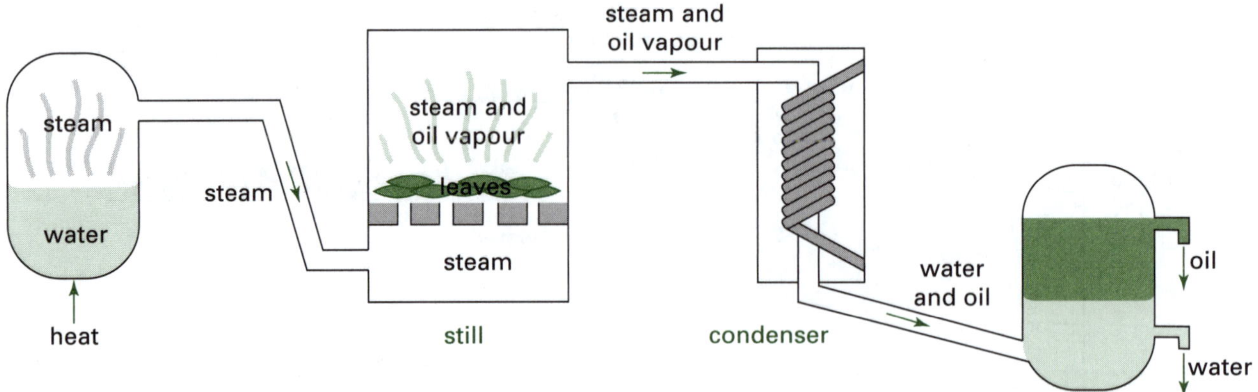

Figure 2.21 Steam distillation of plant oils from leaves

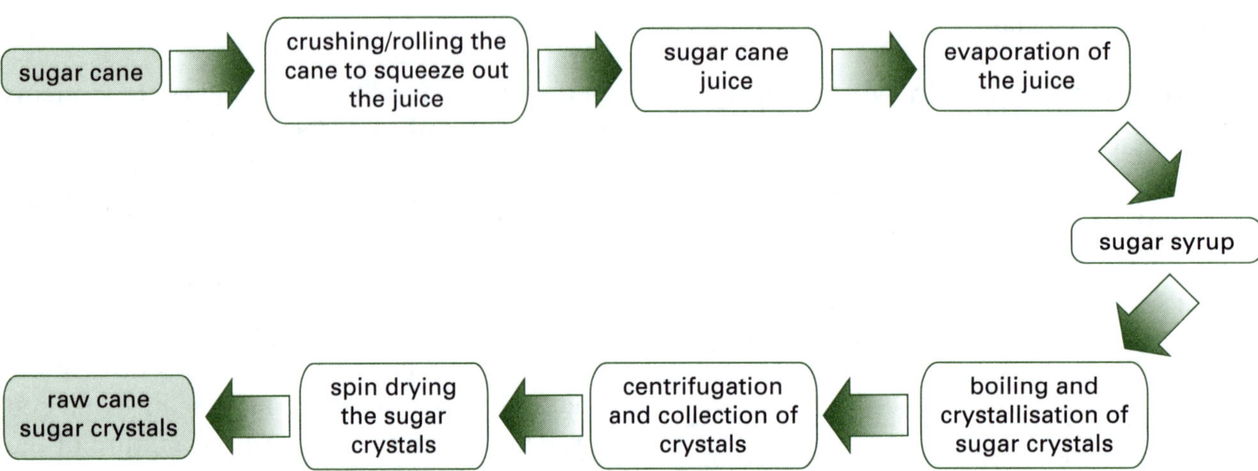

Figure 2.22 Extracting raw sugar crystals from sugar cane

and sorting centre. The process starts when the waste is emptied onto a conveyor belt. Magnetic separation is an important part of this waste sorting process.

Recycling of metals is important as metals are non-renewable resources and the energy used in recycling metals is much less than mining and processing new metals.

Reducing pollution

Separation techniques can be used to reduce pollution. Some of these methods will now be described.

Electrostatic precipitation of smoke particles

Coal-fired power plants are required by law to reduce smoke emissions from their chimneys. One way of doing this is to use electrostatic precipitation. The smoky emissions are passed through a pipe containing a wire that has a high negative charge. The smoke particles gain some of this negative charge as they move past the wire. Further along the pipe are a series of positively charged plates which attract the negatively charged smoke particles. The smoke particles stick to these plates. The emissions minus the smoke particles can then be emitted out of the chimney. The metal plates are periodically cleaned.

Water purifiers in the home

Some homes use a variety of specialised filters to further remove unwanted pollutants from their tap water. Such systems rely on a series of filters arranged in a cartridge that is connected to the kitchen tap. The steps involved in cleaning the water are as follows.

1. Tap water passes through a ceramic filter with a pore size less than one micron (one

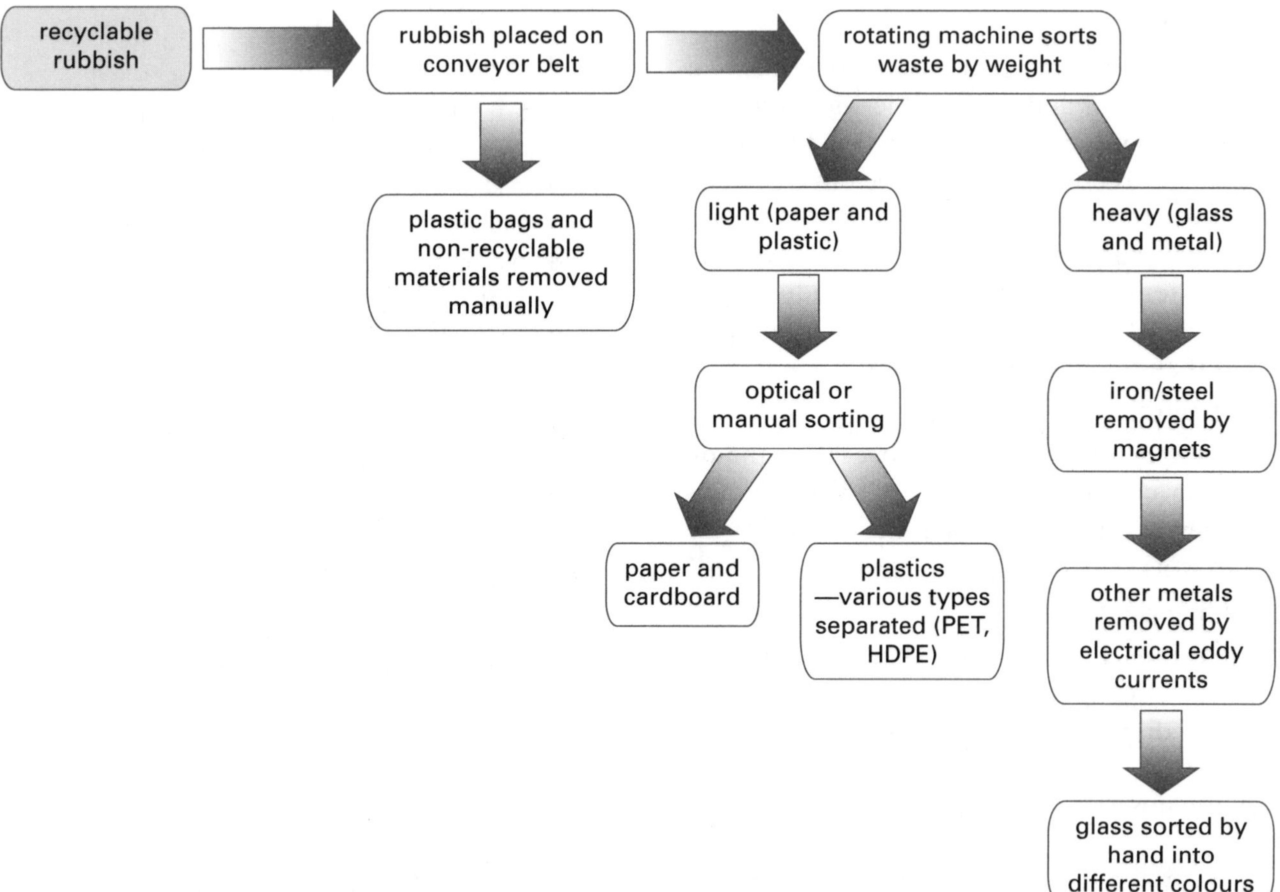

Figure 2.23 Sorting recyclable waste

thousandth of a millimetre). These fine pores will remove bacteria, cysts and fine colloidal matter.

2. The water from the first stage passes into a second ceramic filter containing colloidal silver particles that kill bacteria.
3. Finally, the water passes through a layer of granulated activated carbon that absorbs chlorine and organic pollutants.

These filtration systems are important for people who may be allergic to chlorine, which is used for disinfecting drinking water.

Cleaning up oil spills

Oil spills in the ocean and on coastlines are usually the result of supertankers carrying oil running aground or oil drilling rig accidents. Deliberate discharges of oily industrial wastes and discharges from ships as they clean their ballast tanks also contribute to oil pollution. Sea birds die in their thousands when their feathers become oiled. Fish and seals are also affected. Coral is also killed by oil. The Great Barrier Reef is a very delicate ecosystem that could be devastated by an oil slick.

Cleaning up oil spills is not an easy task. Oil is less dense than water, so it floats on top where winds and waves can move the oil great distances. The following methods have been adopted to clean up oil spills.

Containment and skimming

This method only works if the oil has not had time to spread too far. Long floating booms are used to surround the oil slick and then the contained oil is skimmed from the surface and stored in large containment tanks. Skimming can also involve the use of polyethylene mop-like pads that absorb surface oil.

Dispersants

Dispersants are detergent-based chemicals that mix with the oil and convert the slick into tiny droplets which can then mix with the water and be absorbed into the aquatic environment. Planes are often used to spray these dispersants on the oil slick. While this method works to some extent in the open ocean, studies have shown that this method is more devastating to corals than the untreated oil slick.

In-situ burning

Oil is flammable and on some occasions the oil has been contained with fire-resistant booms and then set alight. The sea must be calm for this to work.

Coastline cleanup

If the oil reaches the beach then machines are used to scoop up the oil and contaminated sand for safe disposal. Absorbent materials called sorbents are also used to soak up the oil. High-pressure hoses are used to wash the oil off rocks.

Test yourself 2

Part A: Knowledge

1. What does decanting involve? *(1 mark)*
   A pouring the mixture through a sieve
   B pouring most of the liquid off the top of a sediment
   C stirring a mixture and pouring it into a filter paper
   D allowing an insoluble solid to settle

2. When a salt water solution is evaporated only the water escapes into the environment because the salt is *(1 mark)*
   A not volatile.
   B insoluble.
   C too dense.
   D not a pure substance.

3. A solution of sugar in water is distilled. The liquid that collects in the flask is called a *(1 mark)*
   A solute.
   B filtrate.
   C solution.
   D distillate.

4. When whole blood is centrifuged, the lowest layer in the centrifuge tube is *(1 mark)*
   A plasma.
   B white cells.
   C platelets.
   D red cells.

5. Eucalyptus oil is extracted from eucalyptus leaves using *(1 mark)*
   A steam distillation.
   B froth flotation.
   C crystallisation.
   D chromatography.

6. Complete the following restricted-response questions using the appropriate word. *(1 mark for each part)*
   a) Recycling of metals is important as metals are .... resources.
   b) Electrostatic precipitation is a method used to remove .... particles from the air emitted from chimneys.
   c) In wine making, sediment and yeast are .... from the fermented mixture to produce a clear white wine.
   d) In milk collected from a dairy farm, the cream can be separated more rapidly from whole milk by ....
   e) The solid material that remains behind in a filter paper after filtration is called the ....

7. Use code letters to match the terms or phrases in each column. *(1 mark for each)*

Column 1	Column 2
A evaporation	F centrifuge
B boiling flask	G crystallisation
C minerals	H distillation
D coloured bands	I froth flotation
E honey	J chromatography

Part B: Skills

8. Figure 2.24 shows a glass rod being used during the filtration of a suspension. Explain the purpose of this glass rod. *(1 mark)*

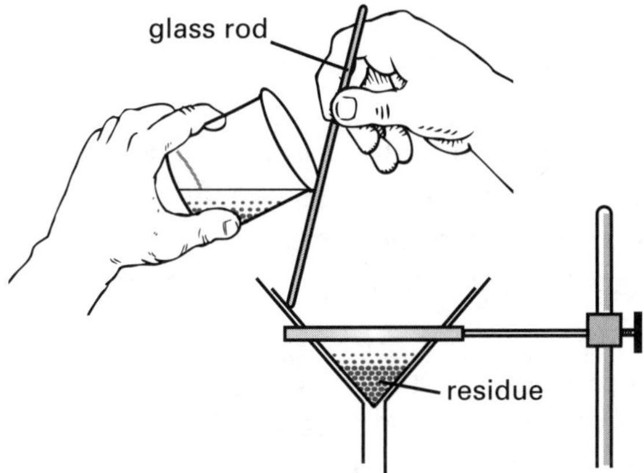

Figure 2.24 Using a glass rod during filtration

9. State one practical industrial application of evaporation as a method of separation. *(1 mark)*

10. A solution is composed of two liquids, X and Y. X has a boiling point of 90 °C; Y has a boiling point of 135 °C. If they are distilled, which one will form the distillate? Explain. *(2 marks)*

11. A sample of a water-based clothing dye was poured onto the top of a chromatographic column and solvent added to flush the dye through the column. As the dye passed down the column, four coloured bands were seen. The colour of these bands, in order from the top, was yellow, green, violet and blue.
    a) Which of these coloured substances is the least soluble in the solvent? *(1 mark)*
    b) Which is the most soluble in the solvent? *(1 mark)*

12. a) The word *centrifuge* is derived from two Latin words: centrum (= centre) and *fugere* (= flee). Using this information, explain what the word centrifuge literally means. *(1 mark)*
    b) Do you think this accurately describes what a centrifuge does? Explain. *(1 mark)*

13. Before froth flotation proceeds, the lumps of mined ore are crushed in a ball mill until they become a fine powder. The balls are made from heavy steel. Why is it important to break up the lumps of ore? *(1 mark)*

14. Most white wines contain 12% alcohol by volume. Calculate the volume of pure alcohol in 750 mL of white wine. *(1 mark)*

Go to pp. 190–191 to check your answers.

Summary

1. Mixtures are materials composed of two or more different substances that are mixed together but not chemically combined.
2. Each component of a mixture retains its own unique properties.
3. Mixtures can be homogeneous or heterogeneous.
4. Solutions are homogeneous mixtures. Some solutions are concentrated and some are dilute, depending on the proportion of solute to solvent.
5. Mixtures can be separated using physical separation techniques including decanting, filtration, evaporation, crystallisation, distillation, chromatography, centrifugation, magnetic separation and froth flotation.
6. Separation techniques can be used to separate the components of a suspension.
7. Separation techniques can be used in the home and in industry.
8. There are various steps in recycling waste.
9. Recycling of metals is important as metals are non-renewable resources and the energy used in recycling metals is much less than mining and processing new metals.
10. Separation techniques can be used to reduce pollution (e.g. electrostatic precipitation of smoke particles, water purifiers, cleaning up oil spills).

Syllabus checklist

Are you able to answer every syllabus question in this chapter? Tick each question as you go through the list if you are able to answer it. If you cannot answer it, turn to the appropriate page in the guide as is listed in the column to find the answer.

	For a complete understanding of this topic	Page no.	✓
1	Can I recall the definition of a mixture?	31	
2	Can I explain why the components of a mixture retain their own unique properties?	31–32	
3	Can I explain the difference between homogeneous and heterogeneous mixtures?	32–33	
4	Can I describe the different types of solutions and can I distinguish between the terms concentrated and dilute with reference to solutions?	34–35	
5	Can I recall the physical separation techniques used to separate the components of a suspension?	37–38	
6	Can I describe common examples of the physical separation techniques used to separate the components of a solution?	37–38	
7	Can I recall practical applications of separation techniques in the home and in industry?	41–46	
8	Can I recall the steps involved in sorting recyclable waste?	43–44	

	For a complete understanding of this topic	Page no.	✓
9	Can I explain why recycling metals is important in terms of energy expended?	44	

	For a complete understanding of this topic	Page no.	✓
10	Can I describe the methods used to reduce pollution?	44–45	

Chapter test

Go to p. v for *Tips for tests and examinations* 80 MIN

Part A: Multiple-choice questions

(1 mark for each)

1. Sports drinks contain dissolved mineral salts as well as dissolved gases. How could you demonstrate that a glass of flat sports drink is a mixture?
   - **A** Filter the drink to collect the residue.
   - **B** Evaporate the drink and look for salts in the basin.
   - **C** Pour the drink through a chromatography column.
   - **D** Centrifuge the drink and collect the sediment.
2. Which of the following substances is a mixture?
   - **A** silver
   - **B** oxygen
   - **C** milk
   - **D** carbon dioxide
3. Freshly squeezed orange juice is best described as a
   - **A** suspension.
   - **B** pure substance.
   - **C** solution.
   - **D** sediment.
4. Red ink is a solution of a solid in a liquid. The red solid is the
   - **A** solvent.
   - **B** sediment.
   - **C** solution.
   - **D** solute.
5. Look at Figure 2.25, which is a graph of the solubility of sugar in a beaker of water at different temperatures.

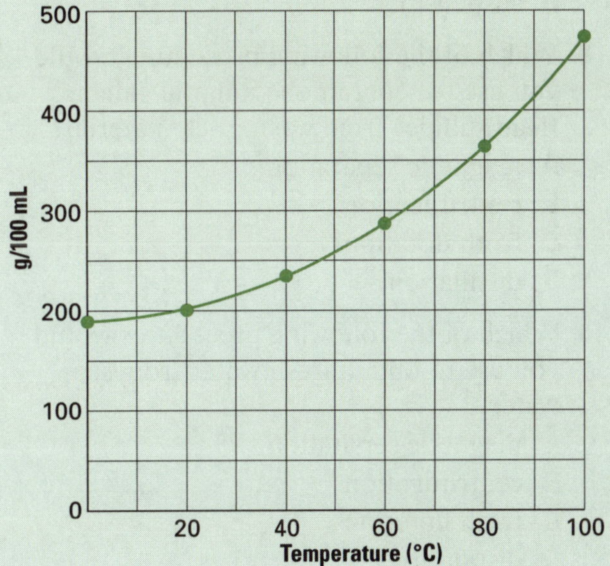

Figure 2.25 Solubility of sugar in water

   From this graph, how much sugar will dissolve at 50 °C?
   - **A** 205 g
   - **B** 360 g
   - **C** 260 g
   - **D** 300 g
6. When making wine, the alcohol has to be separated from the grape residue. Why is evaporation not a good way to do this?
   - **A** The wine will decompose.
   - **B** The alcohol will vapourise and be lost.
   - **C** Grape crystals will contaminate the alcohol.
   - **D** It will become too concentrated.

7. Which of the following substances is a mixture?
   - **A** carbon dioxide
   - **B** tap water
   - **C** oxygen
   - **D** lead

8. If you dissolve solid zinc in hot, liquefied copper to form bronze, the copper is which of the following?
   - **A** solution
   - **B** solute
   - **C** solvent
   - **D** suspension

9. Which of the following procedures would you use to separate the mineral galena (lead sulfide) from waste rock material?
   - **A** magnetic separation
   - **B** centrifugation
   - **C** froth flotation
   - **D** distillation

10. Which of the following procedures would you use to obtain fresh water from sea water?
   - **A** magnetic separation
   - **B** centrifugation
   - **C** froth flotation
   - **D** distillation

Part B: Short-answer questions

11. Complete the following restricted-response questions using the appropriate word. *(1 mark for each part)*
   - **a)** The solid that collects on the bottom of a vessel containing a .... is called the sediment.
   - **b)** The solution that contains very little .... and a great amount of solvent is said to be dilute.
   - **c)** Cream can be separated rapidly from whole milk by ....
   - **d)** .... is a method of separating a liquid from a solution based on differences in boiling point.
   - **e)** Solid–liquid solutions are .... whereas suspensions are cloudy.

12. Use the code letters to match the terms or phrases in each column. *(1 mark for each part)*

Column 1	Column 2
A magnetic separation	F heterogeneous
B muddy water	G petrol
C red blood cells	H centrifugation
D copper minerals	I car bodies
E crude oil	J froth flotation

13. The following words can be used to answer the questions below: solution; suspension; sediment.
   - **a)** Jenna mixed iodine crystals and water in a container which was shaken for several minutes. At the end of this time the crystals collected at the bottom of the container and the water remained clear and colourless. Which word best describes the iodine crystals at the bottom of the container? *(1 mark)*
   - **b)** Craig mixed shavings of candle wax and kerosene and stirred the mixture for 3 minutes. At the end of this time, he noticed that no wax shavings remained and the mixture was clear. Which word best describes the final mixture? *(1 mark)*

14. Chalk is insoluble in water and carbon dioxide is slightly soluble in water. A mixture of chalk powder and soda water was placed in a test tube. The contents were allowed to stand after thorough mixing. Draw a particle diagram of the final appearance of the contents of the test tube. Label your diagram. *(2 marks)*

15. In a washing machine, clothes are spun very quickly just before the cycle is finished. Why do you think this is done? *(1 mark)*

16. In the Australian outback, bore water is often impossible to drink due to the presence of mineral salts. One method of purifying this bore water is to construct a solar distillation plant. In this apparatus, the Sun's energy is used to evaporate the water to produce pure water vapour. The water vapour is condensed back into liquid water which is free of the mineral salts.

a) Design a solar distillation plant using only bowls, sheets of clear perspex and any necessary supports and clamps. Draw a labelled diagram of your final proposal. *(3 marks)*

b) Explain how your apparatus works. *(2 marks)*

17. Alphonse was given the task of extracting a perfume from flower petals. The jumbled method steps for his chosen procedure are given below.

a) Unjumble these steps and write them in the correct order (use the code numbers). *(1 mark)*

b) Draw a labelled diagram of the equipment Alphonse used based on the information in the experimental steps. (Note: an air condenser is a glass tube that does not have a water cooling jacket.) *(3 marks)*

c) What would Alphonse expect to notice about the smell of his collected distillate? *(1 mark)*

d) Explain the purpose of the ice bath. *(1 mark)*

Jumbled method

1. Boil the mixture of shredded petals and water in the round-bottom flask using a Bunsen burner.
2. Use a measuring cylinder to place 30 mL of water into a round-bottom flask.
3. Decant the distillate into a labelled specimen tube.
4. Construct an air condenser from a 1-m length of glass tubing bent into an L shape.
5. Shred the supplied petals and place them in the flask.
6. Connect the short end of the 'L' tube to the flask using a bored stopper.
7. Clamp the flask and air condenser to a retort stand and support the flask over a tripod and gauze.
8. Use a large test tube in an ice bath to collect the distillate from the end of the air condenser.

18. In some traditional societies winnowing is a process in which the husks (chaff) can be removed from the grain kernels of wheat. Before this can be done, the husks are freed from the grains by beating with sticks (threshing). The chaff and grains are then thrown high into the air using pitch forks. A lot of the chaff is removed by this procedure. The last step involves shaking the grain in tambourine-shaped sieves to sift out the remaining chaff.

a) Why does tossing the threshed grain into the air remove a lot of the chaff? *(1 mark)*

b) What remains in the sieve in the final step? *(1 mark)*

19. The solubility of solids X and Y in four different solvents at 20 °C is given below.

Solvent	Solubility (g/100 mL)	Solubility (g/100 mL)
	X	Y
water	0	10
glycerine	0	6
olive oil	4	1
petrol	8	0

a) Two grams of X and 2 g of Y were mixed together and powdered using a mortar and pestle. Samples of this mixture were placed in two separate large stopped tubes (1 and 2).

Tube 1: 100 mL of petrol was added to this tube and the mixture shaken.

Tube 2: 100 mL of glycerine was added to this tube and the mixture shaken.

i) Describe what would be observed in each tube. *(2 marks)*

ii) How could Y be separated from X in tube 1? *(1 mark)*

iii) What would happen if the solid mixture was shaken with olive oil? *(1 mark)*

b) Can water be used to separate X and Y? Explain. *(2 marks)*

20. The following is an advertisement for a product called Strapmag.

Strapmag is an adjustable, reusable magnetic strap. Just wrap it around the oil filter of your car and it turns the filter into a 'super filter'. The powerful magnet in the Strapmag traps very fine metal particles in the oil, holding them against the inside wall of the filter. When the filter is replaced, the trapped particles are thrown away too.

Normal filters can't catch the very fine metal particles; they just pass through the filter and can cause wear and damage to your vehicle. But with Strapmag you have cleaner oil which means longer machine life.

a) Which two separation processes are discussed here? *(2 marks)*

b) How could you find out whether the Strapmag really traps metal fragments? *(1 mark)*

21. Twenty mL of olive oil and 20 mL of water are poured into a beaker. The mixture forms two separate layers. Water is more dense than olive oil.

a) Identify which material will form the upper layer. *(1 mark)*

b) Explain how the separating funnel shown in Figure 2.26 could be used to separate the two liquids. *(2 marks)*

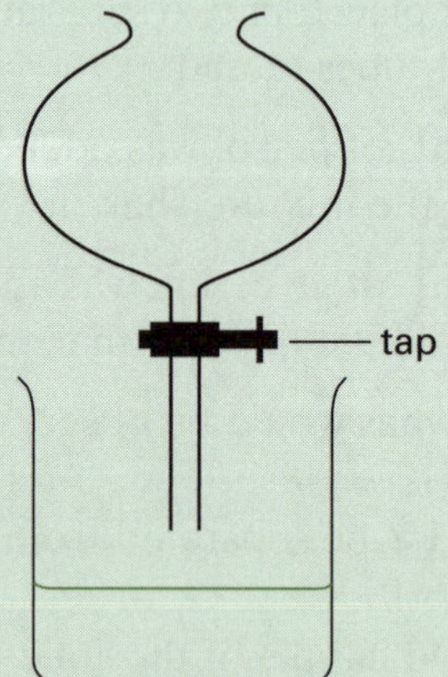

Figure 2.26 Using a separating funnel

22. Iron and brass nails are all painted black and then they are mixed together in a bowl. Suggest an easy method of separating the nails. *(1 mark)*

23. The diagrams in Figure 2.27 are jumbled. They represent the stages in removal of smoke particles from an emission steam using electrostatic precipitation. List the diagrams in the correct order. *(1 mark)*

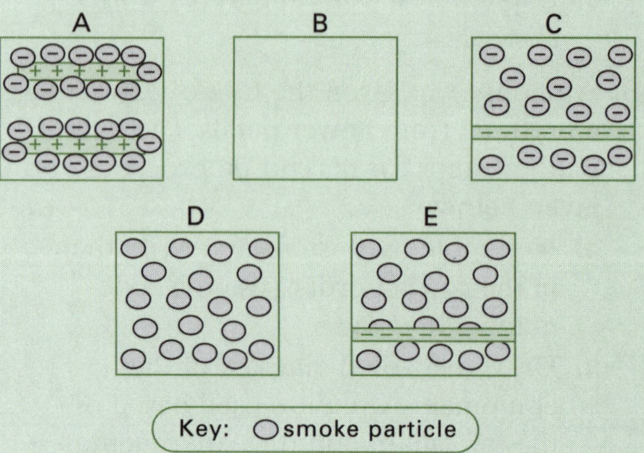

Figure 2.27 Stages in the removal of smoke particles

24. Some whole blood is transfused into patients while other blood undergoes centrifugation. Explain why some is centrifuged. *(1 mark)*

25. a) How does the condenser in a distillation apparatus work? *(1 mark)*

b) Why do you think it is important that the water enters the condenser at the lower end and leaves near the top? *(1 mark)*

26. Tincture of iodine is made by dissolving iodine crystals in alcohol. It is used as an antiseptic. In this example, name the solute, solvent and solution. *(3 marks)*

27. Rewrite each of the following sentences starting with the underlined words.

a) Grease on clothes can be removed by using <u>eucalyptus oil</u> as a solvent. *(1 mark)*

b) In general, a greater quantity of a solute will dissolve in a <u>hot solvent</u> than in a cold solvent. *(1 mark)*

28. Indicate whether the following statements are true or false. *(1 mark for each)*

a) Olive oil is a good solvent for salt water.

b) Some solutions are coloured.

c) A sediment forms when a suspensions settles out on standing.

d) Nail polish remover (acetone) is a solvent for nail polish.
e) Oxygen gas dissolves slightly in water.
f) Light rays are scattered by the particles in a solution.
g) Sterling silver represents a solid solution.

29. When you wish to physically separate a mixture containing two substances, what information do you need to know before you start? *(1 mark)*

30. Melissa folded some filter paper into a cone and used it to filter a muddy suspension she had collected at the local creek. She was amazed to find that the liquid that passed through the filter paper was not completely clear.
    a) Can you explain why this has happened? *(1 mark)*
    b) She found a different packet of filter paper and this time the liquid that came through the filter paper was clear. What do you think was the difference between the two packets of filter paper? *(1 mark)*

31. Explain why solutions cannot be separated by filtration or centrifugation. *(2 marks)*

32. There are many different types of chromatography, yet all rely on one main idea. Describe the main idea behind chromatographic separation. Use a diagram as part of your answer. *(2 marks)*

33. *Chromatography* is a word derived from two Greek words: *khroma*, meaning 'colour', and *graphos*, meaning 'write'.
    a) What does chromatography seem to mean from these Greek words? *(1 mark)*
    b) From your knowledge of separation involving chromatography, explain how the Greek words relate to this process. *(1 mark)*

34. Use the following words in one sentence to describe how a centrifuge works: centrifuge; heavier; lighter; settle; spinning. *(1 mark)*

35. Cleaning up oil spills is not an easy task. Discuss three common methods used to clean up large oil spills caused by the release of oil from supertankers that have run aground. *(3 marks)*

36. Identify the steps in sorting and separating recyclable waste at the collection and sorting centre. *(4 marks)*

37. A mixture of mud, silt, clay and sand is placed in a beaker and water is added. The mixture is stirred and allowed to stand. Three layers of sediment form and one component remains suspended. Draw a labelled diagram of the layers that form and the suspended material. *(4 marks)*

38. Name the process you would use to obtain:
    a) a solid from a mixture of dissolved solids *(1 mark)*
    b) a liquid from a solution *(1 mark)*
    c) a substance dissolved in a liquid *(1 mark)*
    d) an undissolved solid from a liquid. *(1 mark)*

39. A student was separating iron filings from sand. He put a small pile of the mixture on a piece of cardboard on his desk. Next he put a sheet of paper over this before moving the magnet around on the paper. He lifted both the paper and magnet together and removed the iron filings. Why did he use the paper? Why not just pass the magnet through the mixture? *(3 marks)*

40. Magnetite is an ore of iron and is found scattered through igneous rocks in small quantities. It is strongly magnetic. Describe a way of separating the magnetite from the igneous rocks. *(2 marks)*

Go to pp. 191–194 to check your answers.

CHAPTER 3

Astronomy, resources and water

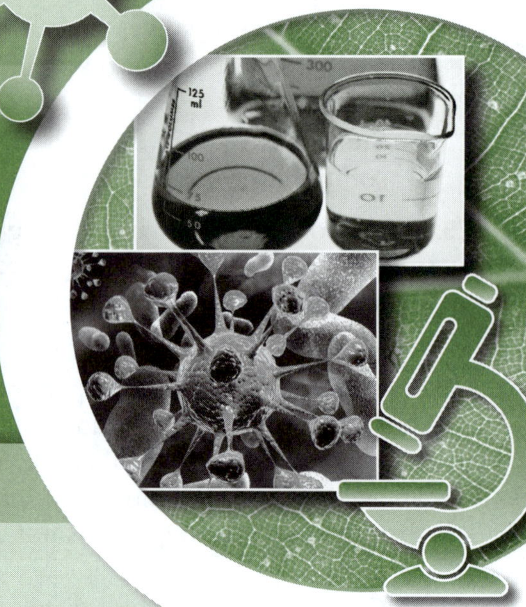

Overview

In this chapter you will learn about:

- eclipses and moon phases
- the rotation of the Earth on its axis and its rotation around the Sun
- seasons and their importance to different cultures
- models to explain the position and motion of the Earth, Moon and Sun
- the way telescopes and space probes have increased our knowledge of the universe
- the Earth-centred and Sun-centred models of the solar system
- gravity and how it holds the planets and the Moon in their orbits
- resources that have been used by humans throughout the ages
- the difference between natural and synthetic resources
- resources that are derived from living things
- resources derived from the non-living world
- fossil fuels and their uses
- alternative energy sources including wind, nuclear and solar energy
- collaborative research in Antarctica
- the water cycle and its effect on climate
- the changes in the state of water
- the effect of humans and their activities on the water cycle
- water management including conservation.

Glossary

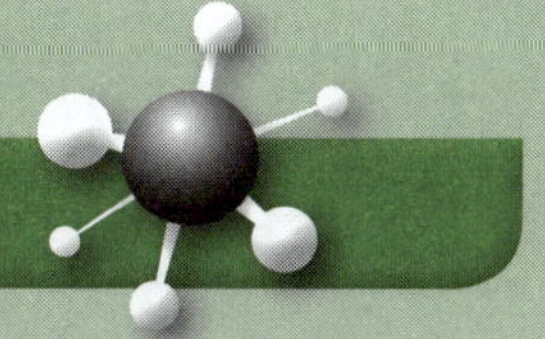

Alloy—a partial or complete solid solution of one or more elements in a metallic matrix. For example, brass is an alloy of zinc and copper.

Antarctica—an extremely cold continent surrounding the South Pole almost entirely below the Antarctic Circle. It is covered by an ice cap that is up to 4000 m deep and about twice the size of Australia.

Aquatic—living in water

Artefact—an item of human manufacture. This term is normally applied only to the products of previous cultures. Examples include bone or stone tools, engraving and paintings.

Astronomical unit—a unit of length used for distances within the solar system, equal to the mean (average) distance between the Earth and the Sun (approximately 149.6 million km)

Biomass—plant materials (mainly cellulose) and animal wastes used as fuel

Blackwater—waste water from a toilet

Condensation—the conversion of a gas or vapour into a liquid state

Eclipse—an astronomical event where one celestial object moves into the shadow of another

Ellipse—a shape that looks like a squashed or distorted circle; a shape that is produced by cutting a cone at an angle

Equinox—either of two times of the year when the tilt of the Earth's axis is inclined neither away from nor towards the Sun, with the Sun being vertically above a point on the Equator. At these times, day and night are of equal length.

Flocculation—the coagulation of fine colloidal particles in water into larger lumps by the addition of chemicals

Fossil fuel—a fuel consisting of the remains of organisms preserved in rocks in the Earth's crust, having high carbon and hydrogen content. The age of these fuels is typically millions of years and includes coal, petroleum and natural gas.

Fusion—the process of melting a solid to form a liquid

Geothermal—relating to the heat energy extracted from reservoirs in the Earth's interior

Gravity—the force of attraction between all objects in the universe, especially the attraction of the Earth's mass for bodies near its surface. On the surface of the Earth, the acceleration due to gravity is $g = 9.8\ m/s^2$.

Greywater—waste water that comes from the kitchen, laundry and shower

Hydrocarbon—an organic compound consisting entirely of hydrogen and carbon. Hydrocarbons form the basis of all petroleum products, and are used as a source of energy.

Industrial Revolution—a period from the 18th to the 19th century where major changes in agriculture, manufacturing, mining and transport had a huge effect on the economic and cultural conditions of people. It started in the United Kingdom and then spread throughout Europe. Machine power replaced human and animal power in the production process, causing a shift from home-based hand manufacturing to large-scale factory production.

Lunar eclipse—when the Moon passes behind the Earth so the Earth blocks the Sun's rays from striking the Moon. This can occur only when the Sun, Earth and Moon are almost or exactly aligned, with the Earth in the middle.

Mass—the amount of matter in an object; the property of an object that causes it to have weight in a gravitational field

Non-renewable resource—a natural resource that cannot be produced, re-grown, regenerated or reused on a scale which can sustain the rate at which it is used. This type of resource often exists in a fixed amount, or is consumed much faster than nature can recreate it. It is formed over very long geological periods. Examples include minerals and fossils.

Nuclear fission—a nuclear reaction where a massive nucleus splits into smaller nuclei with the simultaneous release of energy

Ore—an impure mineral containing metal that is valuable enough to be mined. Ores are mined and then refined to extract the valuable element(s).

Penumbra—a fringe region of partial shadow surrounding an umbra

Glossary

Phases of the Moon—the appearance of the illuminated portion of the Moon as seen from some point on Earth. These phases vary cyclically as the Moon orbits the Earth, and results from the changing relative positions of the Moon, Earth and Sun

Photovoltaic—arrays of electric cells containing materials that generate electricity directly using natural sunlight

Precipitation—rain, snow or hail falling down onto the ground

Renewable resources—a natural resource is renewable if it can be replaced by natural processes at a rate comparable to or faster than humans can consume it (e.g. solar radiation, trees, air, tides, winds and hydro-electricity)

Retrograde motion—the apparent motion of a planet (as viewed from the Earth) in which it seems to moves backwards for a time and then continues to move forwards in its orbit

Revolution—the movement of the Earth around the Sun, or the Moon around the Earth; from the Latin *revolutio* (roll, turn)

Rotation—a single complete turn of an object on its axis; from the Latin *rotatio* (turn around)

Turbine—a machine that uses fluids, such as water or air, to turn a wheel or cylinder transforming the kinetic energy of the moving fluid to other forms of energy, usually electricity

Solar eclipse—when the Moon passes between the Earth and the Sun, and it fully or partially covers the Sun, as viewed from some point on Earth

Solstice—either of two times of the year when the tilt of the Earth's axis is most inclined towards or away from the Sun. This causes the Sun's apparent position in the sky to reach its northernmost or southernmost extreme.

Space probe—a rocket-propelled guided and unmanned space vehicle that can escape the Earth's atmosphere. It makes, and sends back, observations of the solar system (and beyond) that cannot be made as well (or at all) from Earth

Transpiration—the loss of water from the pores in the leaves of plants

Umbra—a region of complete shadow where light is totally obstructed

Weight—the force exerted by a mass as a result of gravity

3.1 Eclipses and moon phases

Children have been singing the following words for generations:

Twinkle, twinkle, little star
How I wonder what you are

But humans have been observing and wondering about the heavens for many thousands of years and invented explanations for their observations. These explanations have changed over the centuries as knowledge and instruments (telescopes and space probes) have improved.

Eclipses

Eclipses happen when the Earth, Moon and Sun are all lined up. When the Earth passes between the Moon and the Sun, a lunar eclipse occurs (see Figure 3.1). When the Moon passes between the Earth and the Sun, a solar eclipse occurs (see Figure 3.2). These create full shadows (umbra) or partial shadows (penumbra).

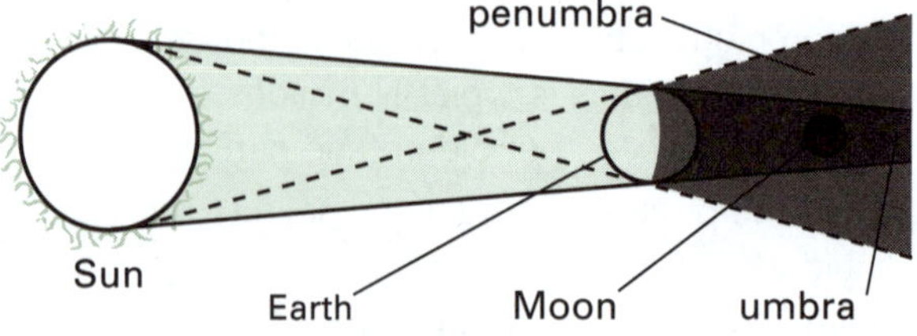

Figure 3.1 Lunar eclipse (not to scale)

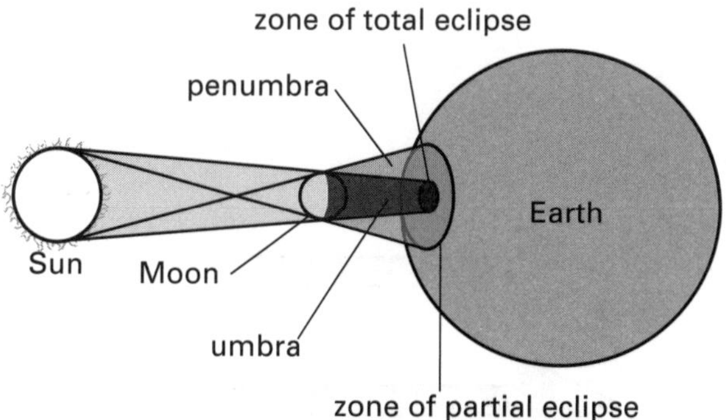

Figure 3.2 Solar eclipse (not to scale)

Because there is a 5° difference between the plane of the Earth's orbit and the plane of the Moon's orbit, eclipses do not occur every month. Occasionally the three bodies do line up and a total or partial eclipse occurs.

A partial eclipse occurs when the Sun and Moon are not exactly in line, and the Moon only partially hides the Sun. This is often seen from a large part of the Earth outside of the track of a total eclipse (see Figure 3.3).

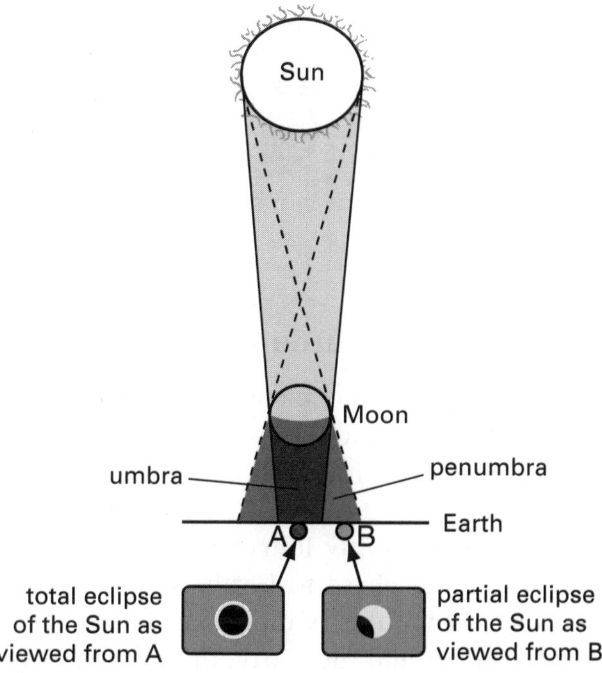

Figure 3.3 What you see during a total and partial eclipse of the sun

Phases of the moon

It takes around 29.5 days from one full moon to the next. Like the planets, the Moon produces no light of its own and shines only by reflecting the rays of the Sun.

The phases of the Moon are caused by the different angles from which we see the bright part of the Moon's surface. As the Moon revolves around the Earth, it appears as if it is changing shape in the sky. As early as 500 BC the Greeks realised that the phase of the Moon depended on the Sun–Earth–Moon angle. They worked out that this implies the Moon is a solid sphere, in orbit about the Earth, half of which is always illuminated by the Sun.

Figure 3.4 shows a view of the Earth when looking down on the South Pole. As the moon orbits clockwise from this view, the different phases of the moon we see depend on the angle between the Sun's rays, the Moon and the Earth.

3.2 Rotations and orbits

From early times people noticed that the Sun rises in the east and sets in the west. Actually, it's not the Sun that moves; rather, it is the rotation of the Earth from west to east. The Earth's rotation is the rotation of the solid Earth around its own axis.

The rotation of the Earth

For many thousands of years people believed that the Sun moved as they didn't feel any movement on Earth. But the movement is relative. Sometimes when sitting in a train that is just starting to move, you get the impression that the platform is moving backwards, not you (and the train) moving forwards. But you know that cannot be.

The Earth rotates around its rotation axis once in 24 hours as shown in Figure 3.5. The axis of rotation is an imaginary line that passes through the North and South Poles of the planet. The time it takes for the Earth to rotate completely around once is what we call a day. It's Earth's rotation that gives us night and day.

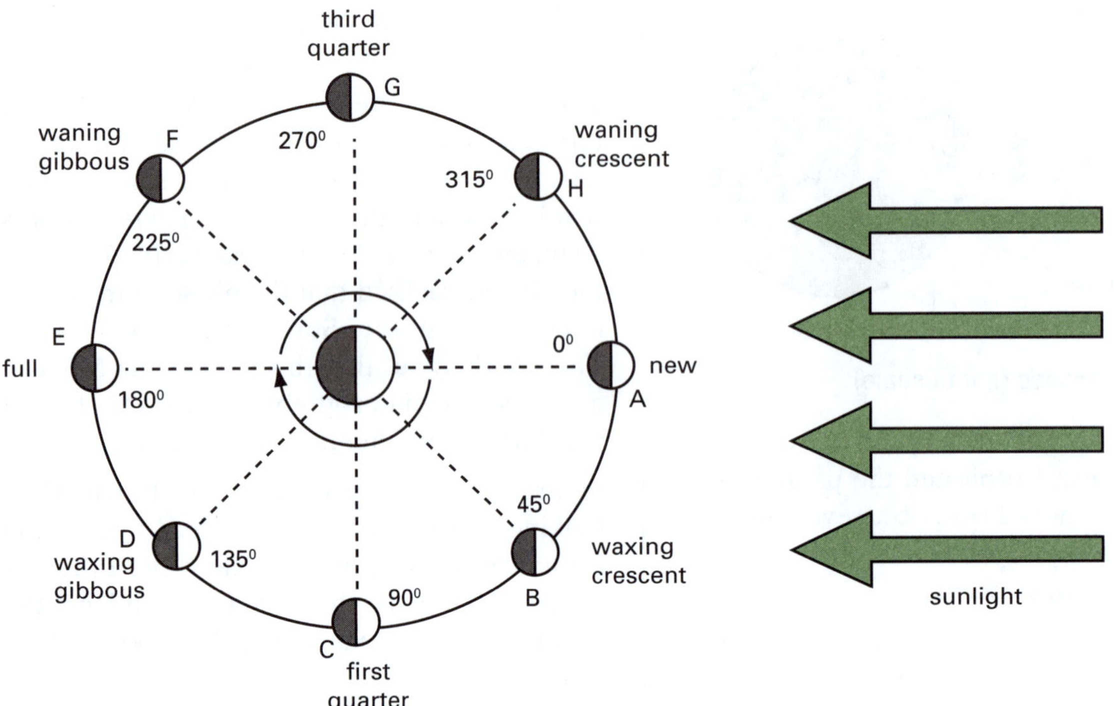

Figure 3.4 (a) Looking down on the Earth's South Pole

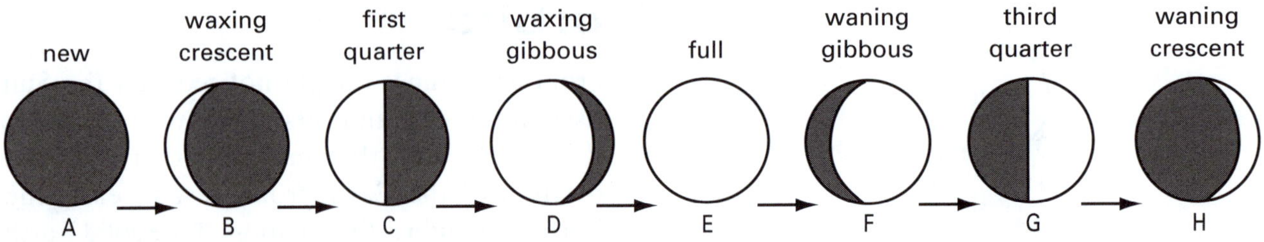

(b) Different phases of the Moon seen depending on the angles in (a)

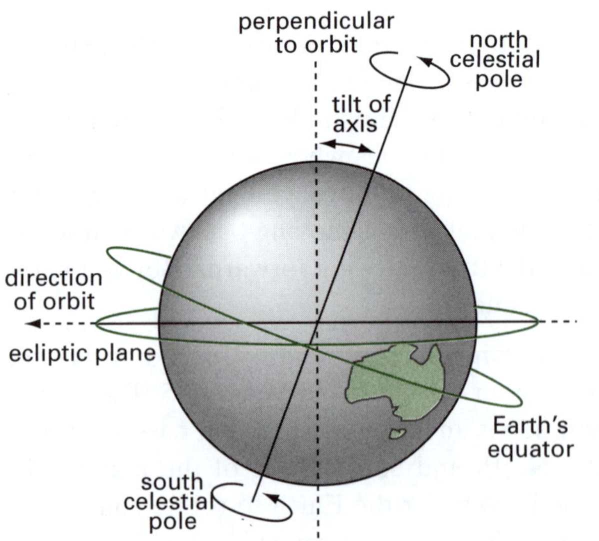

Figure 3.5 The Earth's axis is tilted with respect to its orbit plane around the Sun.

The moon orbits around Earth

The Moon makes a complete orbit around the Earth with respect to the fixed stars about once every 27.3 days. However, since the Earth is moving in its orbit about the Sun at the same time, it takes slightly longer for the Moon to show the same phase to Earth, which is about 29.5 days. Since this is the same time it takes to go once around the Earth, the Moon always presents the same hemisphere (or face) to us, as shown in Figure 3.6.

The path followed by the Moon is not a circle but an ellipse (oval). This means it moves closer to Earth during one part of its cycle, and further away during another. But the path is very close to being circular, and is shown exaggerated in Figure 3.6.

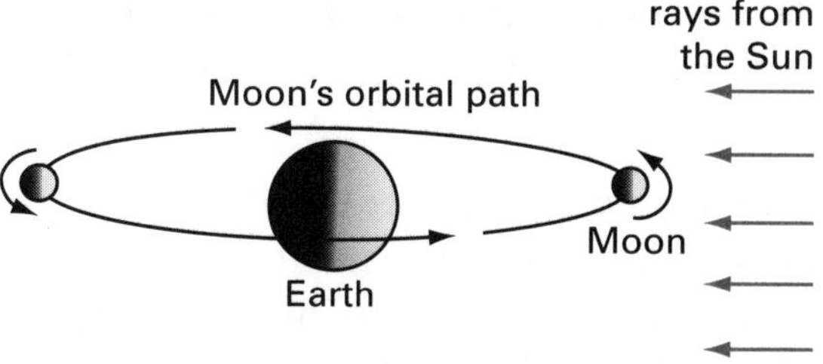

Figure 3.6 The Moon's orbit around Earth

The closest the moon gets to Earth is about 364 000 km; the furthest is 407 000 km. These distances are still very large, being about 60 times the radius of the Earth.

Travelling at over a kilometre each second, it takes the moon a full month to orbit once around the Earth, having travelled some 2.5 million km! This shows the huge distances involved in space.

Earth's orbit around the Sun

The Moon orbits around the Earth and the Earth orbits around the Sun in our solar system. The Earth, too, follows an elliptical path around the Sun, getting as close as 147.3 million km and as far away as 152.1 million km (see Figure 3.7). The orbit of the Earth around the Sun is called an Earth revolution. This celestial motion takes 365.24 days (one year) to complete one cycle. The Earth travels around the Sun at over 100 000 km per hour.

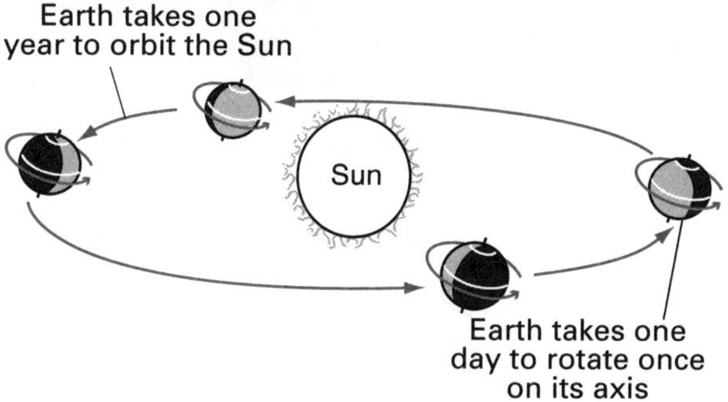

Figure 3.7 Earth's orbit around the Sun

3.3 Seasons

When prehistoric humans observed the periodic changes in the seasons, they tried to find ways to mark the passage of time so they could predict the return of the seasons. For example, the river Nile flooded once a year bringing with it fertile silt, so it was important to know when this would occur so crops could be planted.

One way to monitor the seasons was to observe the phases of the moon. These repeated roughly every 29 days. They found it took about 12 of these lunar cycles for all the seasons to cycle around. Using these time periods for a month ('moonth') and for the year was alright at first, but soon it was out of sync. (Remember, 12 × 29 = 348 days, not nearly close enough to the length of the year.) Different cultures found ways to compensate for this. These early peoples didn't know what caused the seasons. But for many that was no concern; as long as they knew when to hunt and when to sow.

Ancient Egyptians

The ancient Egyptians had 12 months, each with three 10-day weeks. The months were named mostly after their gods. They added another 5 days at the beginning of the year for feasting. (12 × 3 × 10 + 5 = 365 days)

Australian Aborigines

While Australian Aboriginal peoples did not divide the year up into months, many had systems for dividing the year into seasons. In some cases there were as many as eight seasons. These were often named after the animals or plants that were abundant at that time of year, or the weather conditions. The Yolgnu people of the Northern Territory, for example, divided the year into six seasons, each with several phases. Each of these seasons described a pattern, and what people could expect at that time of year.

The Inuit

The Inuit (Eskimos) of far North America divide their year into 12 or 13 lunar months.

Each month is named after something related to a season or an event in the sky.

The reason for the seasons

The seasons are due to changes in temperature on Earth and are a direct result of the amount of solar radiation that hits our planet at a particular time of year. The distance between the Earth and the Sun has nothing to do with it. That is a misconception.

The Earth's seasons are caused by the rotation axis of the Earth not being perpendicular to its orbital plane. As Figure 3.8 shows, the Earth's axis is tilted at an angle of about 23.5° from the orbital plane. So for half of the year (i.e. from around 20 March to around 22 September), the southern hemisphere tips away from the Sun, with the maximum distance reached around 21 June. For the other half of the year, the southern hemisphere tips towards the Sun, with the maximum around 21 December.

The two instances when the Sun is directly overhead at the Equator are called the equinoxes. The word *equinox* comes from the Latin *aequus* (equal) and *nox* (night), because around the equinox, the night and day are approximately equally long.

The March (autumnal) equinox will occur around 21 or 22 March. This marks the beginning of autumn in the southern hemisphere. At the same time, it is spring in the northern hemisphere (from an astronomical viewpoint). The September (vernal or spring) equinox happens around 22 or 23 September. This marks the beginning of spring in the southern hemisphere. At the same time, it is autumn in the northern hemisphere (from an astronomical viewpoint).

A solstice is an astronomical event that happens twice each year. This is when the Sun's apparent position in the sky reaches its northernmost or southernmost extremes. The word *solstice* comes

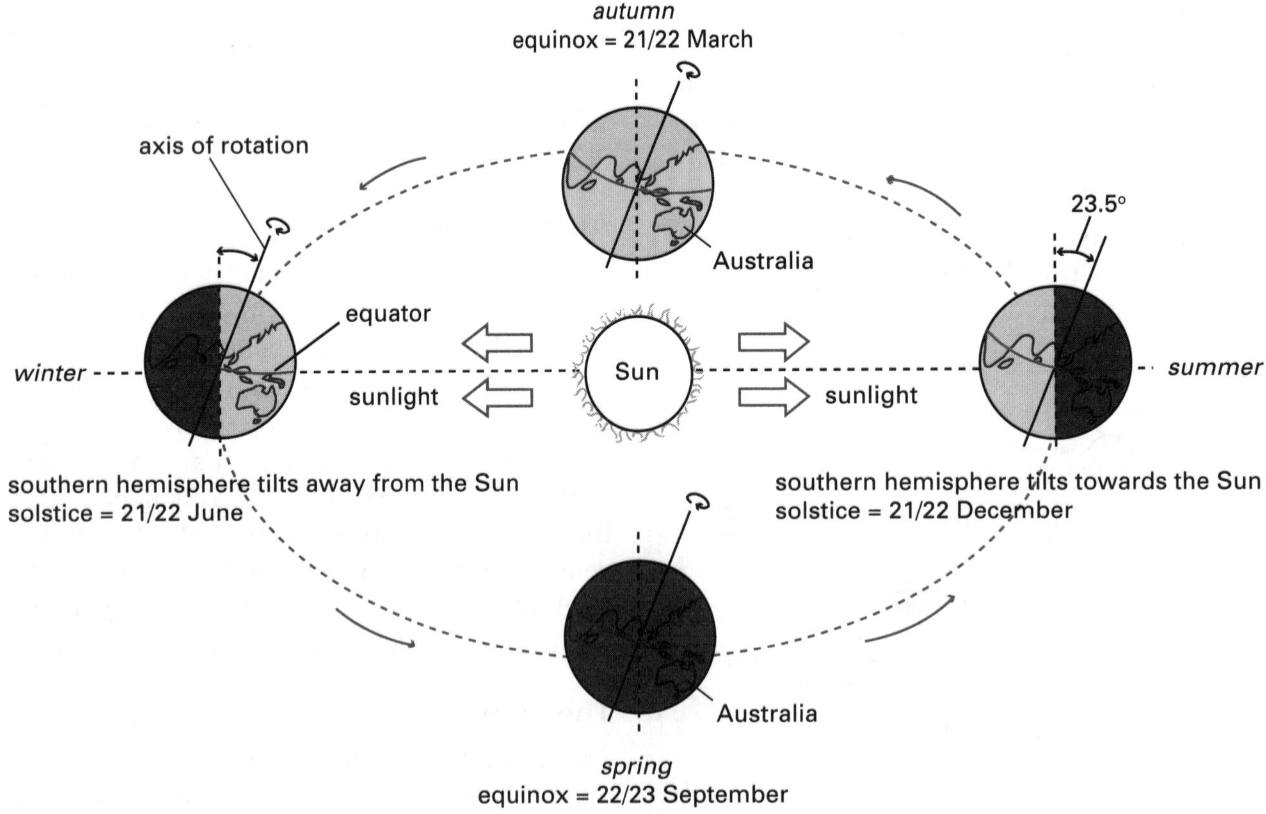

Figure 3.8 The seasons as observed in the southern hemisphere

from the Latin *sol* (sun) and *sistere* (to stand still). This is because at the solstices, the Sun stops moving north or south, being about to reverse direction.

You will be able to see in Figure 3.9 that the southern hemisphere is facing away from the Sun in winter and towards the Sun in summer. But what effect does this have on the amount of sunlight reaching our planet? When the southern hemisphere is facing the Sun the rays falling on it cover a certain area. In the northern hemisphere these rays fall at a greater angle and must heat a greater land mass, so the temperature of the air and ground is not as great as in the southern hemisphere at this time.

Australian seasons begin on the first day of the month. In most other countries seasons begin on the 21st day of the month. The 21st days of December and June are the longest and shortest days for us respectively.

3.4 Models of the Earth, Moon and Sun

Can you imagine the actual size of our solar system? You may have seen a diagram of the Sun and planets in a book. Or you may have seen a revolving model in a museum. But even the largest of such models are far too small. The fact is that the planets are very small and the distances between them are almost ridiculously large.

Modelling the Sun–Earth–Moon system

Here is one way to make a representation that has a scale which is true for sizes and distances in our region of the solar system.

The following numbers are relevant:

- diameter of the Moon: about 3500 km
- diameter of the Earth: about 13 000 km
- diameter of the Sun: about 1 300 000 km
- Earth–Sun distance: about 150 000 000 km
- Earth–Moon distance: about 384 000 km.

1. The aim is to make a scale diagram of the Earth, Moon and Sun and show the relative distances between them.
2. The smallest number in the list is the diameter of the Moon. For convenience let the scale be 3500 km = 1 mm. Thus the Moon has a diameter of 1 mm. By dividing each of the other numbers by 3500, the relative scaled sizes (in millimetres) of the distances and diameters can be calculated, as shown in Table 3.1.

Table 3.1 Scaled measurements for a model of the Sun–Earth–Moon system

Distances	Scaled measurement (mm)
Diameter of the Moon	1.0
Diameter of the Earth	3.7
Diameter of the Sun	371.0
Earth–Sun distance	42 857.0
Earth–Moon distance	110.0

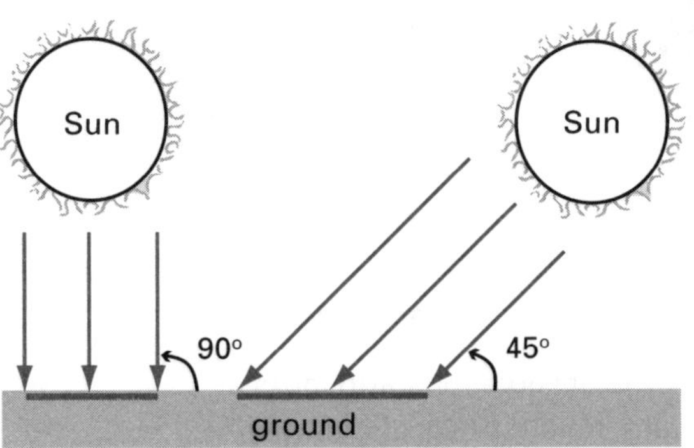

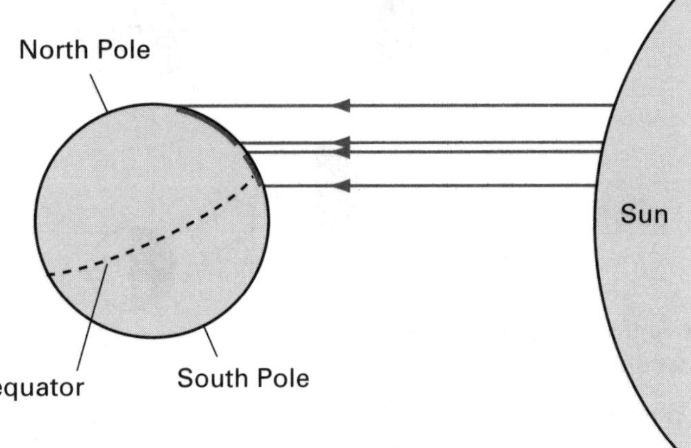

Figure 3.9 The ground is heated more when the Sun's rays are overhead as they concentrate closer together.

3. On this new scale, the Sun would be 371 mm in diameter but its distance from the Earth is about 43 m! This is too large to fit in the classroom.

Astronomers use the average distance between the Sun and the Earth to create a new scale. This distance is called 1 astronomical unit (1 AU). Table 3.2 shows the distances between the Sun and the eight planets of the solar system in astronomical units.

Table 3.2 Distances in the solar system

Planet	Average distance (AU)
Mercury	0.4
Venus	0.7
Earth	1.0
Mars	1.5
Jupiter	5.2
Saturn	9.5
Uranus	19.2
Neptune	30.1

Experiment 1

Modelling the relative movement of the Earth and the Moon

Aim

- To model the phases of the Moon and to model different seasons of the year

Method

Part A: Moon phases

1. Use a basketball and a tennis ball to represent the Earth and the Moon. Use a lamp to represent the Sun.
2. One student holds the basketball and acts as the observer of the Moon. This student will rotate from position 1 through to 8 as shown in Figure 3.10. Another student holds the tennis ball (Moon) and moves in a circle around the basketball without getting in the way of the light.
3. The phases of the Moon should be observed as the tennis ball moves around the Earth-based observer.

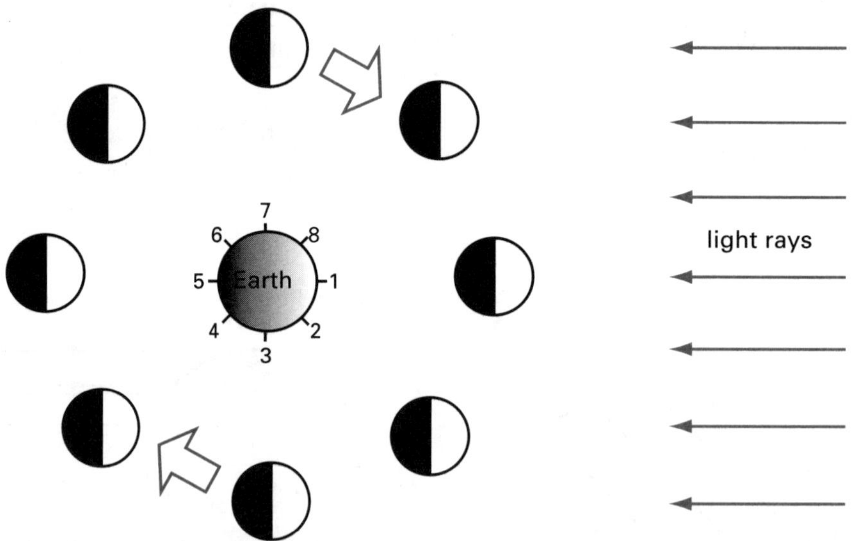

Move the Moon model to different locations around the Earth model.

The observer on Earth looks at the Moon from eight locations.

At position 1, the observer sees the dark side of the ball. This represents the new Moon.
At position 5, the observer sees the bright side of the ball. This represents the full Moon.

Figure 3.10 Modelling the Earth and Moon motion

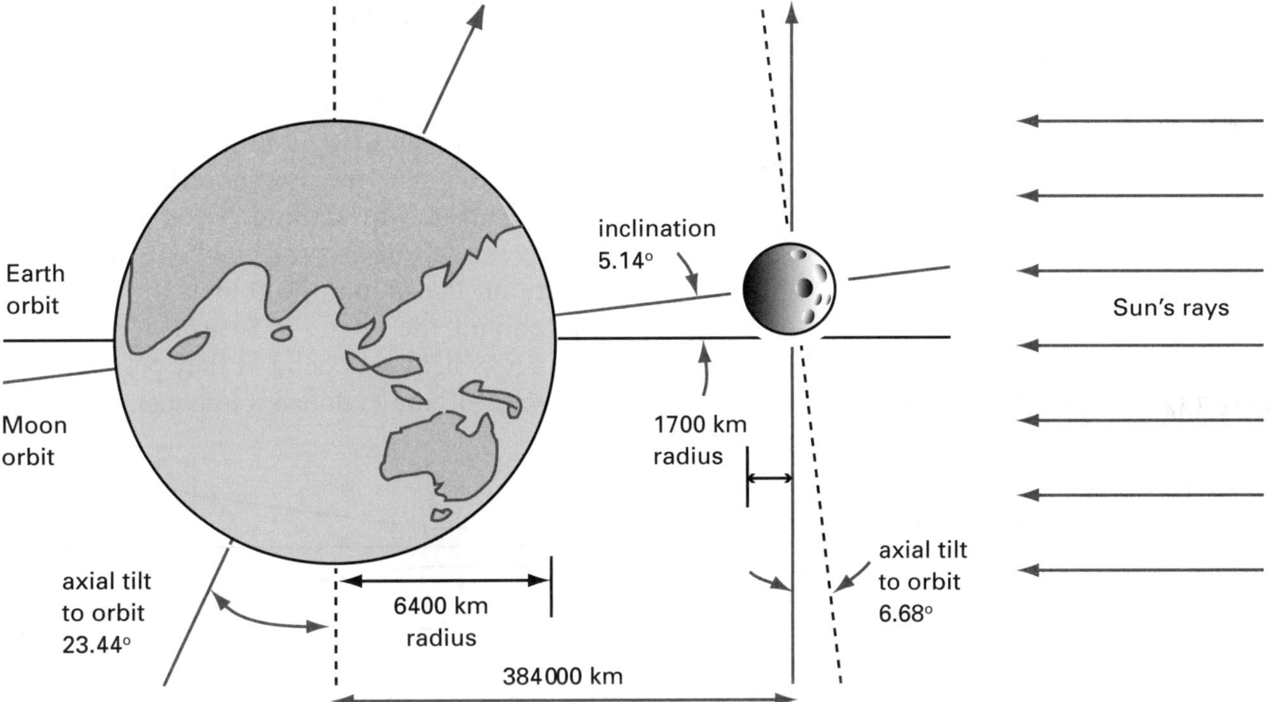

Figure 3.11 The Moon's orbit is inclined at a 5° angle to the Earth's orbit around the Sun.

Note: the Moon's orbit is actually inclined at a small angle to the Earth's orbit, as shown in Figure 3.11. As the Moon revolves around the Earth, it mostly passes either above or below the line of the Earth's orbit.

Occasionally, however, the Moon will line up perfectly with the Earth and Sun. This is when eclipses occur. This will be harder to see as the lamp would need to be just the right size and distance away from the Moon model.

Results

The appearance of the tennis ball at each location (1 to 8) is:

1. dark
2. bright crescent at left
3. half bright at the left
4. dark crescent at right; the rest bright
5. fully bright
6. dark crescent at left
7. half bright at right
8. bright crescent at right.

Analysis

Analyse these results in terms of moon phases.

Go to p. 194 to check your answer.

Conclusion

Write a suitable conclusion.

Go to p. 194 to check your answer.

Method

Part B: Seasons of the Earth

1. The Earth also spins on its axis once each day while it revolves around the Sun once each year. Using a felt-tip pen, place marks on your Earth model to represent the North and South Poles. Rotate the Earth model through this N–S axis while slowly orbiting your lamp Sun. Remember, the N–S axis makes an angle of about 23.5° with the vertical to its plane of revolution (plane of the ecliptic). The axis remains tilted in the same direction towards the stars throughout a year.
2. Observe which hemisphere is pointing towards the lamp and thus how the seasons change. Move your model Earth to the four locations (A–D) shown in Figure 3.12. Which hemisphere is pointing to the Sun at each point?

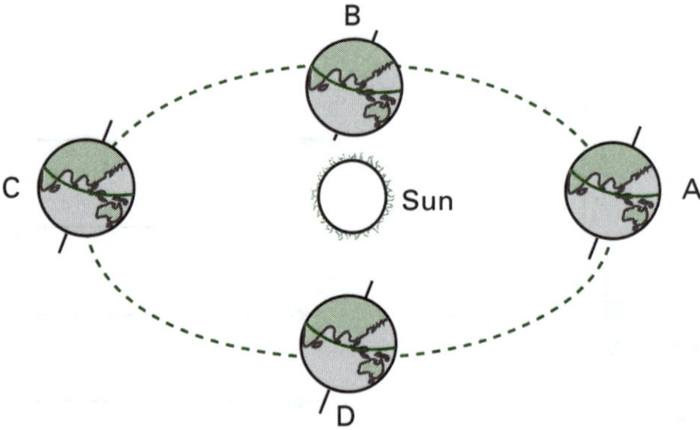

Figure 3.12 Modelling the seasons

Results

The hemisphere pointing towards the lamp (Sun) is:

A southern
B both
C northern
D both.

Analysis

Use the results to identify the seasons for the southern hemisphere at each location.
Go to p. 194 to check your answer.

Conclusion

Write a suitable conclusion.
Go to p. 194 to check your answer.

3.5 Telescopes and space probes

The five planets Mercury, Venus, Mars, Jupiter and Saturn plus the Sun and Moon were visible to the unaided eyes of the ancient astronomers. The planets could be distinguished from stars because through regular observation they were seen to move relative to the background of stars.

Telescopes

Today many scientists make observations of the stars through more and more powerful telescopes. The more powerful they become, the larger the telescope becomes. In order to overcome the distorting effects of the Earth's atmosphere on the final images, telescopes are usually positioned high on mountains away from the glare of city lights.

While he didn't invent the telescope as such, the Italian scientist Galileo Galilei used them in 1609 for his astronomical work. His magnification was limited to only 30 times. However, he discovered sunspots and saw craters on the moon, the four large moons of Jupiter and the rings of Saturn. Figure 3.13 shows how light rays bend as they pass through the lenses of the Galilean telescope.

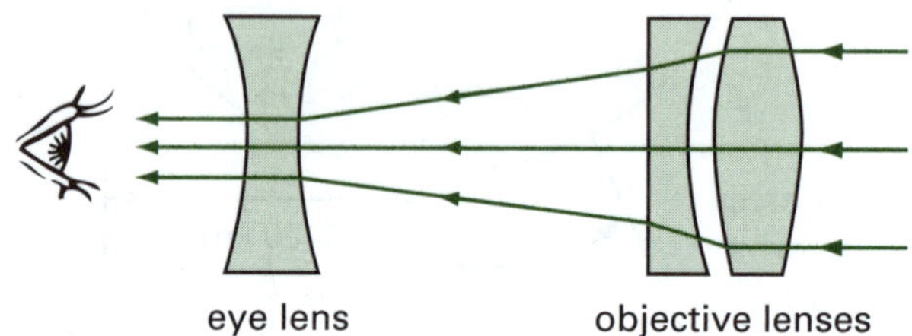

Figure 3.13 The Galilean telescope uses curved lenses to magnify the object.

In 1704 Isaac Newton improved on Galileo's refracting telescope by using a curved mirror instead of glass lenses. The mirror collected the light and focused it onto a central point, as shown in Figure 3.14. This was the dawn of powerful telescopes that could magnify objects millions of times. While Newton's mirror was only about 15 cm in diameter, some modern telescope mirrors exceed 6 m.

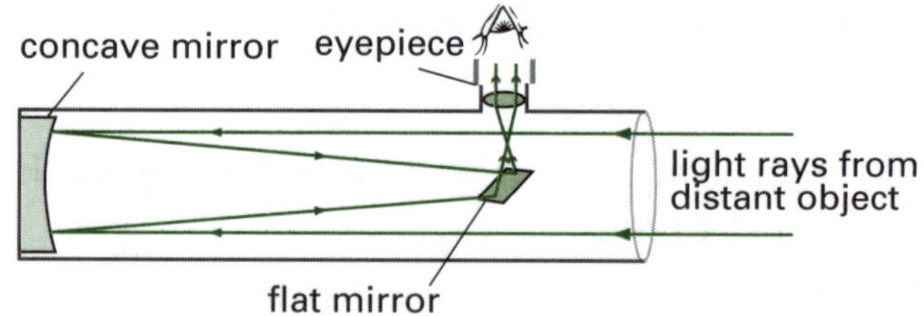

Figure 3.14 In the Newtonian telescope, the primary mirror has a concave (parabolic) shape, which reflects the light to a secondary mirror, and then to an eyepiece.

A boost to the telescope came when the Hubble Space Telescope (see Figure 3.15) was launched on a space shuttle on 25 April 1990. At last, images could be gathered from space using a telescope in orbit around the Earth. Being above the Earth, gases and small particles in the atmosphere did not distort these images. Hubble is one of NASA's most successful and long-lasting science missions. In its time it has beamed hundreds of thousands of images back to Earth. The Hubble Space Telescope is now coming to the end of its productive life and will soon be replaced by even better telescopes with more advanced technology.

Among the many achievements of Hubble are:

- images of the Orion Nebula that confirm the birth of planets around new stars
- images of the Eagle Nebula showing where stars are born
- sending back images that date back more than 10 billion years, showing some 1500 galaxies in various stages of formation.

All of the Hubble Space Telescope's functions are powered by sunlight. The solar arrays convert sunlight directly into electricity. It has a mass over 11 tonnes and orbits 559 km above the Earth every 96 minutes.

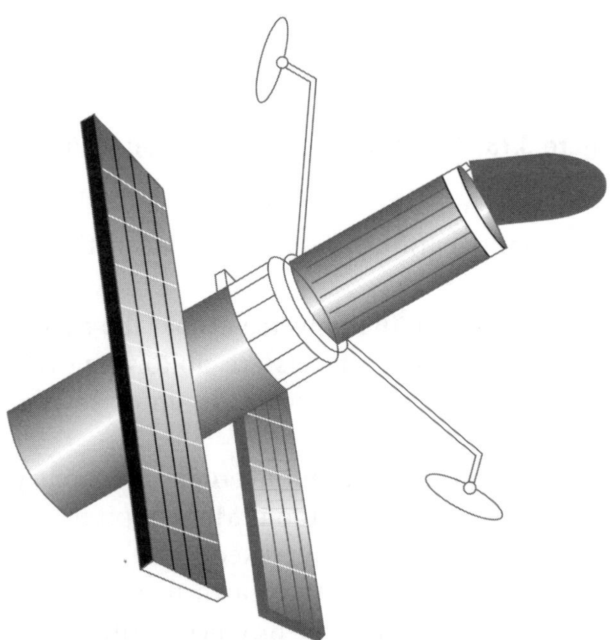

Figure 3.15 The Hubble Space Telescope

In the 20th century telescopes were developed that worked in a wide range of wavelengths from radio to gamma rays, not just visible light. Since then, a huge variety of complex astronomical instruments have been developed.

Space probes

Space probes are unmanned space exploration vessels that leave the Earth's gravity and approach planets, moons, comets or asteroids. A number of countries have launched probes to study objects up close. Probes send data back to Earth for scientists to study. Sputnik 1 was the first probe to go into space, being launched in 1957.

One of the most famous probes is Voyager 1, launched into space in 1977. It has travelled further in space than any human-made object. Voyager 1 flew past Jupiter and Saturn and then headed for the edge of our solar system. This spacecraft is currently some 20 billion km from Earth.

3.6 Historical aspects of astronomy

Astronomy is probably the oldest of sciences. Humans, with their innate curiosity and intelligence, looked up and wondered about what they saw in the sky since prehistoric times. Early cultures identified heavenly objects with gods and spirits, and incorporated their observations into their creation myths and religions. They related the movement of these objects to natural phenomena such as rain, drought, seasons and tides. It is very probable that the first 'professional' astronomers were priests and that their understanding of the night sky was seen as being somehow divine; hence astronomy's ancient connection to what is now called astrology. But astrology, telling your fortune and future by the stars, is not science.

Anaximander

The ancient Greek Ionian school (6th to 4th century BC) developed a natural, mechanical view of the universe. Rather than a supernatural or mythological explanation for what they saw, they asked questions that could then be attempted to be answered using reason and observation, and applying geometry.

Anaximander proposed a model where a cylindrical Earth was at rest in the centre of the universe (see Figure 3.16). This was surrounded by air, followed by one or more spherical shells with holes in them. The holes appeared as stars as there was a rim of fire beyond the solid sphere.

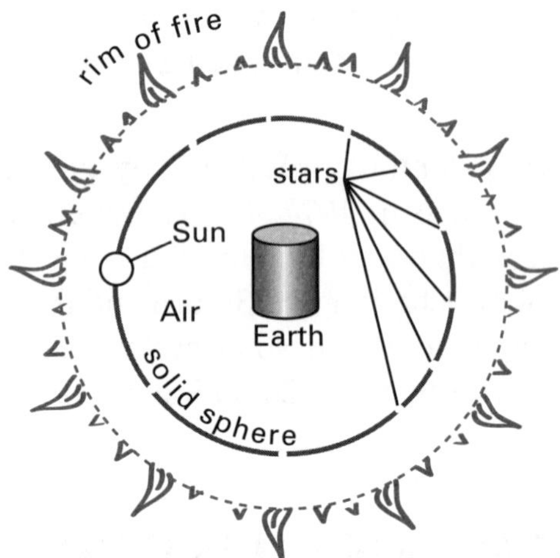

Figure 3.16 Anaximander's model of the universe

Aristarchus

Many ancient models had the Earth at the centre, but Aristarchus of Samos (c. 310–230 BC) proposed a model that placed the Sun at the centre (see Figure 3.17). All the planets, including Earth, revolved around a fixed Sun in circular orbits. The Earth rotated once a day on its axis and the Moon revolved about the Earth. But this model wasn't widely accepted at the time, and for centuries astronomers used a model of an Earth-centred universe.

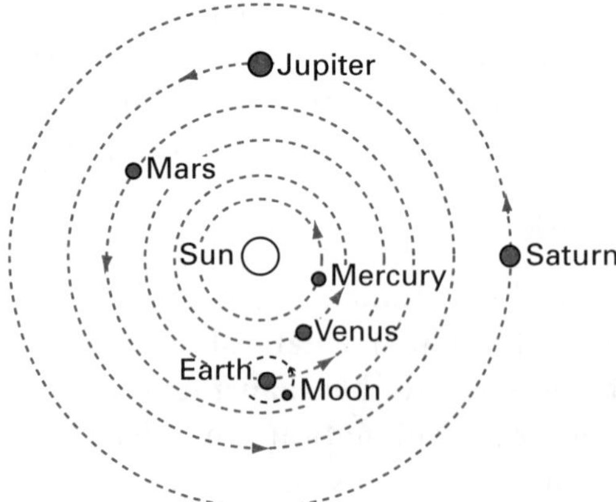

Figure 3.17 Aristarchus placed the Sun at the centre of the universe.

Aristarchus was able to calculate the distance from the Earth to the Moon, and the radius of the Moon. He also roughly estimated the Earth's distance to the Sun.

Copernicus

Nicolaus Copernicus (1473—1543) assumed that the Sun was motionless near the centre of the universe, with Earth and the other planets rotating around it in circular orbits and at constant speeds (see Figure 3.18).

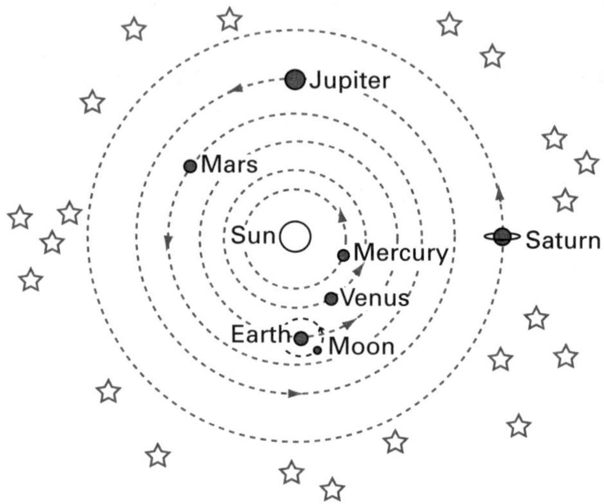

Figure 3.18 The Copernican Sun-centred model of the universe

The major features of the Copernican theory are as follows.

- The centre of the universe is near the Sun.
- The motions of objects are uniform, never-ending, and circular or made up of several circles.
- Circling around the Sun, in order, are Mercury, Venus, Earth and Moon, Mars, Jupiter and Saturn, followed by the fixed stars.
- The distance from the Earth to the Sun is small compared to the distance to the stars.
- The Earth has three motions: daily rotation, annual revolution, and annual tilting of its axis.
- Retrograde motion of the planets is explained by the Earth's motion.

Retrograde motion is the apparent change in direction of planets compared to the background of stars. Astronomers had observed this with planets over millennia, and were looking for ways to explain it. For example, the Earth moves in its orbit faster than Mars and so overtakes Mars about every 26 months. This causes the apparent motion of Mars across the background stars to actually change direction for a short period of time, as shown in Figure 3.19. A straight line from the Earth through Mars also traces a retrograde path against the background of stars, as shown in Figure 3.20.

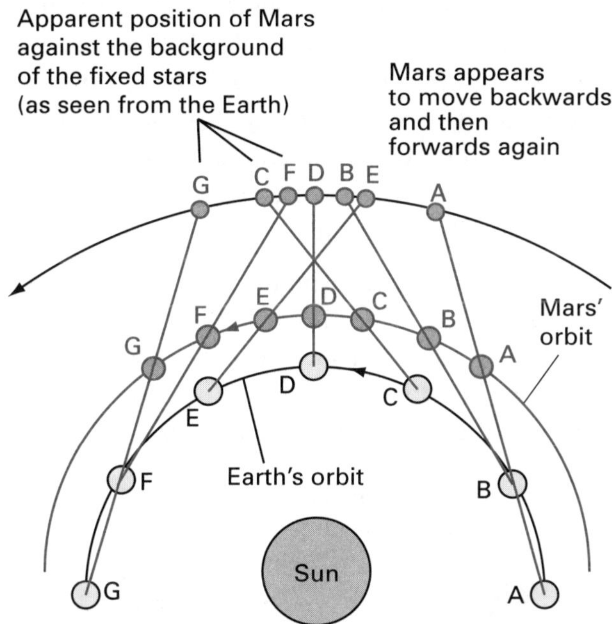

Figure 3.19 Apparent retrograde motion of Mars as seen from Earth

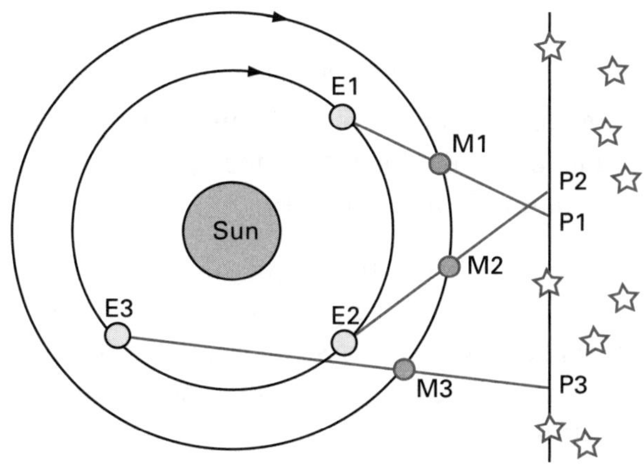

Figure 3.20 Apparent retrograde motion of Earth

Copernicus was able to estimate the distances of the planets to the Sun, compared to the Earth's distance to the Sun. He correctly gave the order of the planets in distance to the Sun.

Galileo

Galileo Galilei (1564–1642) was an Italian physicist, mathematician, astronomer and philosopher whose achievements include improving the telescope and consequent astronomical observations. Galileo used a refracting telescope to observe the Moon, Jupiter and the Milky Way. These and subsequent observations, and his interpretations of them, eventually led to the overthrow of the Earth-centred model of the universe and the adoption of a Sun-centric model as proposed by Copernicus.

Galileo made several key discoveries using telescopes.

- The Moon was not perfectly round and smooth, as people previously thought. He found it to be craggy, with mountains and valleys.
- Venus went through phases similar to the Moon. This could not be explained with Earth-centred models but could be accounted for using the Sun-centred Copernican model.
- He observed the four largest moons of Jupiter, and saw that they moved around the planet.
- He observed the rings of Saturn. These disappeared then later reappeared.
- Galileo observed dark regions (sunspots) that appeared to move across the surface of the Sun. He used this movement to show that the Sun was rotating.

Figure 3.21 shows four of the moons of Jupiter, as observed from Earth at one point in time. (With a good pair of binoculars you can also make similar observations to Galileo's.)

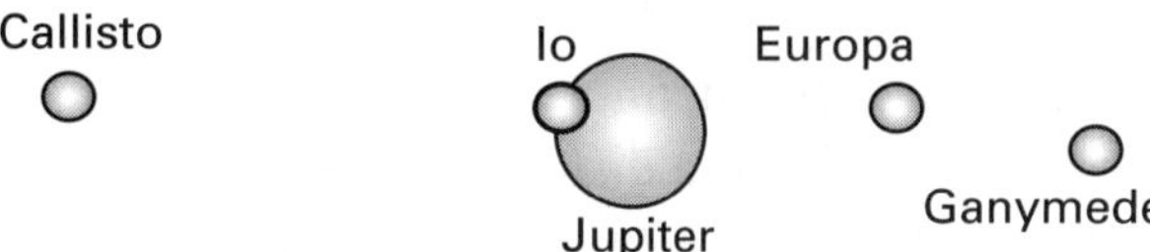

Figure 3.21 Four of the moons of Jupiter

al-Battñ

Muhammad al-Battñ (c. 858–929) was an Arab astronomer, astrologer and mathematician. He determined the length of the solar year as 365 days, 5 hours, 46 minutes and 24 seconds. Compare this to the currently accepted year of 365 days, 5 hours, 48 minutes and 46 seconds. He compiled new and more accurate tables of the Sun and Moon. Some of Al-Battñ's many astronomical measurements were even much more accurate than ones taken by Copernicus many centuries later.

Al-Battñ was the best known of Arab astronomers in Europe during the Middle Ages. His work is important in the development of science for several reasons, especially the large influence his work had on scientists such as Tycho Brahe, Johannes Kepler, Galileo and Copernicus.

Khayyám

Omar Khayyám (1048–1131) was a Persian mathematician, philosopher, astronomer, physician and poet. In 1073, he was invited to build an observatory. With other scientists, he used this to measure the length of the solar year as 365.24219858156 days and so helped reform the ancient Muslim calendar. Khayyám's calendar was more accurate than the Gregorian calendar that came 500 years later. Khayyám's calendar had an error of one day in 3770 years compared to the Georgian calendar, which had an error of 1 day in 3330 years.

Khayyám also outlined the beginnings of a Sun-centred theory, which Copernicus later developed and expanded. This theory challenged the ideas of the time, which included the thought that the Earth was at the centre of a revolving universe.

3.7 Gravity

Forces can act on an object either through direct contact or at a distance. In order to move a desk, you need to be in contact with it. Not all forces, however, are contact forces. Gravity is an example of a force that can exert a force on an object without being in contact with it. Forces like this are called field forces. A field is a space where an object may experience a force. Another example is a magnetic field. A magnet on the outside of a test tube containing iron filings causes them to move around even though the magnet is not in contact with the filings.

Gravity on Earth

Gravity is at work all around us. Any time you drop an object, jump in the air or toss a ball, it is gravity that pulls down. Without gravity we, and everything else on Earth, would float off this planet. It is such a constant presence in our lives we often take it for granted. Isaac Newton defined gravity as a force, one that attracts all objects to all other objects.

Mass is the amount of matter in an object. This mass does not change regardless of whether the object is on Earth, in space or on the Moon (see Figure 3.22). Mass is measured in kilograms (or grams, milligrams, tonnes and so on).

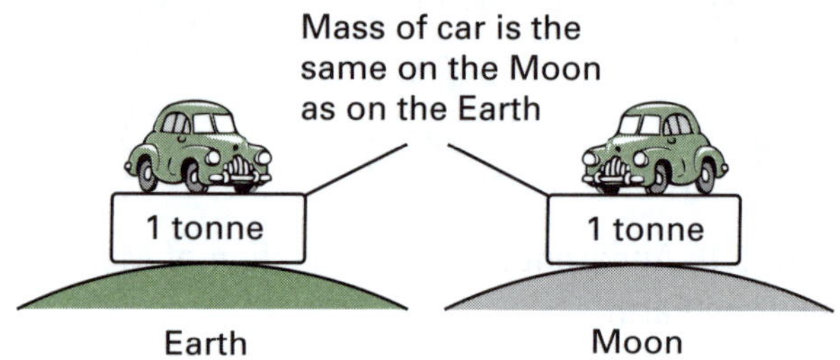

Figure 3.22 A 1-tonne car on Earth still has a mass of 1 tonne on the Moon, even though gravity there is only one-sixth that of Earth's.

The force of gravity acting on an object is also that object's weight. When you step on a bathroom scale, the scale reads how much gravity is acting on your body. The formula to determine weight is:

$$\text{weight} = \text{mass} \times \text{acceleration of gravity}$$
$$w = m \times g$$

Weight is measured in newtons (N) and mass in kilograms (kg). The acceleration due to gravity on Earth is 9.8 m/s^2 and it never changes, regardless of the object's mass.

While scientists distinguish between mass and weight, most people take mass and weight to mean the same thing. It is not uncommon to hear in the playground: 'What is your weight?' with the answer 'Oh, about 50 kilograms'.

Galileo noticed that light and heavy objects fell at the same rate. So if you were to drop a pebble and a large brick off a roof, they would hit the ground at the same time. But the force on the brick (its weight) is larger. It would make a greater dent in the ground. At the time of Galileo, it was difficult for scientists to measure speed or acceleration with any accuracy. Many people believed in the idea, predicted by the ancient Greek Aristotle, that a heavier object falls faster than a lighter one. Galileo clearly shattered this notion.

While the actual weight of a person is determined by their mass and the acceleration of gravity, one's perceived or effective weight comes from the fact that they are supported by the floor, chair or ground. If all support is removed from a person and they begin to fall freely, such as in a falling elevator, they will suddenly feel weightless. Weightlessness refers to a state of being in freefall, in which no support is perceived.

Plants also respond to gravity, as shown in Figure 3.23. Stems and leaves grow in the direction of light but also away from the force of gravity. Roots grow downwards towards the pull of gravity.

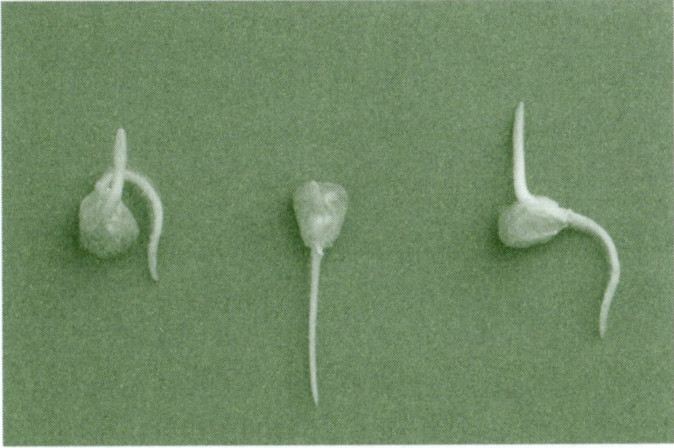

Figure 3.23 No matter how you plant a seed, the stem grows upwards against gravity while the roots grow downwards in the direction of gravity.

Gravity in space

Gravity acts on all matter in the universe. The strength of the gravitational force depends on both mass and distance. So the further apart the objects are, and the less massive those objects, the less the gravitational force. The gravity of the Sun is huge in comparison to everything else in the solar system and it keeps all the planets trapped around it in orbit, as shown in Figure 3.24.

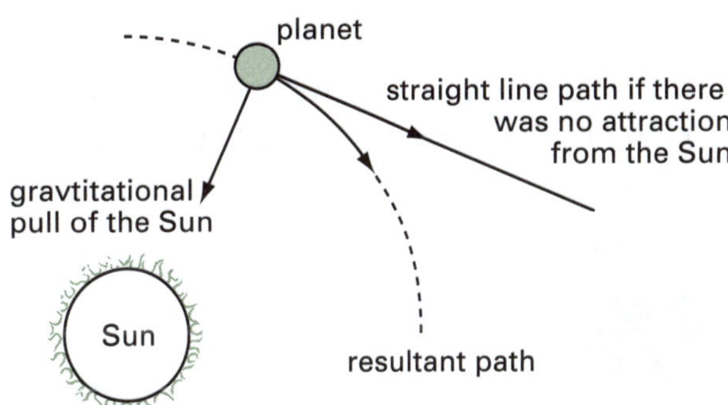

Figure 3.24 The gravity of the Sun keeps all the planets trapped around it in orbit.

Gravity is responsible for the formation of our Earth and all other planets. It is also responsible for the movement of all heavenly bodies. It is gravity that makes our planet revolve around the Sun, and the Moon revolve around the Earth.

The solar system was formed when a cloud of gas and dust in space was disturbed and squeezed. Squeezing made the cloud start to collapse, as gravity pulled the gas and dust together, forming a solar nebula. Eventually particles began to stick together and form clumps. Some clumps got bigger, eventually forming planets or moons. It is this same gravity which now keeps the planets, moons and asteroids in motion around the Sun.

Test yourself 1

Part A: Knowledge

1. Eclipses occur when the *(1 mark)*
   A Earth, Moon and Sun are lined up.
   B Earth is at right angles to the Sun and Moon.
   C Sun passes between the Moon and the Earth.
   D Moon passes between the Earth and the Sun.
2. Which diagram will correctly produce the phase of the moon shown in Figure 3.25? *(1 mark)*

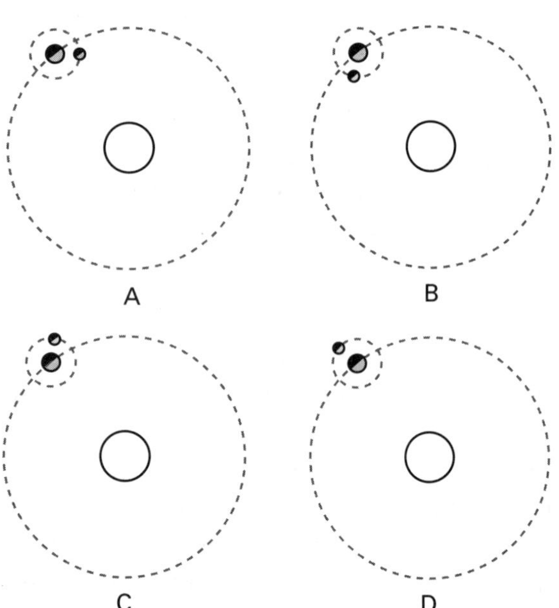

Figure 3.25 Phases of the moon

3. The name given for the two days of the year which exhibit the longest and shortest daylight is *(1 mark)*
   A autumnal.
   B vernal.
   C solstice.
   D equinox.
4. How long is each season in Australia? *(1 mark)*
   A 2 months
   B 3 months
   C 4 months
   D 6 months
5. The cause of the Earth's seasons is *(1 mark)*
   A the Sun.
   B the Moon.
   C the tilt of the Earth's axis.
   D the Earth's orbit.
6. Complete the following restricted-response questions using the appropriate word. *(1 mark for each)*
   a) A .... eclipse happens when the Moon passes between the Earth and the Sun.
   b) It takes one .... for the Earth to complete one rotation on its axis.
   c) .... is the attraction of one object by another object.
   d) One of the first scientists to point a telescope towards the universe was ....
   e) The Earth's only natural satellite is the ....
   f) The weight of a person is determined by gravity and the person's ....
7. Match the word or phrase in the first column with the word or phrase in the second column. *(1 mark for each)*

Column 1	Column 2
A retrograde	F equinox
B season	G moon
C solstice	H summer
D eclipse	I planet
E Galileo	J telescope

Part B: Skills

8. Figure 3.26 shows snapshots of the position of the Earth and a planet further from the Sun.

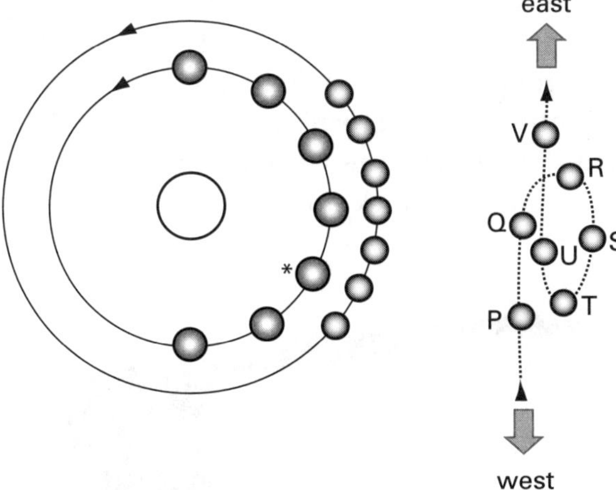

Figure 3.26 Position of the Earth and a planet further from the Sun

a) When the Earth is in the position marked with an asterisk (*), which letter shows the apparent position of the planet against the background of stars? *(1 mark)*

b) Which is moving faster: the Earth or the planet? How do you know? *(2 marks)*

c) What is retrograde motion? How does this diagram help to explain retrograde motion? *(2 marks)*

9. a) Jim planted a seed upside down in the ground. Mary predicted that it wouldn't grow and that he would have to dig it up and place it the right way. Is she correct? Explain. *(3 marks)*

b) Geoff thought the reason was that roots grow away from light, while shoots grow towards the light. Suggest an experiment Jim could perform to show that Geoff is wrong. *(2 marks)*

10. a) Many people believe heavier objects fall faster than lighter objects. Devise a simple experiment you could do to show objects fall at the same rate under gravity. *(3 marks)*

b) A scrunched up piece of paper falls faster than the same flat sheet. Does this fact contradict your experiment? Explain. *(2 marks)*

11. Outline four observations Galileo made with a telescope. *(4 marks)*

12. Identify each of these seasons from its description. *(1 mark for each)*

a) The weather is colder and, in some places, very cold and dark with snow covering the ground. People ski and play in this snow. The plants do not grow as fast. Some animals hibernate through the winter, while others migrate to places where it is warmer.

b) There are blossoms on some trees and many flowers grow. The weather is warmer, but may change considerably. Birds make nests and lay eggs, and many animals have young.

c) The leaves of some trees turn red, yellow or brown, and may then fall off the trees.

d) The weather gets hotter, and people and animals find ways of keeping cool. Many kinds of fruit are ready to pick. Young birds and other animals grow bigger.

13. a) What season in the southern hemisphere does Figure 3.27 show? Explain why. *(3 marks)*

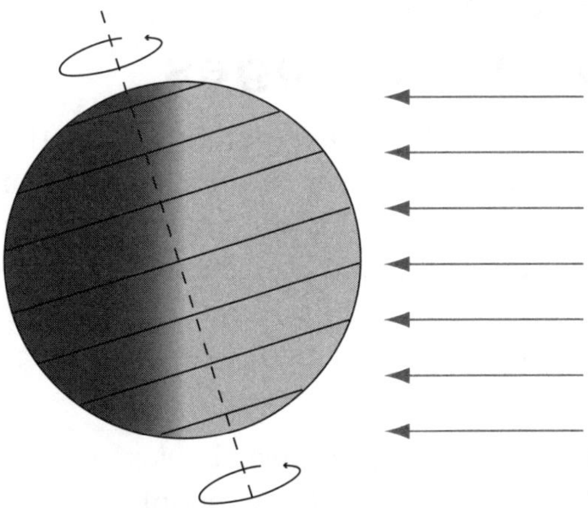

Figure 3.27 Season in the southern hemisphere

b) Is this season the same for the northern hemisphere? Why? *(4 marks)*

14. Identify which parts of Figure 3.28 (A, B or C) correspond to the following times and seasons in Australia: *(3 marks)*

- winter night
- summer night
- spring day.

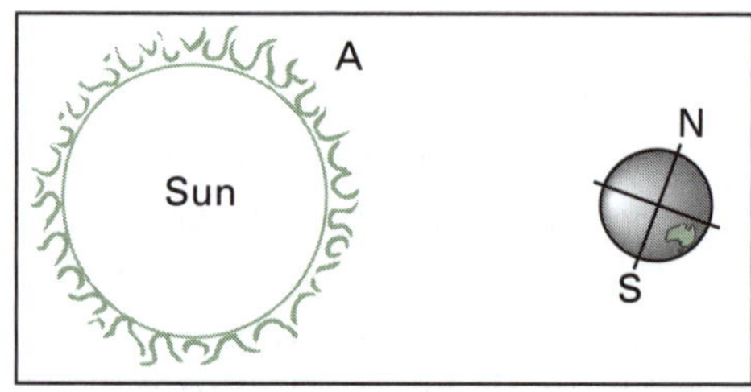

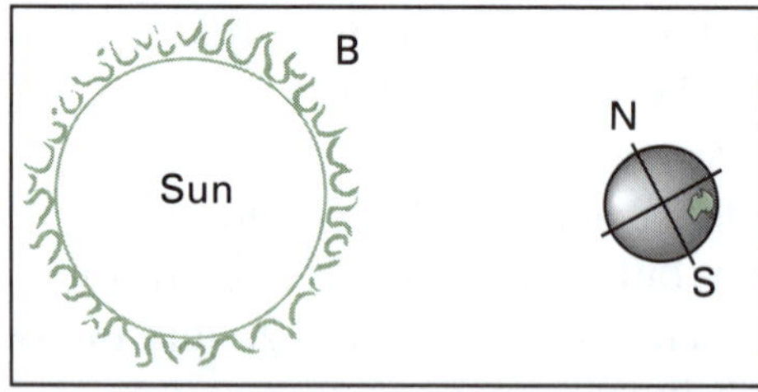

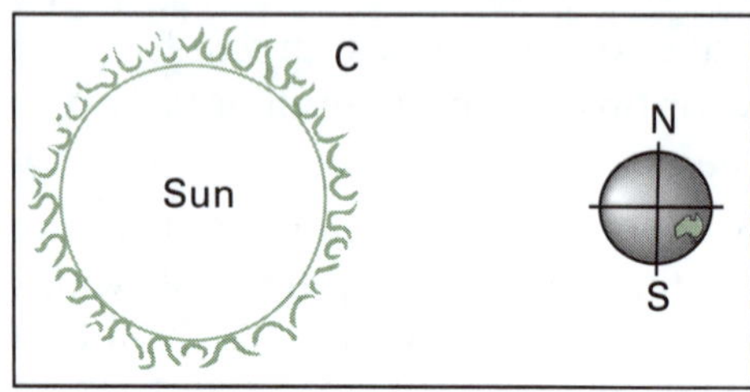

Figure 3.28 Times and seasons in Australia

Go to pp. 195–196 to check your answers.

3.8 Resources through the ages

All the things we use are derived from the earth, air and oceans. Some of these come from living things, such as plants and animals. Others come from the non-living environment, such as minerals and water. While some of these are synthetic, they still have a natural origin. Plastics are a good example.

The materials used by primitive peoples satisfied their basic needs in life. These people needed food, clothing, shelter and fire. Their needs were met simply from their immediate environment. The food most humans ate depended on where they lived. Aboriginal peoples living along the shoreline had diets rich in fish, turtles and shellfish. Those living further inland hunted animals such as kangaroos and wallabies. In the colder parts of Australia they used the hides of these animals to make clothing. Other foods included fruits, grains and the roots of plants.

Many Aboriginal groups used the boomerang (see Figure 3.29) as a tool mainly for hunting but also in religious ceremonies. Boomerangs used for hunting (non-returning) are heavy sticks with a slight curve. The weapon can easily kill a small animal or knock down a larger one.

Figure 3.29 The boomerang was invented by Aborigines

All the implements made by Aboriginal people were uniquely fashioned to use in their immediate environment. They developed and refined these over many generations and made them from readily available bush materials such as stone, wood, bark, lengths of vine, plant fibre, resin and shells. They also used the hair, sinew, bones and teeth of animals. Important materials not locally available (such as certain types of stone or even finished artefacts) were obtained through trade.

One essential factor in human development is the ability to discover and use a wide variety of materials. As humans learned more about how to obtain, extract, refine, alter and combine materials, society became more sophisticated. The discovery of extraction methods for metals such as copper and iron were important milestones in human progress.

Beginning in the 18th century, the Industrial Revolution produced a great variety of machines and led to the manufacture of an increasingly wide array of materials. Today science and technology have enabled us to live in a society that is quite different from that of even a century ago.

However, one of the problems of more sophisticated and advanced societies is that we use far more resources than our primitive cousins could. Not only are there more people on Earth today (currently more than 6.8 billion and growing by 83 million people per year) but we also demand a greater proportion of material and energy resources.

Where will we be in a century or two from now? For example, quill pens lasted for only about a week before it was necessary to replace them. It took some time to fashion from the flight feathers of large birds using a pen knife, and when used they needed to be constantly dipped in ink (see Figure 3.30). It wasn't until the 20th century that ballpoint pens would consign most earlier writing implements to history. Today much writing is done using computer technology (word processing, emails, SMS, twitter, Facebook), so will pens become a thing of the past at some time in the future?

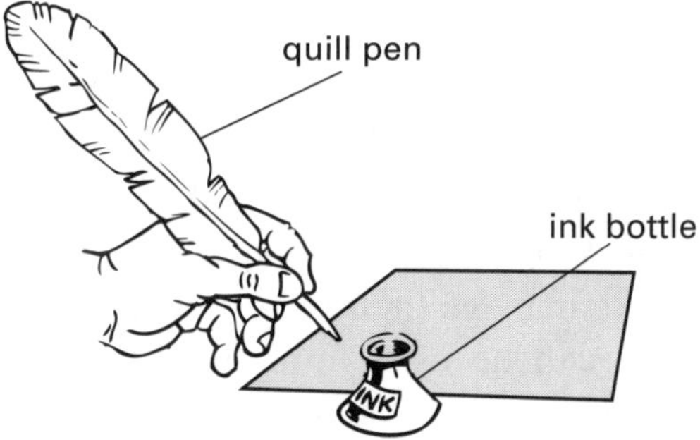

Figure 3.30 A quill pen

3.9 Natural and synthetic resources

Our basic foods (meat, fish, vegetables, fruit), building materials (wood, bricks and cement), and many fibres used for clothing (cotton, wool) are derived directly from the natural environment. These are referred to as natural resources.

However, humans have learned to modify or create alternatives to some of these natural resources. These are called synthetic resources. Some reasons for making them are:

- to provide variety and choice for consumers
- that man-made resources are often better suited to the purpose they will be used for than are the currently available natural resources
- that man-made resources are cheaper to produce than are natural resources.

While clothing is still produced from cotton and wool, a number of artificial fibres have added variety to these natural fibres. Polyester, dacron, rayon and nylon are some of the artificial fibres now available. Not only have these extended the variety, texture and durability of clothes, but many of these fibres are cheaper to produce, thereby decreasing the cost.

Foods, too, have been modified. Packaged foods are convenient and longer lasting than fresh food. Human ingenuity and the addition of man-made materials have made this possible. Artificial colours, flavours and preservatives are common additions to such foods. Some foods that are advertised as 'chicken flavoured' or 'pineapple flavoured' may not contain any chicken or any pineapple; rather, the flavour may be a chemical having this smell or taste concocted cheaply in the laboratory.

Human-made resources are those that have been altered from their original form by human ingenuity and technology. All resources we use are either renewable or non-renewable.

- Renewable resources are those that can be produced within a short period of time, usually no longer than a few decades. These include fish, wood, vegetables and drinking water. They are produced and maintained by the processes and functions of the ecosystem. Forests are planted for their wood and as each mature forest is cut down, another one is planted in its place.
- Non-renewable resources are those which cannot be replaced, or take too long to do so. It took millions of years to produce coal, oil and natural gas. As they are removed from the ground, they are not being replaced. The same is true of the many minerals we rely on to make metals.

For a long time, humans thought of the world's resources as infinite and took freely whatever they wanted. While populations were small, and had limited technology, this did not present much of a problem to the environment. But as demand and human population increased, more damage was done to the environment especially as machinery became available. This continues to occur.

- Cutting down trees. The early settlers in Australia cut down trees for use in building and firewood.
  - Some forests were turned into unproductive wastelands.
  - Many animals and other plants were deprived of their habitat and food sources.
  - In many developing countries, cutting down trees has exposed the soil to the elements and it is no longer stable. Consequently, mudslides during wet weather have killed many people and buried entire villages.
- Overfishing. Fish are considered a renewable resource.
  - Overfishing has wiped out many fisheries and elsewhere fish numbers have been drastically reduced. Many are close to extinction.
  - Commercial fishers often catch as much as they can, as quickly as possible. With large boats and nets, they can plunder huge tracts of ocean in a short period.

Many people now realise that resources must be conserved, managed and recycled if we are to use them wisely and make them last. By increasing wastes we are polluting the planet and making some resources scarce. As supplies dwindle, prices increase.

3.10 Using resources

The resources we use come from both the living and non-living world. While some of these are renewable, they must be used wisely and allowed to regenerate.

The wide range of resources which are used by humans are produced by an equally wide range of methods.

- Some resources can be used directly: fruit, vegetables, fish, sand, clay and granite.
- Some require simple methods of extraction before they can be used: salt from seawater, sugar from sugar cane, gas from oil or natural gas, and coal from mines.
- Some require more complex methods of extraction and chemical reactions are used to change naturally occurring materials: aluminium from bauxite, iron from iron ore and petrol from petroleum.
- Some need synthesising: plastics and nylon are made from simpler raw materials. Most plastics are built up entirely from materials obtained from petroleum or coal.

3.11 Resources from living things

Humans have used plants and animals for food, fibre, clothing and building materials for thousands of years. The book you are reading is made from wood chips, one result of forest management. When humans became farmers, they learned to manage their plants and animals so that some can be used immediately, some can be sold to others, and some can be used to generate or store for next season.

Farmers have devised methods over the years to grow plants and animals which have resulted in specimens that are:

- bigger and hardier
- disease resistant
- quick growing and give the farmer a greater yield.

For example, there are cows which are bred because they produce high quality milk. Other cows are raised for their meat. There are cattle that have been crossed with Asian cattle and are able to withstand the heat, biting insects and dryness of northern Australia. There are chickens that after only about a month or so have grown enough for meat usage. Other varieties of chicken are smaller but they are very efficient at turning feed into eggs.

Some farmers have turned to breeding fish in artificial fish farms. This way they control their breeding cycles and maintain numbers from season to season. Oyster farmers raise oysters for their meat and/or pearls. These farmers have learned a lot about the animals they are breeding and their conservation and preservation. Fishermen harvest fish (a resource) from the natural world. Many of these realise the importance of maintaining stocks for future fishing.

One controversial example of wildlife exploitation is whaling. While no species is yet extinct, many have been reduced to commercial extinction. This means their numbers are so few that it is no longer worthwhile hunting and catching them. Many local populations have been wiped out. For over 1500 years many species of whale have been hunted: right whales, bowhead whales, baleen whales and sperm whales. In some cultures whale meat is a delicacy and this provides the incentive to hunt whales. Many of the materials once only available from whales have been replaced by man-made alternatives. Why does whaling continue?

3.12 Resources from the non-living world

The Earth contains a fixed amount of minerals and fuels. Once they have been mined or pumped out they cannot be regenerated. These non-renewable resources can be mined only once. Once coal or petroleum is burnt, it has gone. Some metals and other non-renewable materials can be recycled. This will allow our supplies to last longer but at each stage some material is lost. In the long run they are permanently consumed.

Most of the mineral and fuel resources occur in the Earth's crust. Some are present only in small amounts and are not evenly distributed around the world. Sometimes they are concentrated in particular areas or rock types. Local concentrations like this are called ore bodies. Geologists who work for mining companies are employed to find new sources of ore to replenish those which are near depletion.

Three of the most common metals used in modern society are iron, aluminium and copper.

Iron

About 5% of the Earth's crust contains iron but most of it is in low concentrations, or it is chemically combined in rocks which makes extraction difficult and therefore expensive. Four ores, all oxides of iron, are important as iron ores: haematite, magnetite, limonite and siderite.

Today we use some 20 times more iron (in the form of steel) than all other metals together. The world's potential sources of iron are vast. It is not anticipated we will run out of iron in the near future.

Iron is used in all sorts of industries. In the form of steel (iron mixed with a small percentage of other metals and carbon, called alloys), it is the most useful metal available to humans. Table 3.3 outlines information about some different types of steel.

Table 3.3 Some types of steel

Alloy	Special properties	Uses
stainless steel	does not rust, strong	medical instruments, cutlery, vessels
silicon steel	strong, electrically resistant	electric motors, generators, transformers
tungsten steel	very hard	cutting tools, high-speed drills
manganese steel	hard wearing and tough	excavators, armour plates
vanadium chrome steel	shock resistant, hard wearing	axles, grinding machines, gear wheels

Depending on the percentage of the other alloys, it is possible to make steels for a wide range of uses. Examples include knives, railway lines, girders, cars, trains, food cans, tools, supports for buildings, casings for computers and other appliances, and so on. The list is endless.

Figure 3.31 shows the major mining sites for iron ore in Australia. Most iron ore is mined in Western Australia.

Figure 3.31 Operating iron ore mines in Australia

Aluminium

About 8% of the Earth's crust consists of aluminium. While it is the most abundant metal in the Earth's crust, other metals such as gold, copper, iron and zinc have been used far longer than aluminium. Humans only learned to purify aluminium a century ago.

Bauxite, which is a mineral consisting of aluminium hydroxides, is the ore mined for aluminium. Australia is one of the most important bauxite-mining countries in the world, mining some 40% of the world's bauxite. Aluminium is the second most used metal after iron, as it can be alloyed with a wide variety of other metals to produce materials with a whole range of useful properties. It is very strong, yet lightweight and resistant to rusting. For most purposes aluminium is used as an alloy (see Table 3.4). Small quantities of other metals or silicon make it harder and stronger than pure aluminium, which is soft and ductile.

Table 3.4 Alloys of aluminium

Alloy: aluminium +	Special properties	Uses
silicon	low melting point	welding fittings, castings
copper	stronger than steel but not as resistant to corrosion as other alloys	high-strength structures
zinc	can be heated very strongly	mobile structures, aircraft
magnesium	good for welding, high strength, resistant to corrosion especially in salt water	tanks, ships
manganese	easy to work with, very strong	sign boards, cooking utensils, tools

Figure 3.32 shows the major mining sites for bauxite in Australia.

Figure 3.32 Operating bauxite mines in Australia

Copper

The United States and Russia between them produce over one-third of the world's copper. The main sources of copper are the ores chalcopyrite, chalcocite, bornite and malachite. Copper is used both as a metal and as an alloy. Its most obvious use is in electrical wires since copper is a better conductor of electricity than many other metals. Salts of copper are used in industrial processes, textile dyeing, fungicides (prevents timber rotting) and disinfectants.

Copper was the first metal used by humans. Around 9000 BC, Neolithic peoples found that it was easier to shape than stone. About 4000 BC, it was discovered that by mixing it with a small amount of tin, the harder alloy bronze was formed. It was then used to make not only decorations, but also weapons, armour and tools. (The ancient wonder of the world, the Colossus of Rhodes, was made of bronze and reinforced with iron. The Statue of Liberty in New York Harbour contains more than 37 000 tonnes of copper.)

Figure 3.33 shows the major mining sites for lead, zinc, silver, nickel and copper mines in Australia. Most nickel is mined in Western Australia.

Figure 3.33 Operating lead, silver, zinc, copper and nickel mines in Australia

Our so-called silver coins (5¢, 10¢, 20¢, and 50¢ pieces) are alloys of 75% copper and 25% nickel, while the $1 and $2 coins consist of 92% copper mixed with aluminium and nickel. Table 3.5 outlines information about some different types of copper alloys.

Table 3.5 Alloys of copper

Alloy	Special properties	Uses
bronze (copper + tin)	resistant to corrosion, good for casting	marine applications e.g. propellers, statues, bearings in car engines and heavy machinery
cupro-nickel (copper + nickel)	resistant to corrosion, ductile	coins, turbines
brass (copper + zinc)	resistant to corrosion, easy to machine, good appearance	fittings and valves, hulls of sailing boats, musical instruments, decorative pieces (e.g. light fittings and taps), astronomical instruments
phosphor bronze (copper + tin + zinc + phosphorus)	resistant to corrosion, softens when heated	bearings

Other mineral resources

Humans mine and use other non-metallic resources as well. Here are some of these together with their uses.

- Building stone. Once very popular in buildings, many varieties of stone are no longer used. Bricks and mortar have taken over, partly because of changing fashions and partly because of the cost of digging out and working with building stone.
- Sand, crushed rock and gravel. These are still very popular in building roads, dams, constructing breakwaters and dams. Sand is also a major component of cement.
- Clay. There are many types of clay used to produce a variety of materials: china, drainage pipes, bricks, tiles, stoneware, abrasives, cement and glass to name a few.

- Cement. This is made from limestone and used daily for a wide variety of purposes like building and coverings.
- Fertilisers. While phosphorus, nitrogen, sulfur and potassium are usually available in the soil, farming can seriously deplete the soil of these necessary elements. Farmers add appropriate fertilisers to the soil to make good the depletion and to increase crop yields. For example, nitrogen is usually added to the soil in the form of ammonium sulfate. This is produced as a by-product of coke at steelworks.

3.13 Fossil fuels

Any source of energy can be considered to be a fuel. Until recent times, and even now in many parts of the world, humans used wood, peat and charcoal. Today's fossil fuels include coal, petroleum and natural gas. These substances are called fossil fuels as they were formed from the remains of living organisms millions of years ago.

The distribution of energy supplies around the world, especially fossil fuels, is very uneven. The largest producers of these fuels are not necessarily the largest consumers. The movement of energy supplies around the world is therefore an important component of international trade. Currently world movements of oil alone exceed 2000 million tonnes a year.

As the finite reserves of fossil fuels are used up, renewable energy resources such as solar, wind, tidal and biomass will become increasingly important energy sources to people on this planet.

Petroleum

Millions of years ago, primitive marine life died and fell into the silt beds on the ocean floor, where they then decomposed. These remains underwent complex chemical and physical changes, in an environment deficient in oxygen, as the sediments were compressed into rock. Bacteria assisted these changes. As the source rock was compacted, oil and water was forced out, and these gradually migrated to porous reservoir rocks (see Figure 3.34).

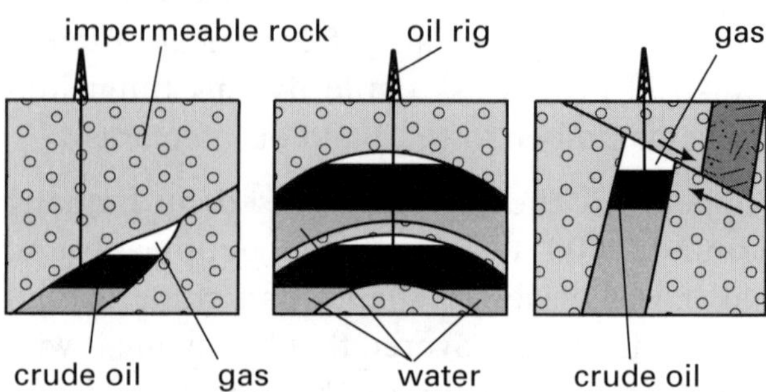

Figure 3.34 Some ways petroleum (crude oil) and gas may be trapped underground

Once oil is found, it is transported to oil refineries using tankers or pipelines. There it is separated into its various components using fractional distillation.

A number of products are made from petroleum, including bitumen, petrol, diesel, kerosene, aviation fuel and heating oil. While many of these products are well known, there are other less obvious ones. For example, Vaseline is a jelly used for lubrication and for soothing sore skin; it's made by distilling petroleum. Almost all man-made plastics are derived from petrochemicals.

Petroleum, sometimes called crude oil, is composed almost entirely of hydrocarbons. These are molecules consisting of long chains of carbon with hydrogen atoms. Petroleum is a mixture of solids, liquids and gases. Underground petroleum reservoirs, often found under the sea but now also found on dry land, are largely liquids and gases trapped in porous rocks like sandstone or limestone.

While geologists can recognise the Earth's underground features that might hold petroleum and can test for these, the only sure way of knowing is to do a test drill. This is a costly process, so oil companies do not want to embark on expensive drilling unless they are fairly certain they will find oil. The first oil well was drilled in the United States in 1859 but since then technology has allowed oil companies to drill wells to over 5 km depth and below 150 m of water.

The major oil-producing countries in the world are Saudi Arabia, Kuwait, Iran, Iraq and the United Arab Emirates. The major consumers are the highly industrialised Western countries such as the United States, Canada, Australia and Europe.

Petrochemicals are chemicals that come from petroleum and natural gas. Polymers, detergents, solvents and nitrogen fertilisers are just some of these.

Natural gas

Natural gas is a mixture of gaseous hydrocarbons found in reservoirs of porous rocks such as sandstone. Other impervious rocks, which cap these porous ones, trap the gas. Gas and petroleum are often found together but some wells, such as those in the Marlin field in the Bass Strait, produce only gas.

Natural gas is formed the same way as petroleum. It is mainly methane (CH_4), ethane (C_2H_6), with some propane (C_3H_8) and butane (C_4H_{10}). These last two gases are separated and can be used as bottled gas or LPG (liquefied petroleum gas) and are used as fuel by campers and for tractors, buses and cars. Figure 3.35 shows ball-and-stick models of the molecules of some simple hydrocarbons found in natural gas.

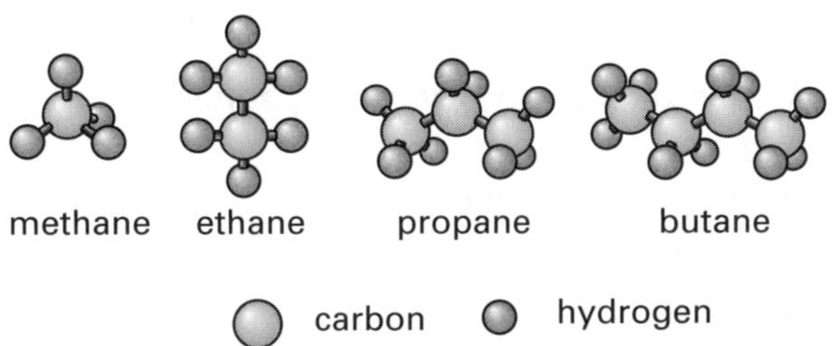

Figure 3.35 Ball-and-stick models of some simple hydrocarbons

Natural gas is used as an industrial and domestic fuel. The gas used in gas stoves and heaters has come from natural gas fields. Other uses include manufacturing fertilisers, generating electricity, and making hydrogen for oil refining.

Natural gas may also contain other hydrocarbons. The gases oxygen, nitrogen, carbon dioxide, hydrogen sulfide and (sometimes) helium can all be separated out from natural gas and used.

Across Australia, large pipelines channel natural gas to refineries. They are also transported as a liquid, in refrigerated tankers. The gas mined in Bass Strait is pumped to the Victorian coast using undersea pipes.

As a fuel, natural gas is versatile and clean; it is easily contained and transported. When petroleum and natural gas are burned, the greenhouse gases carbon dioxide and water are produced.

Coal

Coal is a hard, black or brown mineral often burned as a fuel. As Figure 3.36 shows, it consists of the compressed remains of plants that grew in tropical and subtropical regions about 250 million years ago. When these plants died, they fell into swamps and partly decomposed to form layers of peat. A lack of oxygen prevented bacteria from fully decaying these plants. As sea levels rose (or the land fell) these layers were buried under marine sediments. The weight of these sediments compressed the peat and, under high temperature and pressure, it was transformed to coal. The higher the pressure, the harder the coal became.

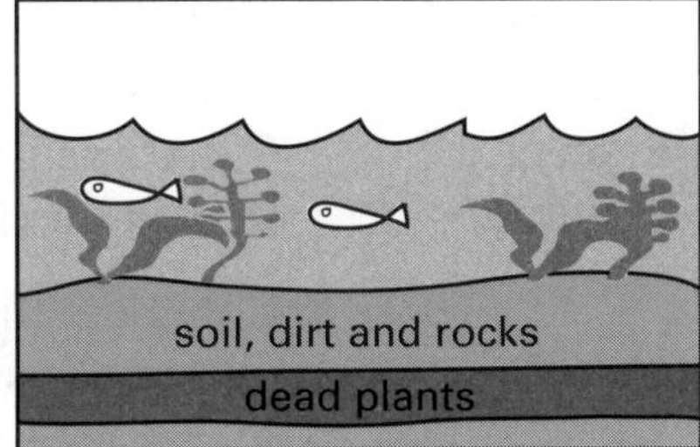

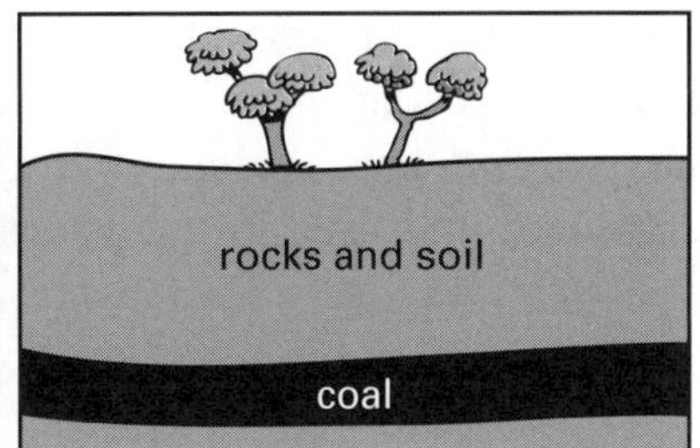

Figure 3.36 Steps in the formation of coal

Coal is mainly carbon and is ranked according to the proportion of carbon it contains. Lignite, or brown coal, is mined in Victoria but is a poor fuel as it contains up to 40% contaminants. New South Wales (38% of Australian total) and Queensland (58%) have large reserves of quality coal and these are near the surface (see Figure 3.37).

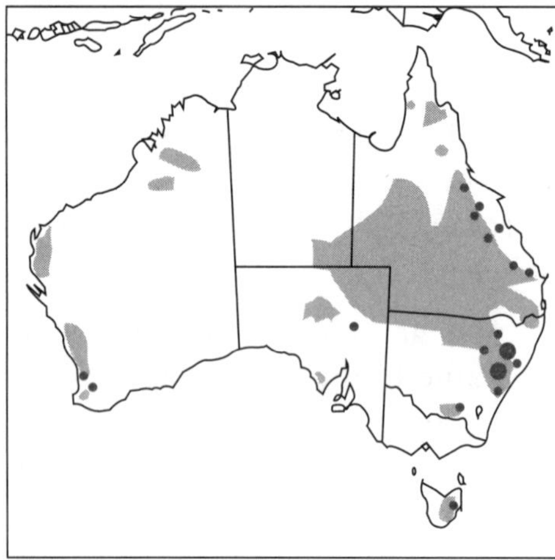

Figure 3.37 Known black coal areas in Australia are shaded; circular dots show coal-producing areas.

About one-third of Australia's main energy comes from coal. In Australia coal is mined for two main reasons: for use as a fuel in electricity power stations and to produce coke for the iron and steel industries. Coke is what remains when quality coal is heated in special furnaces to boil off the volatile components. This coke is strong, porous and relatively pure. The gas which boils off, coal gas, was once used as a fuel (e.g. it was used for street lighting in Sydney) but it is rarely used today.

Coal tar is a thick black liquid that is also produced from coal. This is used in a wide variety of industries. Examples of substances made from coal tar include creosote (a wood preservative), pitch, solvents, preservatives, lubricants, disinfectants and soap. Many chemicals such as benzene, toluene, naphthalene and anthracene can be isolated and used in the pharmaceutical industry.

Test yourself 2

Part A: Knowledge

1. Synthetic resources are good alternatives to natural resources for a variety of reasons. Which of the following is not a correct reason for choosing a synthetic resource? *(1 mark)*

A They increase variety and choice.
B They are better suited for the purpose for which they will be used.
C They are cheaper to produce than natural resources.
D They are made from renewable resources.

2. A resource that cannot be replaced in a reasonably short time is usually referred to as *(1 mark)*

A renewable.
B non-renewable.
C natural.
D synthetic (man-made).

3. Silk, used for clothing, is obtained from silk worms. This material would be a *(1 mark)*

A renewable, natural resource.
B non-renewable, natural resource.
C renewable, synthetic resource.
D non-renewable, synthetic resource.

4. Which of the following is a common use of clay? *(1 mark)*

A brick, tiles and china
B fertiliser and building stone
C tiles, dams and fertilisers
D ornaments, china and making copper

5. Identify the response that contains two fossil fuels. *(1 mark)*

A wood, coal
B petroleum, copper
C natural gas, hydrogen
D coal, crude oil

6. Complete the following restricted-response questions using the appropriate word.
(1 mark for each part)

a) Our silver coins are alloys of copper and ....
b) .... is the mineral from which aluminium is extracted.
c) Iron and carbon form an alloy called ....
d) Non-renewable .... are those which cannot be replaced.
e) Coal is a hard, black or brown mineral often burned as a ....

7. Use code letters to match the terms or phrases in each column. *(1 mark for each)*

Column 1	Column 2
A coal	F girders, axles and nails
B copper	G soil erosion
C steel	H non-renewable resource
D aluminium	I window frames
E tree felling	J electrical wires

Part B: Skills

8. a) Before compact fluorescent light bulbs were made, incandescent light bulbs used a tungsten wire which glowed white hot when electricity passed through it. Suggest why it is a suitable metal to use as the filament in such light bulbs. *(1 mark)*
b) Australia mines a small proportion of the world's tungsten ores. Table 3.6 shows the percentage of tungsten ores mined in Australian states.

Table 3.6 Tungsten ores mined in Australia (%)

State	% of tungsten ore mined
Tasmania	61
Queensland	27
Western Australia	7
Other states	X

The percentage of tungsten ore mined in the states other than Tasmania, Queensland and Western Australia is marked with an X. Calculate a value for X. *(1 mark)*

9. From where are each of the following resources mainly extracted: air, water or earth?
a) oxygen *(1 mark)*
b) salt *(1 mark)*
c) nickel *(1 mark)*
d) nitrogen *(1 mark)*
e) building stone *(1 mark)*

10. Ancient people caught or harvested and ate their food within a short period of time.
a) Why did they eat the food so quickly? Choose *two* possible answers. *(2 marks)*
A They were always very hungry.
B Food sources were scarce.
C They could not preserve the food for very long.
D Food goes bad when it cannot be refrigerated.
b) What developments have enabled humans to lengthen the time between collecting their food and eating it? Choose *two* possible answers. *(2 marks)*
A Freezing or refrigeration
B Cooking and canning
C Non-stick cookware
D Better hunting tools
c) What is the advantage of preserving food resources? Choose *two* possible answers. *(2 marks)*
A Particular foods are available out of their normal growing season.
B Global warming will ultimately destroy our food resources.
C Packaged, preserved goods are readily transported overseas.
D Minerals and vitamins do not break down.

11. Bill: 'So what if fishermen catch a lot of fish? Fish is a renewable resource and as they say, "there are plenty of fish in the sea".'
Jane: 'We need to conserve even our renewable resources. We need to find out what is available and only take out a small number of fish.'
a) Which person is correct? *(1 mark)*

b) Are fish numbers finite or infinite? *(1 mark)*

c) True or false: if fish are not given a chance to reproduce they will become extinct. *(1 mark)*

d) Why should fishing be controlled? *(1 mark)*

12. Cashmere is a type of wool for clothing that comes from the Kashmir goat, originally from Mongolia and China. The down shed by these goats that live in high, dry and windy plateaus is made of coarse outer hair that protects them from the weather. Beneath this is a finer fibre, cashmere, which insulates the animal from the cold. It takes about 4 years to produce enough cashmere from one goat to produce one jumper.

a) Some farmers in Australia keep flocks of Kashmir goats on their farms. Choose *two* reasons for growing cashmere here rather than importing it. *(2 marks)*

A Australian buyers have direct access to the wool without the cost of importing it.

B Kashmir goats prefer the Australian climate.

C Australian farmers can ensure direct quality control and improve the product for Australians.

D It takes a shorter time for the goats to mature in Australia.

b) For what purpose does the goat use this fur? *(1 mark)*

c) For what purpose do humans use this fur? *(1 mark)*

d) Do you think a cashmere jumper would be cheap or expensive? Why? *(2 marks)*

13. Scientists have found that soon after bushfires, or deliberate burning off near the coast, red algal blooms develop in the waters along the shore. An algal bloom is a rapid and excessive growth of algae. This has the effect of smothering other plant and animal life and leads to a serious decline in fish. They noticed the connection by analysing water samples soon after the fires. The coastal waters were found to be high in nitrogen and phosphorus.

a) Suggest how the phosphorus and nitrogen and may have found their way into the water. Choose *two* possible answers. *(2 marks)*

A Winds blow smoke and ashes (containing the nitrogen and phosphorus) out to sea.

B Water birds carry these minerals out to sea.

C Sewage wastes pollute the ocean with nitrogen and phosphorus.

D Rainwater washes the smoke and ashes from the land into the sea.

b) Suggest a hypothesis to explain why high nitrogen and phosphorus levels in the water would cause an algal bloom. *(1 mark)*

c) The algal blooms float on the water surface.

i) When the algal blooms float on the surface, will more or less sunlight be able to penetrate into the water? *(1 mark)*

ii) Will the normal underwater plants receive enough sunlight to carry out photosynthesis to produce food and oxygen? *(1 mark)*

iii) Why will this process lead to a serious decline in fish stocks? *(1 mark)*

d) Tree felling by humans has had a serious impact on the environment. Explain what damage this activity causes. *(1 mark)*

14. Many farmers grow wheat crops in Australia. Their aim is to harvest the maximum amount of wheat to make a profit. This means that they need to manage the land.

a) From the following six methods, identify *four* ways a wheat farmer can get the maximum yield of wheat from using land sustainably. *(4 marks)*

A Add fertiliser to the soil before planting.

B Build dams to collect rainwater.

C Allow cows access to the crop area so their droppings will fertilise the ground.

D Spray the wheat crop with toxic pesticides daily.

E Grow wheat varieties that are disease resistant.

F Use trickle irrigation to conserve water.

b) Some people believe that farming the land and sustainable land management are contradictory ideas. Do you agree? Explain. *(2 marks)*

Go to p. 196 to check your answers.

3.14 Alternative energy resources

The main energy sources considered so far are petroleum, natural gas and coal. Unfortunately, burning these non-renewable energy sources produces carbon dioxide and other waste chemicals. Over time, these create serious environmental consequences through air pollution and warming. There are other sources, many of which are renewable, and will play a more important role in the years ahead.

Hydro-electricity

Hydro-electricity is a renewable energy source that utilises the power of falling water from dams, rivers and waterfalls. Often, power stations are located near dams because storing and releasing water can be controlled. Water moves through a generator and turns the blades of a turbine. The rotating turbine creates electricity in the generator, which is passed onto the electricity grid as shown in Figure 3.38.

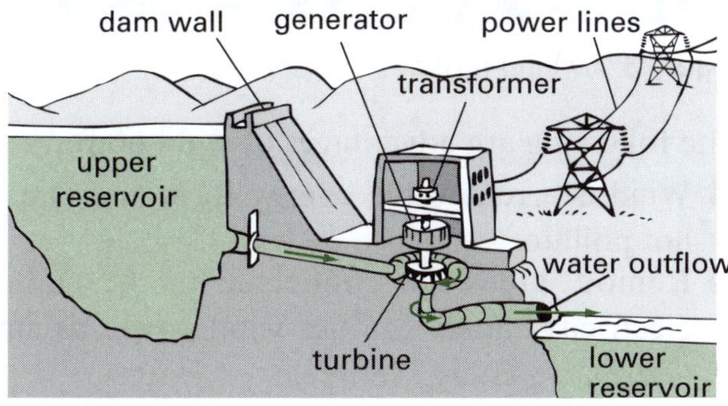

Figure 3.38 Generating hydro-electric power

Australia has a number of large hydro-electric energy sources, the most famous being the Snowy Mountains Hydro-electric Scheme, the largest in the country. Tasmania's Hydro-Electric Corporation generates the second largest amount of hydro-electricity in Australia. Hydro-electricity currently accounts for 97% of Australia's renewable energy consumption. In terms of total electricity generation, hydro-electric power represents 10% of Australian production and 18% of world production.

The following are the advantages of hydro-electricity.

- Water is a renewable resource and does not pollute.
- As water can be stored, the electricity generated can be controlled to meet demand.
- Hydro-electric power stations are relatively inexpensive to operate.
- Irrigation and water supply can share costs.
- It is cheaper and less invasive than mining for fossil fuels.

Hydro-electricity also has the following disadvantages.

- Hydro-electric schemes often require constructing dams which can have significant environmental effects on river flows, water quality, and plant and animal life.
- Hydro-electric schemes are expensive to build.
- Appropriate sites for dams are not always available.
- When water supplies are reduced drastically, such as in droughts, the amount of electricity that can be generated is limited.

Solar power

Solar power is a renewable energy source produced from the sun. While the sun's energy has been used for centuries in several ways, it wasn't until 1954 that scientists were able to utilise its power. There are two main forms of solar power.

- Solar photovoltaic energy. Panels trap the sun's energy causing electrons in the solar panels to move, creating an electric current (see Figure 3.39). This current can be used to run appliances or as excess power sold to an electricity company.

- Solar thermal energy. The sun's heat energy is used to heat other substances such as water. In most homes water heating accounts for up to 60% of the energy used, so installing solar hot water makes sense.

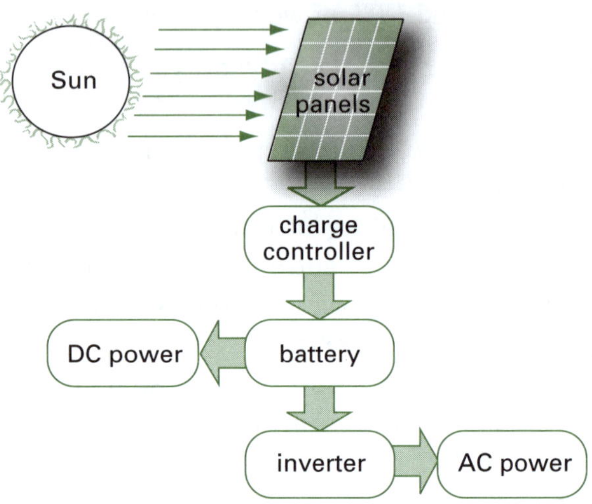

Figure 3.39 Generating electricity using solar panels

The following are advantages of solar power.

- The sun's energy is sustainable. This is a renewable resource.
- Solar panels are cheap to run.
- There is no ongoing maintenance other than keeping the solar panels clean.
- Operating solar cells do not produce greenhouse gases or other emissions.
- Solar cells are long lasting.
- Hot water can be stored until needed, while electricity generated requires storage batteries or can be fed back into the power grid.

Solar power has the following disadvantages.

- Electrical energy or water heating can only be generated during daylight hours.
- Cloudy days significantly reduce the amount of electricity or hot water that can be produced.
- Initial purchase and installation costs can be high.
- Stored excess electricity can result in energy losses.

Solar vehicles, such as bicycles, boats and cars, use solar panels on the surface (generally, the top) of the vehicle where photovoltaic cells convert the sun's energy directly into electrical energy. Many are not practical day-to-day transportation devices at present, but are experimental and engineering exercises.

Wind power

Wind power is a renewable energy source that uses windmills to harness the energy of moving air. The turning blades of the windmill drive a generator which generates electricity. Modern turbines are mounted on top of towers some 40 to 60 m high to maximise their exposure to the strongest winds (see Figure 3.40). While only 1% of all electricity supplied in Australia is through wind power, new wind farms being built around the country should see its popularity increase. The tower of a wind turbine is usually hollow and made of steel. The rotors (blades), curved on one side and flat on the other, are made from fibreglass and polyester.

Figure 3.40 Wind turbines

The following are advantages of wind power.

- Wind is a renewable energy form and does not pollute.
- Remote areas not connected to the main electricity supply can use wind power as an alternative energy source.

- The base of wind turbines only occupy small amounts of land so other uses, such as farming, can occur underneath.
- Wind farms are cheap to operate and have low maintenance levels.

Wind power has the following disadvantages.

- Wind strength is variable, and so it cannot be relied on as the main generator of electricity.
- Turbines are noisy so they are not suitable near residential areas.
- Construction costs can be high.
- Generators require a large area.
- Wind turbines are very large and many people consider them visual pollution.
- Currently wind turbines are not very efficient and many turbines are needed to supply a reasonably sized community.
- Windmills can be dangerous to bird life and need to be built away from flight paths.

Biomass

Biomass is a renewable energy source utilising a variety of animal and plant matter, such as plant waste, wood, saw millings, animal manure and even human sewage, converting it into useable fuels such as ethanol and biogas. Decomposing organic matter produces heat that can be used to produce energy. This natural process can be accelerated with the resulting heat used to drive turbines and so create electricity (see Figure 3.41). Biomass does not produce extra carbon dioxide; it merely hastens a natural process.

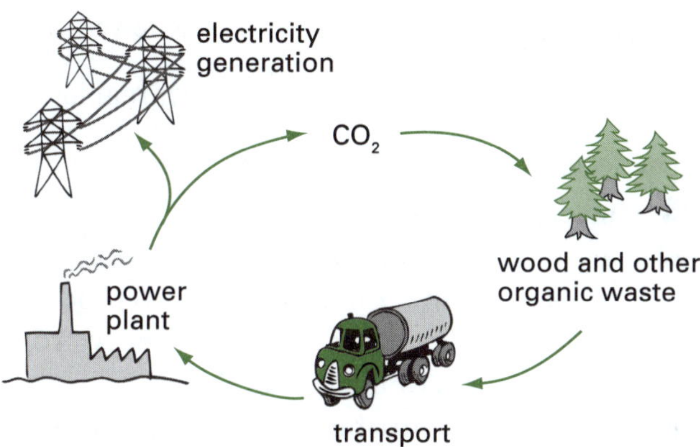

Figure 3.41 Simple stages in using biomass to generate electricity

The following are advantages of using biomass.

- Biomass is a renewable energy form.
- Fuels made from biomass, such as ethanol, can be used as an additive to petrol.
- It re-uses what would have been agricultural wastes.
- It can be made greenhouse neutral.
- It breaks down methane (a greenhouse gas).
- Biomass can be stored and used when needed with minimal energy loss.

Biomass has the following disadvantages.

- Organic wastes are energy poor, so handling and transporting is not economical.
- Large land areas are needed to produce the fuel.
- Waste organic matter might have improved the soil, had it remained.
- The energy processing plant needs to be located close to the source of biomass.
- Biomass creates odours.

Tidal power

Tidal power is a newer source of energy that converts the energy of tides into electricity or other useful forms of power. The tide moves a huge amount of water twice each day, and harnessing it could provide a great deal of energy. But converting it into useful electrical power is not easy.

A huge dam (barrage) is built across a river estuary. When the tide goes in and out, the water flows through tunnels in the dam. This ebb and flow of the tides can be used to turn a turbine as shown in Figure 3.42. Large lock gates allow ships to pass.

The largest tidal power station in the world is in La Rance estuary in northern France, built in 1966. There are no tidal energy facilities currently operating in Australia, although a site near Derby in the north-west of Western Australia is suitable.

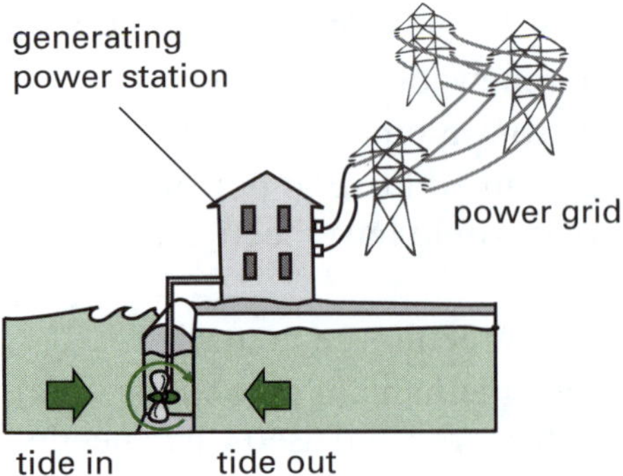

Figure 3.42 The tide moving in and out turns a turbine.

The following are advantages of tidal power.

- Tidal power is the only renewable form of energy derived directly from the relative motions of the Earth–Moon system.
- Tides are more predictable than wind energy or solar power.
- Recent technological developments and improvements, both in design and turbine technology, are making it more competitive.
- The energy supply is reliable and plentiful.
- Once the barrages and turbines are built, tidal power is free, with very little maintenance.
- It does not produce greenhouse gases or other waste.
- No fuel is required.
- Offshore turbines are not expensive to build and don't have a large environmental impact.
- Tidal power can provide secondary benefits such as bridges and roads, which are built over the tidal generators.

Tidal power has the following disadvantages.

- Tidal power is generally more expensive than other energy forms.
- There is a limited availability of sites with sufficiently high tidal ranges or flow velocities.
- A barrage across an estuary is expensive to build, affecting a very wide area.
- The environment is changed upstream and downstream. Many birds rely on the tide uncovering the mud flats so that they can feed, and fish migration is hampered.
- Tidal power changes the way sediments move and the turbidity (clarity) of the water system.
- Power is only available for around 10 hours each day, when the tide is actually moving in or out.

Nuclear energy

Nuclear power produces around 11% of the world's energy needs. Worldwide there are around 500 nuclear power plants operating. As Figure 3.43 shows, a nuclear reactor uses uranium rods as fuel. Heat is generated by nuclear fission where neutrons smash into the nucleus of the uranium atoms, splitting them roughly in half and releasing energy in the form of heat. Carbon dioxide gas or water is pumped through the reactor to take the heat away; this then heats water to make steam. This steam drives turbines that drive electricity generators.

Australia does not currently have any nuclear power reactors. There is one small nuclear reactor at Lucas Heights in Sydney, which is used for research and to provide medical radio-isotopes.

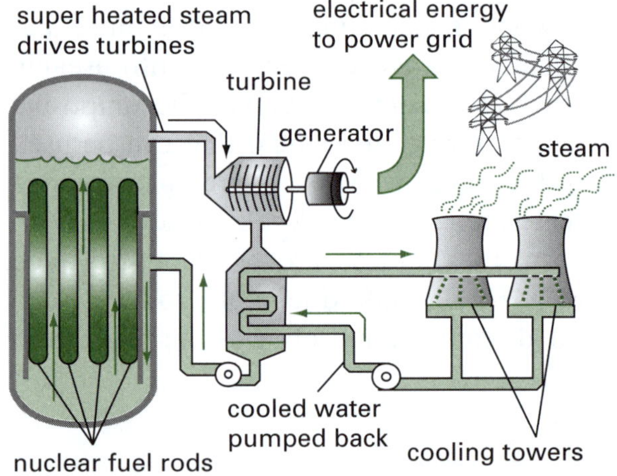

Figure 3.43 Principle of a nuclear power reactor

The following are advantages of nuclear energy.

- Nuclear power is reliable.
- Currently, nuclear power costs about the same as coal, so it's not expensive to produce.
- It produces large amounts of energy from small amounts of nuclear fuel.
- It does not contribute to the greenhouse effect, as it doesn't produce smoke or carbon dioxide.

- It has minimal amounts of waste.

Nuclear power has the following disadvantages.

- Nuclear energy from uranium is not renewable.
- Building a nuclear reactor is expensive.
- Although not much waste is produced, what is produced is very dangerous.
- While nuclear power is reliable, maintaining safety can be expensive.
- Many people in Australia don't want a nuclear power reactor built near them.

The main concern many people have is of an accident at a nuclear power reactor. In March 2011, several reactors were partly damaged when a major 8.9 magnitude earthquake hit Japan, followed soon after by a tsunami. Failure of the cooling systems allowed pressure to build up beyond the design capacity of the reactors. Small amounts of radioactive vapour were released into the atmosphere to prevent damage to the containment systems. Some radioactive material had also seeped outside, with radiation levels nearby some eight times normal. One of the reactors exploded soon afterwards due to a build-up of hydrogen gas formed when cooling water decomposed on contact with the hot core. Such damage can potentially affect many square kilometres for quite a long time, but such accidents are rare since safety is paramount at nuclear power reactors.

Geothermal power

Temperature increases below the Earth's surface. Even a few kilometres down in the ground, it can be over 250 °C where the Earth's crust is thin. Hot rocks underground can be used to heat water to form steam. Holes are drilled down to the hot region; steam comes up, is purified and is used to drive turbines, which drive electric generators (see Figure 3.44). Sometimes it may be necessary to drill more holes and pump water down to the hot rocks.

The Earth's magma and hot dry rocks will more widely provide cheap, clean and almost unlimited energy as soon as the technology is developed to make it cost-competitive with traditional forms of energy.

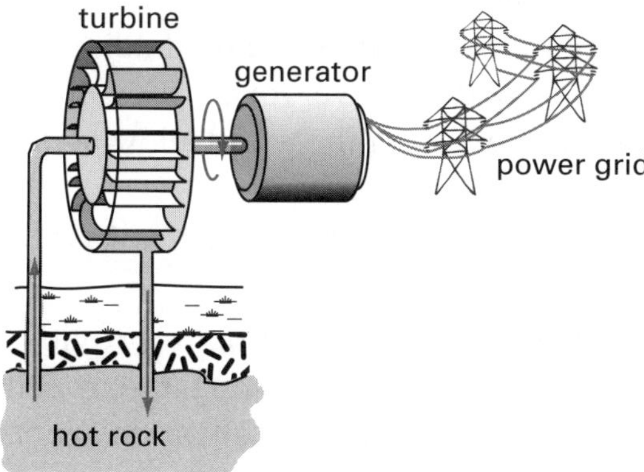

Figure 3.44 Geothermal power production

The following are advantages of geothermal power.

- Geothermal energy does not produce any pollution, and does not contribute to greenhouse warming.
- There is not much impact on the environment as the power stations do not take up much room.
- No fuel is needed.
- Once a geothermal power station is built, the energy is almost free.

Geothermal power has the following disadvantages.

- Sites for locating a geothermal power station are limited.
- Not all hot rocks are suitable or at a depth reasonably close to the surface.
- The overlying rock must be easy to drill through.
- Sometimes a geothermal site may not produce enough steam for quite a period.
- Hazardous gases and minerals may issue from underground that may be difficult to safely dispose.

3.15 Collaborative research in Antarctica

Antarctica is the most untouched and undisturbed region on the planet since it is remote, isolated and frozen all year. This offers scientists many advantages over anywhere else on the planet for study. Antarctica:

- has no borders allowing research findings to be freely available to everyone
- has the cleanest air in the world allowing for reliable air quality monitoring
- is an ideal setting for astronomy as it is the darkest place on Earth.

Being undisturbed, the continent provides scientists with a global baseline against which to measure the damage inflicted by humans on the rest of the planet.

Over 50 years ago the Antarctic Treaty was signed by several nations to protect the pristine condition of the Antarctic environment and the wise use of its resources. There are now 44 countries involved in the treaty. Representatives meet annually to discuss issues as varied as scientific cooperation, measures to protect the environment, the management of tourism and the preservation of historic sites.

Antarctica currently holds 70% of Earth's fresh water, and 91% of its ice. However, the existence of plant fossils and coal beds found in Antarctica clearly proves that the Antarctic was not always an icy desert. But now, except for a few tiny insects, algae, lichens, mosses and microscopic life forms, the Antarctic interior is very barren and has too harsh an environment to support significant plant and animal life.

The Australian Antarctic Division undertakes science programs and research projects contributing to an understanding of Antarctica and the Southern Ocean. It conducts and supports collaborative research programs with other Australian and international organisations. It maintains three permanently manned stations (Casey, Davis and Mawson) on the Antarctic continent, and one on Macquarie Island. Figure 3.45 shows the locations of these bases as well as some other Antarctic research stations. McMurdo and Halley bases are operated by the United Kingdom and Palmer base is operated by the United States.

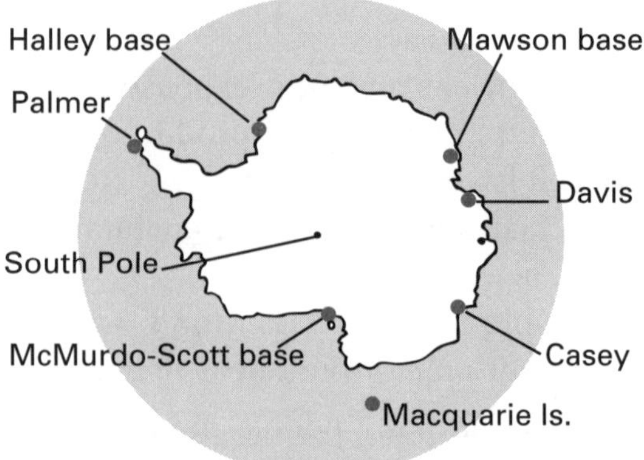

Figure 3.45 Some of the research stations on the Antarctic continent

Some research conducted recently includes the following.

- In the 1960s and 1970s, research into how radio waves move through the Antarctic ionosphere was conducted by researching the electromagnetic properties of the Antarctic ice sheet.
- Glaciology is the study of the movements and formation of the Antarctic ice sheets and how ice forms and accumulates. This study led to the discovery of lakes under the glaciers.
- Past climate and atmospheric changes have been researched by examining Antarctic ice cores and ocean sediment records.
- Sea ice extent and duration, its temperature, salinity and movements, critical to the support of marine food webs, have been researched.
- The process of ice formation, which affects ocean circulation around the globe, has been researched.
- Research was carried out mapping the landscape in Antarctica, both above and below the ice sheets.
- The geographic structure of Adélie penguin populations, and how their numbers vary, has been monitored.
- Stratospheric ozone amounts and cloud profiles have been measured.

- Current animal and plant relationships between biodiversity and ecosystem processes in a cold desert ecosystem have been studied.
- It is the perfect place to study cosmic microwave background radiation, which is believed to be the remaining echo of the big bang.

3.16 The water cycle

The Earth contains vast amounts of water. The oceans cover about 70% of the Earth's surface. Yet there are many places on the Earth (such as deserts) where there is very little water, fresh or otherwise.

While the amount of water found in rivers, oceans, in the ground or in the atmosphere remains fairly constant, water is always moving from one area to another. This never-ending circulation is called the water cycle.

Change of state

The water cycle involves water changing state between solid, liquid and gas.

- In the solid state, water exists as ice or snow.
- Liquid water is present in oceans, lakes, rivers, in porous rocks beneath the surface and inside the bodies of living things.
- Gaseous water is called water vapour and it is mainly present in the atmosphere.

The water cycle is powered using energy from the Sun. The physical processes of evaporation and condensation are important in understanding the water cycle (see Figure 3.46).

- Evaporation: the process of converting a liquid into the gaseous state. The speed of evaporation can be increased by warming the liquid.
- Condensation: the process of converting a gas or vapour into the liquid state. The speed of condensation can be increased by cooling the vapour.

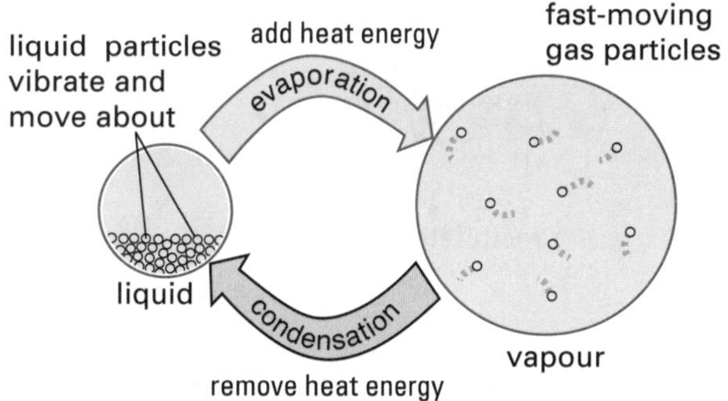

Figure 3.46 Particle model of evaporation and condensation

Water evaporates from:

- oceans, rivers and lakes as well as the land
- the leaves of plants in a process called transpiration
- our skin when we sweat and from our lungs as we exhale.

Water vapour condenses into liquid water in the upper atmosphere. These water droplets form clouds. Some water remains in the atmosphere as water vapour.

Water returns to the earth by a process called precipitation. Sometimes the liquid water in these clouds freezes to form ice or snow. When hail or snow forms, the liquid water is converted to solid water. This physical process is called solidification or freezing. When snow or ice turns back to liquid water, the process is called melting or fusion. This is shown in Figure 3.47.

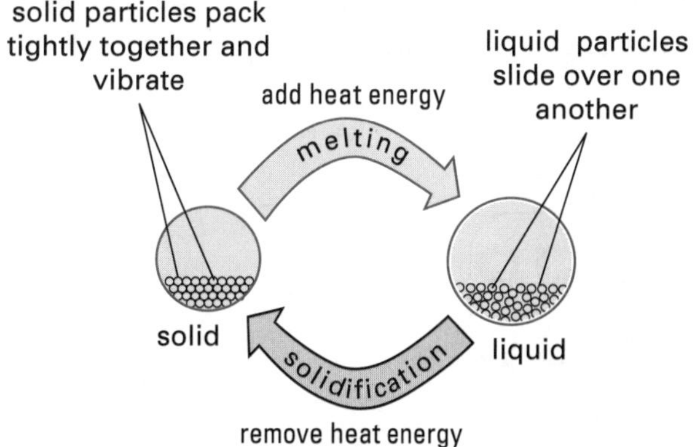

Figure 3.47 Particle model of solidification and melting

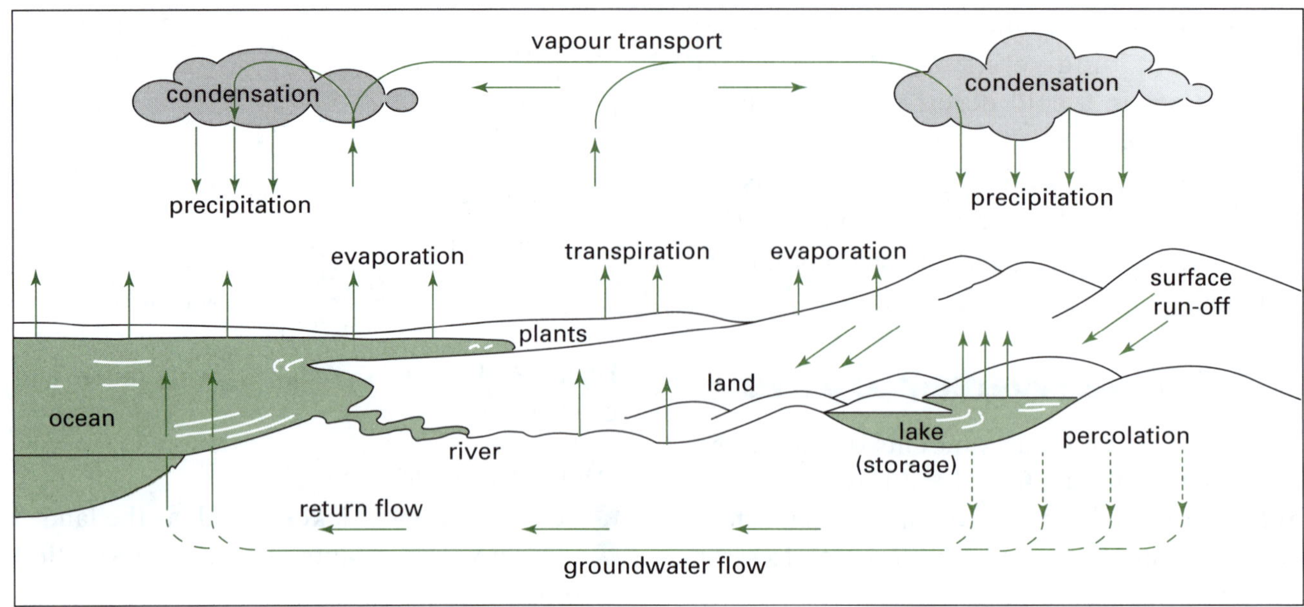

Figure 3.48 The water cycle

A lot of the water falling on land soaks or percolates into the ground. Some water has been trapped inside the ground in porous rocks for thousands of years. The remainder runs off the top in the form of streams and rivers and heads back to the sea. And so the cycle goes on. Other water may be trapped in the form of ice and snow; some of it has been there for millions of years.

The wind and weather play an important role in this cycle. Water evaporating in one area and forming clouds can be blown thousands of kilometres before being dumped. The direction and strength of the wind determines where the water ends up. Figure 3.48 shows a simplified view of the water cycle.

Particle models can be used to understand the processes occurring in the water. Particle models for water undergoing changes of state are shown in Figure 3.49.

Experiment 2

Modelling the water cycle

Aim

- To construct a model that demonstrates part of the water cycle

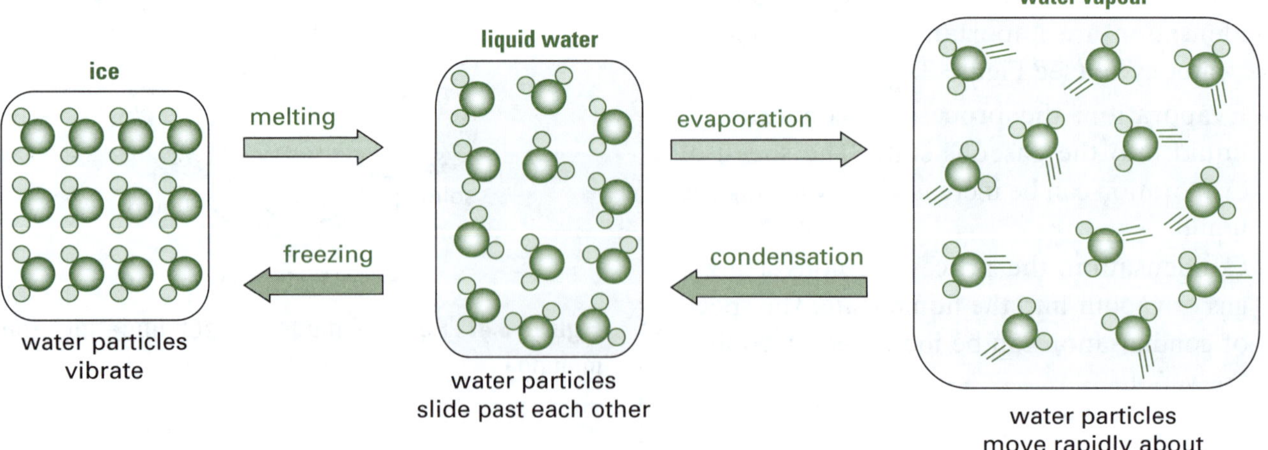

Figure 3.49 Particle models of water changing state

Method

1. Set up the apparatus shown in Figure 3.50. Place some copper sulfate solution in the larger beaker. Insert an empty smaller beaker.
2. Fasten the plastic wrap with a large rubber band. Place a small mass in the centre of the plastic wrap so that it creates a depression.
3. Leave the apparatus in the Sun (in a sunny window, for example) for a day or so. Observe what happens to the water.

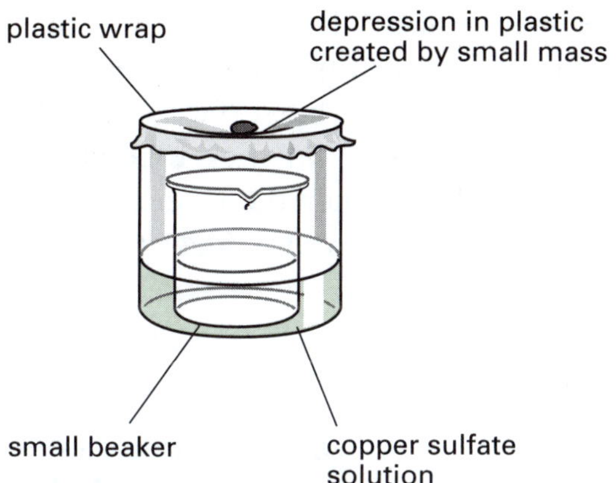

Figure 3.50 Water cycle experiment

Results

The beaker gradually collects clear, fresh water and the copper sulfate solute is left behind in the larger beaker.

Analysis

Analyse this result and write your response.
Go to p. 194 to check your answer.

Conclusion

Write a suitable conclusion.
Go to p. 194 to check your answer.

Clouds and the water cycle

Cloud formation and the water cycle are closely connected. Consider the following examples of ways in which cloud formation is promoted.

Colliding air masses

Warm and cold masses of air contain different amounts of water vapour. As Figure 3.51 shows, a warm air mass that forms over an ocean promotes evaporation and so the warm air mass is quite humid. A cold air mass that forms over ice or snow will contain very little water vapour and is quite dry. When these two air masses collide, the denser cold air pushes the less dense warm air mass upwards where it rapidly cools. The water vapour it contains then condenses and clouds form. This process contributes to the production of rain or snow.

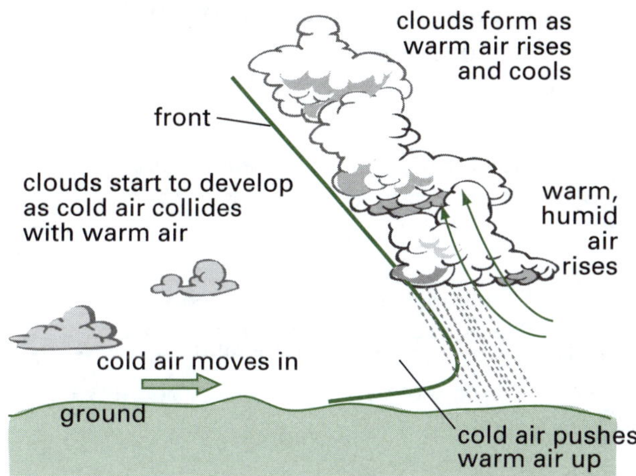

Figure 3.51 Cold air collides with warm, moist air

Mountain ranges

When a warm, humid mass of air moves over the land, it is sometimes forced upwards as it approaches a mountain range. As the warm, moist air moves upwards it cools and cloud formation occurs. Cool air can't hold as much water, thus forming clouds. This process is shown in Figure 3.52.

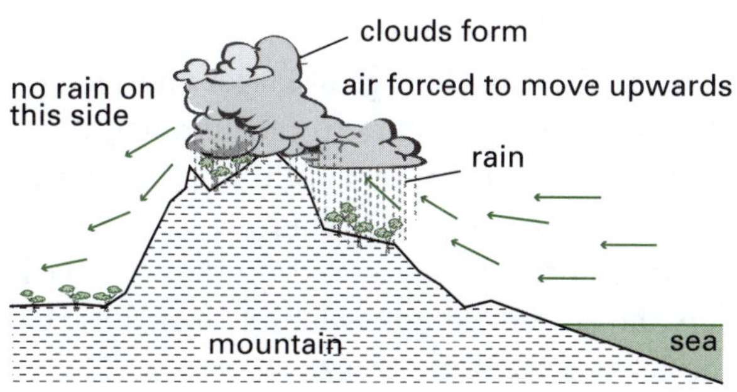

Figure 3.52 Moist air moving over a mountain range

Convection currents

The surface rocks and soil of the Earth become very warm when exposed to the Sun during the day. These hot rocks heat the layer of air nearest the ground and the heated air becomes less dense. This less dense air rises into the upper atmosphere. Eventually this air becomes so cool that any water vapour condenses to form clouds.

At the same time, the convection current is completed as cooler air sinks. These convection currents are responsible for cooling sea breezes along the coast. At night the land cools faster than the sea and so a convection current is established and land breezes form. In this case, clouds form over the sea. Figure 3.53 shows this process.

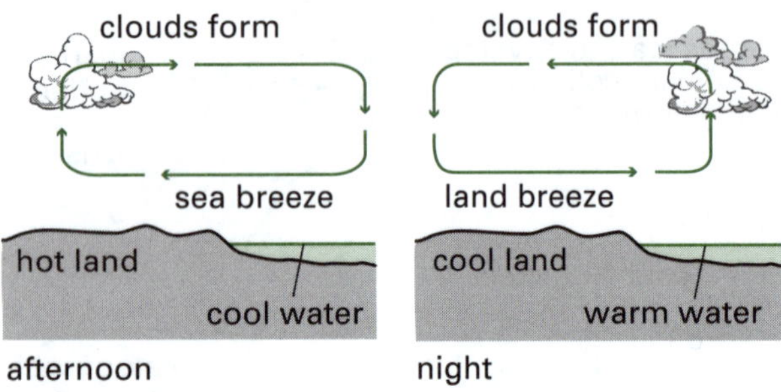

Figure 3.53 Sea breezes and land breezes

Hail and snow

Liquid water drops can freeze to form hail and snow.

Hail

The steps in the formation of hailstones are as follows.

- Upward-moving air inside cumulonimbus clouds pushes water drops higher into freezing conditions where they solidify into very small ice crystals.
- The small ice crystals grow larger and, as they get heavier, they start to fall under gravity but they get caught by another updraught of air and move upwards again.
- More moisture is gathered on the way back up and so the ice grows larger in size as this additional moisture freezes.
- This process is repeated until the weight of the hail exceeds the force of the updraughts and then the hail falls to the ground.

Snow

The steps in the formation of snow crystals are as follows.

- Snow forms in weather systems called extratropical cyclones.
- Individual six-sided ice crystals begin to form as water freezes in the clouds. Once they are large enough, they start to fall.
- Falling ice crystals have six branches and as long as the temperature is low enough they grown bigger by absorbing more moisture.
- As the ice crystals fall further down, they come in contact with warmer air and their surfaces start to melt. This surface water can lead some ice crystals to stick together. If too many stick together then sleet instead of snow falls to the ground. Sleet feels wet whereas snow feels dry on the skin.

Figure 3.54 shows a snowflake and the cumulonimbus clouds which produce snow.

Figure 3.54 Snowflake shape and cumulonimbus clouds

3.17 Human impact on the water cycle

The influence of humans on the water cycle is quite varied.

Population growth and urbanisation

The growth of the human population has led to the construction of cities and homes as well as industries. The original rural landscape is then replaced by an urban landscape. These changes have a great effect on the water cycle. Here are some examples:

- removal of vegetation and reduction in plant transpiration
- loss of small depressions in the ground that temporarily store water
- reduction of natural drainage of rain into the soil and evaporation of water from the soil surface
- more surfaces (e.g. rooftops, concrete driveways, roads) that do not absorb rain and lead to an increase in surface run-off into drains
- large rises in the amount of storm water; this can lead to periods of intense flooding
- changes in patterns of rainfall.

Deforestation

Deforestation is the cutting down of natural forests for timber and paper production. This has an effect on the water cycle.

- There is an increase in the runoff of water after rain as there are no tree roots to absorb the water.
- More water collects in rivers and streams and this can lead to local flooding.
- Weather patterns are changed leading to droughts and crop losses downstream of a deforested area.
- A reduction in plant transpiration into the atmosphere occurs due to the loss of trees.
- Long-term drying out and the formation of deserts can be promoted if deforestation continues.
- Tree removal promotes increased salinity as water tables rise. Groundwater contains dissolved salts and these salts do not reach the surface while trees are growing. With the removal of trees, salty water reaches the surface and evaporates to produce salt. The salt kills the surface grasses.
- In 2000, the Murrumbidgee Irrigation Area (MIA) was listed as a region in which salinity was causing major problems.

The Amazon Basin, South America, is one of the largest forested areas in the world. Figure 3.55 shows the main causes of deforestation over the last decade.

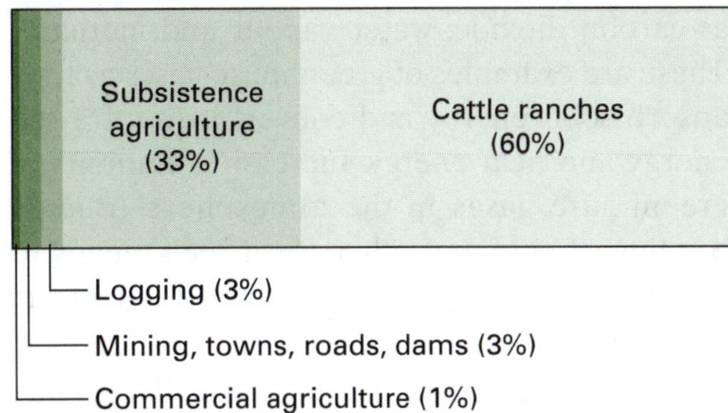

Figure 3.55 Deforestation in the Amazon Basin

Irrigation

Irrigation is the artificial watering of the land to produce crops. Irrigation has increased as the world population has increased. This has had an effect on the water cycle.

- Irrigation water comes from natural sources such as rivers, lakes and underground bores. The water is then pumped through pipes to the farms.
- The large amounts of water that irrigators use can lead to dramatic loss of river, lake and groundwater, and stresses on natural plant and animal habitats.
- Rivers and lakes can become polluted with fertilisers from farms due to irrigation water runoff.
- Irrigation water can leach or dissolve natural minerals in the soil. The loss of minerals requires further fertiliser applications.

Hydro-electricity

One method of generating electricity is to store large masses of water in high dams. This water is then released and rushes under gravity through pipes. The energy of the moving water is used to generate electricity using turbines.

- Rivers have to be dammed to store large masses of water. This reduces natural water flows in the rivers.
- Dammed rivers lead to habitat destruction. Fish cannot migrate through a dam.

Global warming

The atmosphere of the Earth contains gases such as carbon dioxide, water vapour and methane. These are examples of greenhouse gases as they absorb solar energy and convert it to infra-red energy and heat energy. Increasing amounts of greenhouse gases in the atmosphere (due to burning of coal and other fossil fuels) leads to global warming. Global warming will affect the water cycle.

- The polar ice caps and glaciers will melt due to increased global warming.
- Ocean temperatures will increase.
- Increasing rates of water evaporation will occur due to global warming. Increased severe weather events such as cyclones and floods can result.
- Greater rates and intensity of precipitation will occur.
- Higher night-time temperatures will prevent frost formation and prevent the deaths of insect pests.
- Climate change will affect humans and lead to changes in natural habitats.

3.18 Water management

Fresh water is an important resource which must be conserved and managed.

Production of drinking water

Water is collected in dams and this water has to be treated to make it suitable for drinking. Natural waterways contain many contaminants including:

- mineral salts such as soluble iron compounds
- suspended clay particles that make the water cloudy
- animal waste
- fertilisers from farms
- algae
- aquatic animals
- microbes
- detergents washed into the waterway from towns.

After large objects such as tree branches and leaves are screened out of the water, the main steps in this purification are as follows (see also Figure 3.56).

1. Aeration. The water is sprayed into the air to increase its oxygen content and to convert any iron minerals to an insoluble sediment.
2. Flocculation. Chemicals such as alum are added to coagulate or flocculate fine clay particles suspended in the water into larger particles that will sediment.
3. Sedimentation. The suspended particles settle out to the bottom of the sedimentation tank.
4. Filtration. Sand and gravel filters are used to remove the solid sediments.

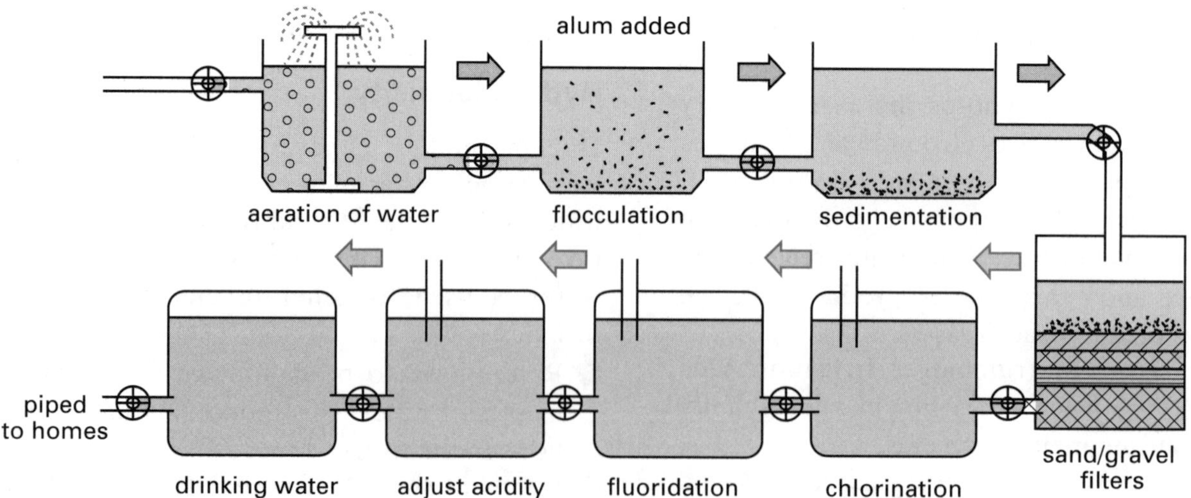

Figure 3.56 Producing drinking water

5. Chlorination. The water is disinfected to kill any dangerous microbes. This is achieved by adding chemicals that release chlorine.
6. Fluoridation. Small amounts of fluoride compounds are added to the water. Fluoridated water strengthens teeth against decay.
7. Acidity adjustment. The water is chemically adjusted so it is close to neutral (i.e. neither too acidic nor basic). Lime (calcium oxide) is added to adjust the pH.

A seawater desalination plant has been built near Botany Bay in Sydney, and began operating in 2010. This plant was built to produce 15% of Sydney's water needs. It has the capacity to produce 30% of Sydney's water should this be required by prolonged drought. The desalination process is very expensive as very fine membranes are required to filter out the salt. The water that is produced, however, is of very high quality.

Australia's first working desalination plant was the Kwinana plant in Perth, and was completed in 2006. A second plant on the Gold Coast began operations in 2009. Other desalination plants are planned and being built around the country.

Treatment of waste water

Waste water is water that comes from your toilet (called blackwater) as well as from your bath, shower, laundry and kitchen (called greywater). This water moves into sewerage pipes where it is pumped to a sewage waste treatment plant.

In some small towns where houses are not connected to the sewage pipes, the waste water is pumped into a septic tank on the householder's property (see Figure 3.57). This septic tank contains bacteria that decompose the waste. A dark sludge is formed and effluent water can be released into the environment via buried pipes where it percolates further into the soil. The organic matter in the effluent helps to fertilise the soil. Periodically the sludge is pumped out into special trucks and transported away for further processing.

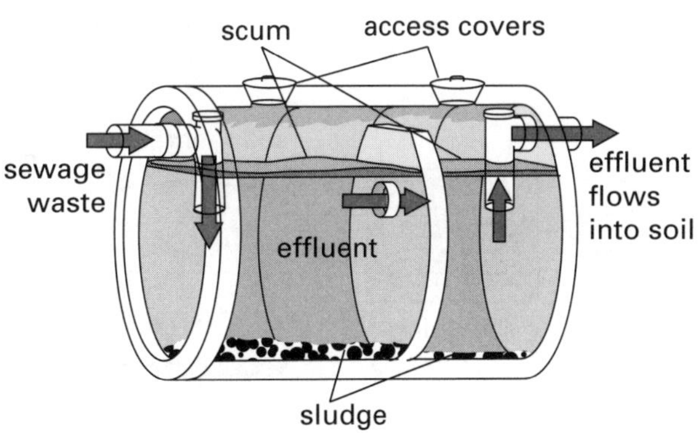

Figure 3.57 Septic tank

A sewage waste treatment plant is designed to treat sewage so it is not dangerous when released into the environment. Sewage treatment plants vary in their design depending on their location and the size of the population in that area. Engineers, environmental chemists, biologists and local government officials are involved in the design and maintenance of the plant.

The steps in a typical sewage waste treatment plant are as follows (see also Figure 3.58).

1. A wire mesh filter is used to screen the sewage to remove large debris.
2. Sewage then flows into a sedimentation tank where suspended solids settle to the bottom of the tank. Flocculants may be added to help sediment fine colloidal material. Materials such as plastic and oil float to the top of the tank and are removed. Once the sludge has settled the remaining water can be chlorinated and released into the environment. In some areas the treatment proceeds to a second phase described below.
3. Aerobic microbes act on the organic sludge in what is called the activated sludge process. Bacteria attack the aerated mixture and some of the organic material is oxidised to carbon dioxide and water. Stable solids form after this process.
4. Solids are finally allowed to settle before the clear water is decanted and chlorinated and pumped out into the environment. Final destinations for this treated water include the ocean and rivers.

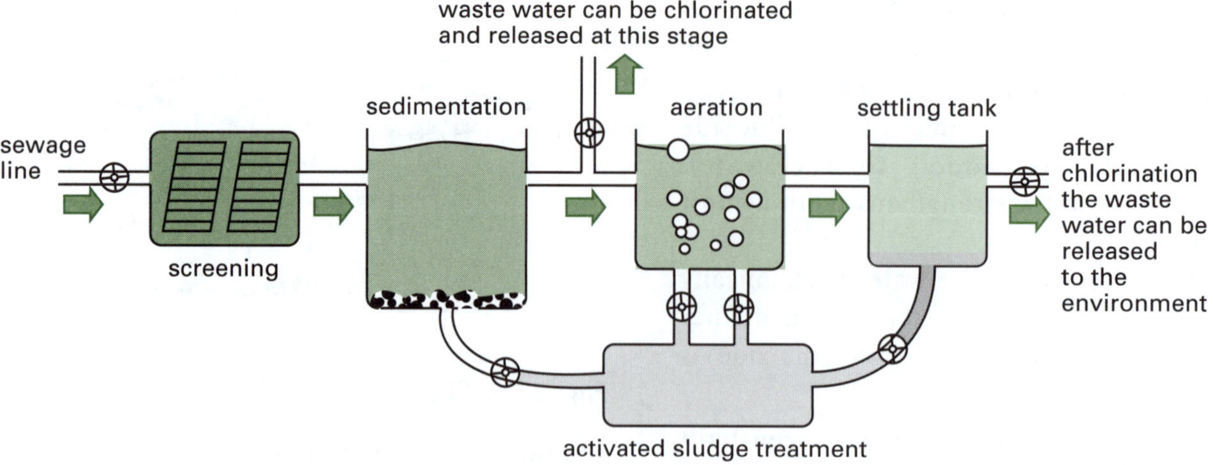

Figure 3.58 Sewage waste treatment plant

Treatment lagoons

There are other methods of dealing with wastewater. One of these is using waste treatment lagoons. These lagoons consist of about 10 pond-like structures that hold and treat the wastewater. They are lined with materials such as clay to prevent leakage into the groundwater. The water gets cleaner as it moves from pond to pond.

The treatment processes used are largely natural. Additional aeration and chemical treatments may also be used. The treatment lagoons are less expensive than the conventional waste treatment works.

The steps of the process are as follows.

1. Sewage flows into the first pond and methane gas is captured as decomposition begins. Methane is used as a fuel to power the treatment plant.
2. Bacteria break down the wastes as the wastewater flows from pond to pond.
3. Aerators dissolve more oxygen in the water and this helps the aerobic bacteria to decompose the waste (see Figure 3.59).
4. After about 35 days the effluent emerges from the pond system and can be discharged into the sea, recycled or used for irrigation.

Figure 3.59 Aerated lagoons for wastewater treatment

Constructed wetlands

In natural wetlands such as marshes and swamps, the plants and microorganisms assist in purifying the water and removing natural waste materials. In many parts of Australia and overseas, constructed wetlands are used to:

- clean wastewater and remove toxic materials such as heavy metals from effluents
- filter out suspended solids and remove dangerous bacteria and viruses
- remove nitrogen and phosphorus compounds which can cause the explosive growth of algae in rivers and lakes
- improve the quality of water released from conventional primary waste water treatment plants
- improve the quality of greywater released from kitchens, bathrooms and laundries.

Marsh plants such as reeds and various freshwater aquatic species that are native to these areas are used.

Water recycling

Australia is a dry continent and water conservation is essential. Recycling of water is an important component of water conservation. It is a waste of freshwater resources when large volumes of stormwater, for example, are allowed to run out to sea.

Recycled water can be treated to various levels of purity using the variety of procedures already discussed. In many locations, people use recycled water for drinking, showering and irrigation. For irrigation the recycled water only needs to be treated to the levels required by the Environmental Protection Authority. Further treatments can produce water suitable for drinking.

Caring for waterways and water management

The traditional hunter-gatherer lifestyle of Australian Aborigines required them to sustainably care for the land, waterways and living things. They used the resources of the land to the minimum extent that allowed them to survive.

It is important not to degrade the land and waterways so that the requirements of living things are destroyed. The farming practices of early European settlers often led to land and waterway degradation as they did not have the knowledge of the land that the Aboriginal people did. Early settlers to Australia considered the natural environment there to be plundered for its resources, while Aboriginal peoples believed in living harmoniously with the land.

Australian governments are now actively involved in caring for waterways and water management. Some important ways include:

- reversing past poor management practices (e.g. over-allocation of river water to irrigators) that led to the degradation of natural waterways
- returning more water to natural waterways and closing down unsustainable agriculture
- supporting Aboriginal water and land management principles and including Aboriginal representatives in water reform initiatives.

Future waterway management in Australia needs to ensure that ecosystems are not damaged. If global warming increases over time then water management will become even more important. Farmers and gardeners will need to develop ways of sustainably using water. This will involve the prevention of evaporation of water from soils by covering the exposed soil with plant mulch and using trickle irrigation rather than overhead spraying. Open water channels will need to be replaced by covered ones.

Experiment 3

Water distribution

Aim

- To demonstrate how water is distributed over the Earth

Method

1. Fill a large measuring cylinder to the 1000 mL mark. This represents the entire supply of water on Earth.
2. Pour 28 mL of this total water into a 50-mL measuring cylinder. This represents the total freshwater supply. The remaining water in the 1000-mL measuring cylinder represents saltwater. This is found mainly in the oceans.
3. Use the 28 mL that represents freshwater and take out the following amounts.
   a) Pour 23 mL into another 50-mL cylinder. This represents freshwater locked up in glaciers and the icecaps.
   b) Pour 4 mL into a 10-mL cylinder. This represents groundwater.
   c) Transfer two drops into a Petri dish with a dropper. This represents surface water.

d) Transfer one drop into another Petri dish. This represents the water in the atmosphere and soil.

Results and analysis

1. Draw a diagram of the results of this modelling experiment.
2. Construct a flow chart to show how water is distributed on Earth. Go to p. 194–195 to check your answer.

Conclusion

Write a suitable conclusion.

Go to p. 194–195 to check your answer.

Test yourself 3

Part A: Knowledge

1. Select the statement that is true of liquid water. *(1 mark)*
   - **A** The particles are tightly packed together and vibrate slightly.
   - **B** Liquid water can be converted to solid water by adding heat.
   - **C** Water vapour condenses to form liquid water when heat energy is removed from it.
   - **D** Liquid water contracts when it is converted into water vapour.
2. Transpiration is the process in which *(1 mark)*
   - **A** water percolates into the ground.
   - **B** water evaporates from the pores in the leaves of plants.
   - **C** groundwater flows towards the sea.
   - **D** condensed water falls to the ground from clouds.
3. Cloud formation is promoted when *(1 mark)*
   - **A** warm, humid air moves upwards over a mountain range.
   - **B** a cold air mass forms over an ocean and promotes rain.
   - **C** a cold air mass forces a warm air mass down close to the ground.
   - **D** cold, moist air rises and becomes warmed by the Sun.
4. How do paved surfaces in cities influence water management? *(1 mark)*
   - **A** The paved surfaces are porous and allow water to become trapped.
   - **B** They reduce the amount of storm water that reaches the sea.
   - **C** They do not absorb rain and there is a large increase in surface run-off into drains.
   - **D** The paved surfaces absorb more heat and cause global warming.
5. Constructed wetlands are useful in water management as *(1 mark)*
   - **A** fish and other aquatic animals can help to remove waste from effluents.
   - **B** they filter out suspended solids and remove dangerous bacteria from wastewater.
   - **C** they generate various gases such as methane and carbon dioxide that can be used to make electricity.
   - **D** they selectively absorb phosphate minerals from blackwater.
6. Complete the following restricted-response questions using the appropriate word. *(1 mark for each)*
   - **a)** Sewage effluent flows into a sedimentation tank where .... solids settle to the bottom of the tank.
   - **b)** Rivers and lakes can become polluted with .... from farms due to irrigation water runoff.
   - **c)** Convection currents are responsible for cooling .... breezes along the coast.
   - **d)** A lot of the water falling on land soaks or .... into the ground.
   - **e)** Removal of vegetation has a significant effect on plant .... and therefore on the water cycle.

7. Use the code letters to match the terms or phrases in each column. *(1 mark for each)*

Column 1	Column 2
A condensation	F change of state
B adding heat energy	G chlorine
C microbes	H lime
D tree removal	I increased salinity
E acidity adjustment	J evaporation

Part B: Skills

8. Figure 3.60 shows three steps in the formation of storm clouds. Match the following three captions with the correct diagram. *(1 mark)*

Figure 3.60 Steps in the formation of storm clouds

i) When the cloud reaches a height where the air temperature is low enough, water droplets merge together growing in size.

ii) Moist air rises and gradually cools, causing water vapour to condense into tiny droplets.

iii) When the droplets of water reach a certain weight, the air can no longer support them and they fall to the ground as rain, hail, snow or sleet.

9. Figure 3.61 shows the moisture capacity of the air at various temperatures.

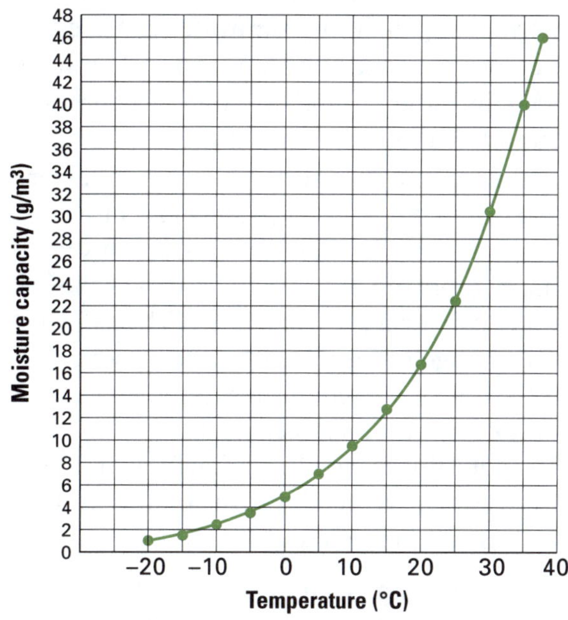

Figure 3.61 Moisture-holding capacity of air versus temperature

a) State one conclusion that can be made about the moisture-holding capacity of air from this graph. *(1 mark)*

b) What is the moisture-holding capacity of air at 25 °C? Answer to the nearest whole number. *(1 mark)*

c) A sample of air has a moisture-holding capacity of 10 g/m^3. What is the temperature of the air sample? *(1 mark)*

d) To what temperature does a sample of air, originally at 15 °C have to be warmed to double its moisture-holding capacity? *(1 mark)*

10. Complete the passage using some of the words supplied (not all words need to be used): condenses; cools; evaporation; fresh; groundwater; heat; moisture; permeable; rain; sun; surface; winds. *(5 marks)*

When water vapour **a)** .... sufficiently in the atmosphere it **b)** .... to form clouds. Eventually the clouds can't hold the **c)** .... and so release it as rain, hail or snow.

d) .... that falls on the ground can seep into it. This depends on how e) .... the ground is. The more permeable, the more rainwater can seep or percolate into it. If it rains faster than it can soak into the ground, the water becomes runoff. Runoff remains on the f) .... and may flow into streams, rivers, lakes or the ocean. g) ...., too, can move towards large bodies of water. As this is happening, the h) .... drives the cycle by causing i) .... This is just the change of liquid water into vapour. The water cycle continues to move water and keep water sources j) .... Without this cycle life on Earth would be impossible. We need it to sustain us and for all our life processes to function.

11. The following is a jumbled list of some of the stages in the production of drinking water. Use the code letters to place them in the correct order. *(1 mark)*
    A. The fine suspended colloidal particles are flocculated into larger particles.
    B. Gravel and sand layers are used to filter the water.
    C. Flocs sediment to form sludge at the bottom of the tank.
    D. Chlorine is added to disinfect the water.
    E. The acidity is adjusted.
    F. Large objects are screened out of the water.

12. Water can be converted to different states as shown in Figure 3.62. Match the code letters to the following processes occurring: melting; evaporation; freezing; condensation. *(2 marks)*

Figure 3.62 Water conversion flow chart

13. The apparatus in Figure 3.63 was used for an experiment involving water.

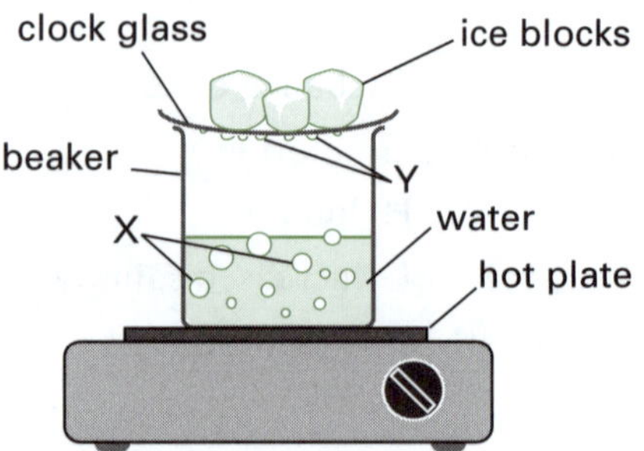

Figure 3.63 Experiment involving water

    a) Large bubbles (X) are seen in the water as the water is heated. What are these bubbles composed of? *(1 mark)*
    b) Droplets of a clear liquid form at Y. What are these droplets? *(1 mark)*
    c) Explain how these droplets (Y) form. *(1 mark)*
    d) Identify where the process of fusion occurs. *(1 mark)*

14. You are given the following percentages of the world energy consumption of various energy sources: oil 40%; coal 27%; natural gas 23%; nuclear 7%; hydro-electricity 3%. Draw a labelled sector (pie) graph for this information. *(5 marks)*

Go to p. 197 to check your answers.

Summary

1. Eclipses happen when the Earth, Moon and Sun are lined up.
2. The phases of the moon are caused by the different angles from which we see the bright part of the Moon's surface.
3. The Earth rotates around its rotation axis once in 24 hours.
4. The Moon makes a complete orbit around the Earth with respect to the fixed stars about once every 27.3 days.
5. The orbit of the Earth around the Sun is called an Earth revolution. This takes 365.24 days (one year) to complete one cycle.
6. The seasons are due to changes in temperature on Earth and are a direct result of the amount of solar radiation that hits our planet at a particular time of year.
7. The Earth's seasons are caused by the rotation axis of the Earth not being perpendicular to its orbital plane. The Earth's axis is tilted at an angle of about 23.5° from the orbital plane.
8. Five planets (Mercury, Venus, Mars, Jupiter and Saturn) plus the Sun and Moon can be seen without a telescope.
9. Telescopes are usually positioned high on mountains away from the glare of city lights and the polluted air. Space telescopes are not hampered by the distortion of images caused by the atmosphere.
10. Aristarchus and Copernicus proposed models of the universe in which the Sun was at the centre and the planets and stars revolved around it.
11. Galileo used a telescope to observe the craters on the Moon and the moons of Jupiter. He supported the Copernican model.
12. Bodies experience a force in a gravitational field. This force is the weight of the body.
13. Resources can be classified as natural or synthetic.
14. Resources can be classified as renewable or non-renewable.
15. Resources can be obtained from the living or non-living world.
16. Metals are important materials that are extracted from ores.
17. Fossil fuels include petroleum, natural gas and coal.
18. The various forms of alternative energy include hydro-electricity, solar power, wind power, tidal power, biomass, nuclear energy and geothermal power.
19. Antarctica has many natural resources but countries need to decide how to make use of these without damaging the environment.
20. Water can exist as a solid (e.g. ice), a liquid (e.g. liquid water) or a gas (e.g. water vapour).
21. The water cycle involves water changing state between solid, liquid and gas.
22. Cloud formation and the water cycle are closely connected. Rising, moist warm air cools down and water vapour condenses to form clouds.
23. Human activities affect the water cycle. These activities include urbanisation, population growth, deforestation, irrigation, hydro-electricity production and global warming.
24. Fresh water is an important resource which must be conserved and managed.
25. Water is collected in dams and this water has to be treated to make it suitable for drinking.

Summary

26. The job of the sewage waste treatment plant is to treat it so it is not dangerous when released into the environment.
27. Australia is a dry continent and water conservation is essential. Recycling of water is an important component of water conservation.
28. Australian governments are now actively involved in caring for waterways and water management.

Syllabus checklist

Are you able to answer every syllabus question in this chapter? Tick each question as you go through the list if you are able to answer it. If you cannot answer it, turn to the appropriate page in the guide as is listed in the column to find the answer.

For a complete understanding of this topic		Page no.	✓
1	Can I recall what causes eclipses?	56–57	
2	Can I state what causes us to observe different phases of the Moon?	57	
3	Can I recall how long it takes the Earth to rotate on its axis?	57	
4	Can I recall how long it takes the Moon to complete an orbit of the Earth?	58	
5	Can I recall how long it takes the Earth to revolve around the Sun?	59	
6	Can I explain what causes the different seasons throughout the year?	60–61	
7	Can I recall how the tilt of the Earth's axis gives rise to different seasons?	60–61	
8	Can I recall the five planets that can be seen without a telescope?	64	
9	Can I explain why telescopes are positioned on high mountains and why space telescopes produce much clearer images?	64–65	
10	Can I recall who produced Sun-centred models of the universe?	66–67	

For a complete understanding of this topic		Page no.	✓
11	Can I recall the experiments Galileo performed to support the Copernican model of the universe?	67	
12	Can I explain why all bodies experience a force in a gravitational field?	68	
13	Can I classify resources as natural or synthetic?	73	
14	Can I classify resources as renewable or non-renewable?	73	
15	Can I name some resources that can be obtained from the living or non-living world?	74–78	
16	Can I recall from what substance metals can be extracted?	72–78	
17	Can I recall some types of fossil fuels?	78–80	
18	Can I name the various types of alternative energy?	83–87	
19	Can I explain why Antarctica is regarded as a pristine environment where scientists can conduct physical, biological and environmental research?	87–89	
20	Can I name the three states in which water can exist?	89	

For a complete understanding of this topic		Page no.	✓
21	Can I give examples of the changes of state for water in the water cycle?	89	
22	Can I recall how clouds are formed?	91–92	
23	Can I recall which human activities influence the water cycle?	92–94	
24	Can I give reasons why fresh water must be managed and conserved?	97	

For a complete understanding of this topic		Page no.	✓
25	Can I recall the process behind supplying water for drinking?	94	
26	Can I recall what must happen to waste water before it is released into the environment?	95–96	
27	Can I explain the importance of water recycling?	97	
28	Can I explain why Australian governments are actively involved in caring for waterways and water management?	97	

Chapter test

Go to p. v for *Tips for tests and examinations*

80 MIN

Part A: Multiple-choice questions

(1 mark for each)

1. Equinoxes are

A the first day of spring and autumn.
B the two days in the year when the Sun shines directly overhead on the equator.
C June 22 and December 22.
D the dates when the Sun is the furthest north or south of the equator.

2. The moon travels at a speed of 1.023 km/s as it revolves around Earth. The distance it travels in one hour is

A 3683 km.
B 61 km.
C 368 km.
D 88 387 km.

3. In Australia the 3 months of autumn are

A December, January, February.
B September, October, November.
C June, July, August.
D March, April, May.

4. Which of these is not a fossil fuel?

A coal
B petroleum
C natural gas
D uranium

5. Which of these is not a renewable energy resource?

A solar power
B wind power
C tidal power
D nuclear power

6. Which of these would be the most important in locating a tidal power plant?

A There needs to be a great difference between high and low tides.
B There needs to be a short distance between one side of the bay and the other.
C There must be at least one high and low tide during the day.
D A city must be located nearby.

7. Which one of the following statements about irrigation water is true?

A Irrigation water can leach or dissolve natural minerals in the soil.
B Irrigation water comes from underground bores only.
C Irrigation is beneficial to natural plant and animal habitats.
D Irrigation keeps natural rivers and lakes free of pollution.

8. Ice crystals turn into hail stones when
   - **A** large ice crystals melt to form rounded hail stones in a series of cycles in cumulonimbus clouds.
   - **B** downward-moving air inside cumulonimbus clouds pushes water drops lower into freezing conditions where it solidifies into very small ice crystals.
   - **C** moisture is gathered as ice crystals are forced back up inside a cumulonimbus cloud and the ice grows larger in size as this additional moisture freezes.
   - **D** cycles of freezing and melting of water occur in cumulus clouds.

9. The process of converting liquid water to solid water is called
   - **A** melting.
   - **B** solidification.
   - **C** condensation.
   - **D** evaporation.

10. Which one of the following statements about events in the water cycle is true?
    - **A** As warm, moist air moves upwards (e.g. over a mountain range) it cools and cloud formation occurs.
    - **B** Warm and cold masses of air contain the same amounts of water vapour.
    - **C** Snow forms when individual five-sided ice crystals begin to form as water freezes in the clouds.
    - **D** In the night the land cools more slowly than the sea and so a moist convection current is established and sea breezes form.

Part B: Short-answer questions

11. It is all right to look at the Moon during a lunar eclipse, but dangerous to look at it during a solar eclipse. Explain. *(2 marks)*

12. Figure 3.64 shows the shadows that were cast during a solar eclipse. A person is standing at point X on the Earth's surface. Which of the shadow patterns would they observe? *(1 mark)*

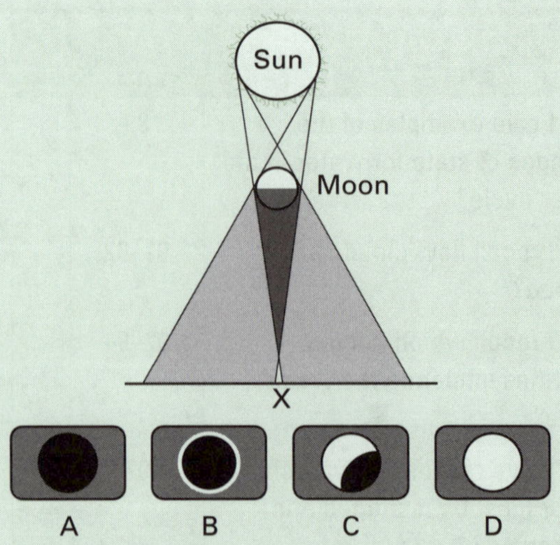

Figure 3.64 Solar eclipse shadows

13. **a)** At which phase of the Moon is the only time to see a lunar eclipse? *(1 mark)*
    b) Will a lunar eclipse always be seen at that time? Explain. *(2 marks)*

14. In which month of the year is it possible to not get a full moon? *(1 mark)*

15. Viewed from the North Pole, the Earth rotates anticlockwise on its axis. Looking down at the South Pole from space, would the Earth's rotation appear clockwise or anticlockwise? *(1 mark)*

16. The circumference of the Earth at the Equator is about 40 000 km. The time for the Earth to complete one rotation equals 24 hours. Use the formula

    $\text{speed} = \frac{\text{distance covered}}{\text{time taken}}$ to calculate the speed of rotation, in km/h, at the equator. *(1 mark)*

17. Four of Jupiter's moons (Io, Callisto, Ganymede and Europa) are readily visible using a simple telescope. Figure 3.65 shows the positions of the moons around Jupiter over a period of 20 days.

Earth days	Moon positions
1	G E I (J) C
2	(J) I C
3	I (J) E G C
4	(J) I G C
5	E (J) G C
6	G I (J) C
7	G (J) I C
8	G E I (J) C
9	C E (J) I
10	C I (J) E G
11	C (J) I G
12	C E (J) G
13	C G (J) E
14	C G (J) IE
15	CG EI (J)
16	C E (J) I
17	(J) EG
18	(J) I ECG
19	E (J) G C
20	G I (J) E C

Key: C = Calisto E = Europa G = Ganymede I = Io (J) = Jupiter

Figure 3.65 Positions of the moons around Jupiter over a period of 20 days

a) Use this information to determine the number of Earth days that it takes each moon to orbit Jupiter. *(4 marks)*

b) Why can we not see all four moons on each of the nights of observation? *(1 mark)*

c) List the moons in order of increasing distance from the planet. *(1 mark)*

18. Use the code letters (A to G) to arrange the phases of the Moon in Figure 3.66 in the correct order, starting with the full moon as viewed from Australia. *(2 marks)*

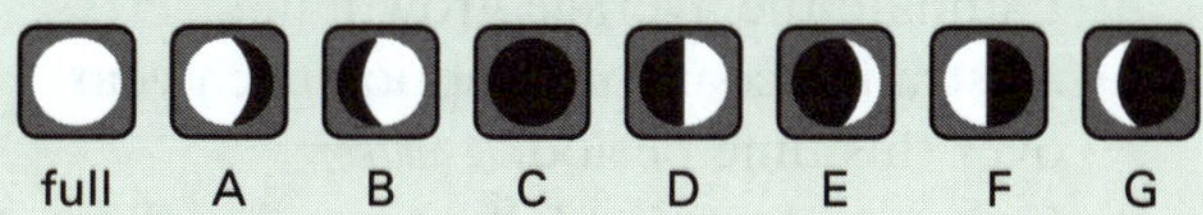

Figure 3.66 Phases of the moon

19. Figure 3.67 shows a diagram drawn by Copernicus in the 16th century.

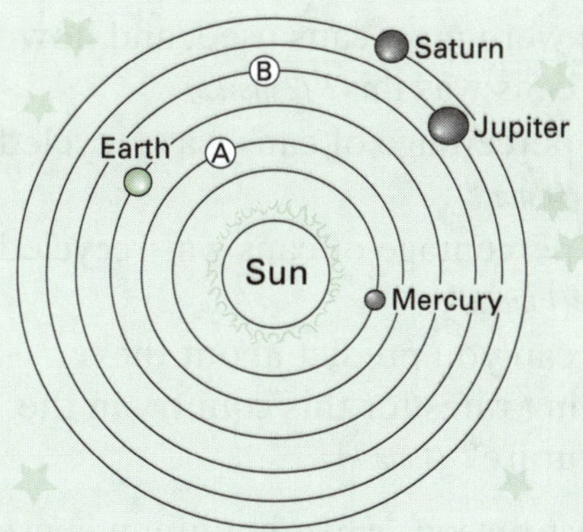

Figure 3.67 The universe according to Copernicus

a) Why isn't Uranus shown in this drawing? *(1 mark)*

b) Name planets A and B. *(2 marks)*

20. State an advantage of using telescopes to explore the heavens. *(1 mark)*

21. List five resources that were not available to humans 200 years ago. *(5 marks)*

22. Using examples, explain the differences between renewable and non-renewable resources. *(4 marks)*

23. Identify and describe three steps that can be taken by individuals to conserve non-renewable resources. *(3 marks)*

24. Figure 3.68 shows the numbers of aluminium cans used, and recycled, in a country between 1999 and 2011.

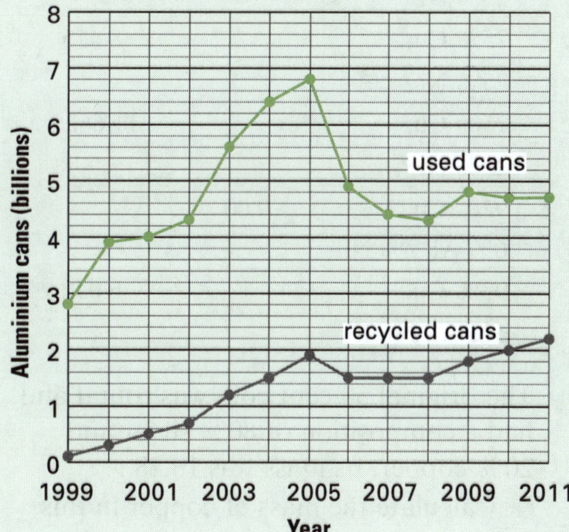

Figure 3.68 Number of aluminium cans used and recycled, 1999 to 2011

a) When were most cans used, and how many cans was this? *(2 marks)*

b) What percentage of cans was recycled in 2001? *(1 mark)*

c) What percentage of cans was recycled in 2010? *(1 mark)*

d) What can you predict about the recycling rates for this country in the near future? *(1 mark)*

25. a) Suggest reasons why aluminium is used so extensively worldwide. *(3 marks)*

b) Around 24 million tonnes of aluminium is used worldwide each year. The global demand for aluminium is growing at about 3% per annum. By how many tonnes is the aluminium demand growing each year? *(1 mark)*

26. a) Table 3.7 gives some information about Australian coins. Fill in the missing information. *(6 marks)*

Table 3.7 Composition and mass of Australian coins (Cu = copper; Ni = nickel; Al = aluminium)

Coin	Metal composition	Mass (g)	Mass of copper present (g)
5 ¢	75% Cu, 25% Ni	2.83	
10 ¢	75% Cu, 25% Ni	5.65	
20 ¢	75% Cu, 25% Ni		8.475
50 ¢	75% Cu, 25% Ni		11.6625
$1	92% Cu, 6% Al, 2% Ni	9.00	
$2	92% Cu, 6% Al, 2% Ni		6.072

b) The original 50-cent coin was round and had a composition of 80% silver and 20% copper. Its mass was 13.28 g.

i) Calculate the mass of copper in this coin. *(1 mark)*

ii) Why is silver no longer used in 50-cent coins? *(1 mark)*

27. The World Nuclear Association (WNA) has analysed Australian sources of energy for electricity generation. They state in their online report (www.world-nuclear.org/info/inf64.html):

In 2008–09 the electricity was produced from 56 gigawatts (GWe) capacity, of which 30.3 GWe (54.1%) was coal-fired, 14.7 GWe (26%) gas or multi-fuel, 1.35 (2.4%) oil, 7.1 GWe (13%) hydro and 2.5 GWe (4.5%) other renewables.

a) Convert this data into a pie chart. *(5 marks)*

b) The WNA also states:

Australia is heavily dependent on coal for electricity, more so than any other developed country except Denmark and Greece.

How does your pie chart show that this statement is true? *(1 mark)*

28. Figure 3.69 shows coal consumption, in terms of energy units, between 1965 and 2010 inclusive.

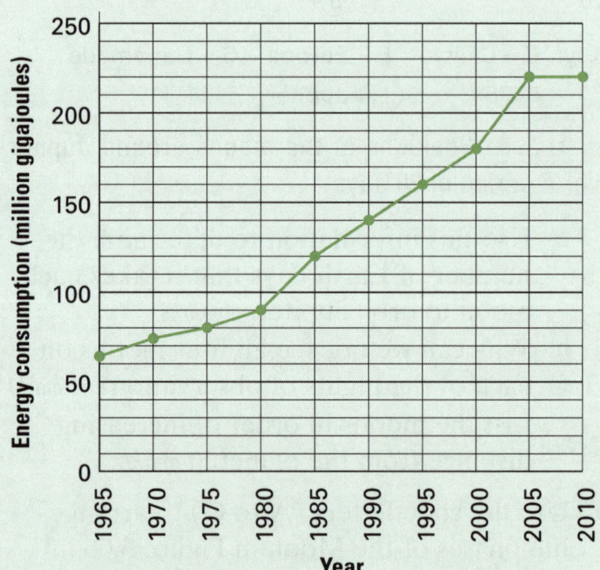

Figure 3.69 Coal consumption, 1965 to 2010

a) Estimate the average growth in Australian coal consumption each year over this time period. *(2 marks)*

b) Given current trends, how many millions of gigajoules of energy would coal in Australia be providing in 2030? *(1 mark)*

c) How accurate do you think our answer to b) is? *(1 mark)*

29. Match each of the following statements to an energy source. *(1 mark for each)*
   a) The amount of energy available depends on the amount of water flowing and the height of the surface of the water above the turbine.
   b) This power plant is simply a very large and expensive device for generating electricity using the heat generated by atoms splitting.
   c) Large dams bring considerable environmental and social change. People living in the area to be flooded are displaced.
   d) Tubes are attached to a black absorber plate. As heat builds up in the collector, it heats the water passing through the tubes. The storage tank then holds the heated water, ready for use.
   e) Turbine blades are shaped a lot like aeroplane wings, having an airfoil design. In an airfoil, one surface of the blade is somewhat rounded, while the other is relatively flat.
   f) When sunlight hits the surface of a photovoltaic panel part of the light passes through the panel, starting a flow of electricity.
   g) The higher the turbine is placed, the more energy it can capture because air speeds increase as elevation increases. Ground friction and ground-level objects interrupt the flow of the wind.
   h) In volcanic areas, molten rock can be very close to the surface. Sometimes we can use that heat.
   i) A wind turbine works the opposite of a fan. Instead of using electricity to make wind, a turbine uses wind to make electricity.
   j) When tides come into the shore, the water can be trapped in reservoirs behind dams created by using sluice gates.
   k) This heat energy has been used for thousands of years in some countries for cooking and heating, especially by native peoples, who bury food to be cooked in the ground.

30. Consider the map in Figure 3.70 showing where wind farms are suitable.

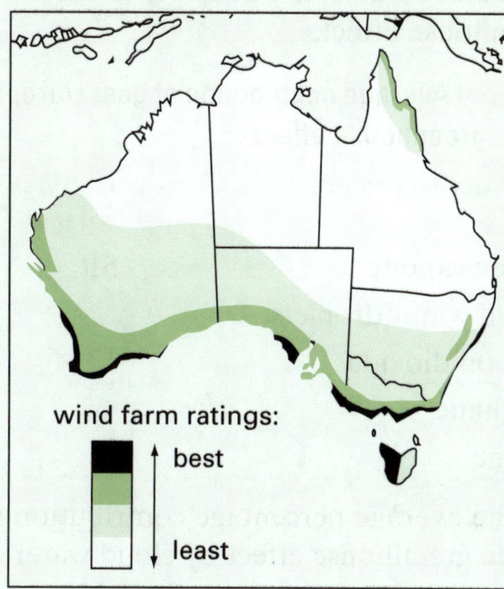

Figure 3.70 Suitable locations for wind farms

   a) Describe where it is most suitable to locate wind farms in Australia. *(2 marks)*
   b) Explain why these areas are most suitable. *(2 marks)*
   c) What might be some criticisms of locating wind farms in these areas? *(2 marks)*

31. Does a liquid like water need to be heated to its boiling point (100 °C) for it to evaporate? *(2 marks)*

32. Plants transpire water through pores in their leaves. Where does the plant obtain the water that is inside its body? *(2 marks)*

33. Figure 3.71 shows some steam coming from a boiling kettle of water. Is steam the same as water vapour? *(2 marks)*

Figure 3.71 Steam from a boiling kettle

34. Table 3.8 contains data on the average relative contribution of various gases as well as cloud water droplets to the greenhouse effect.

Table 3.8 Average contribution of gases/droplets to the greenhouse effect

Greenhouse gas/droplet	Average contribution (percentage)
water vapour	51
cloud water droplets	
carbon dioxide	17
methane	6
ozone	5

a) The average percentage contribution to the greenhouse effect by cloud water droplets is missing from the table. Calculate this value. *(1 mark)*

b) Present this data as a column graph. *(5 marks)*

35. To flocculate means to form into small clumps or masses. Figure 3.72 represents four stages in the flocculation of colloidal particles when certain chemicals, such as alum, are added to the waste water. The steps are jumbled. What is the correct order? *(4 marks)*

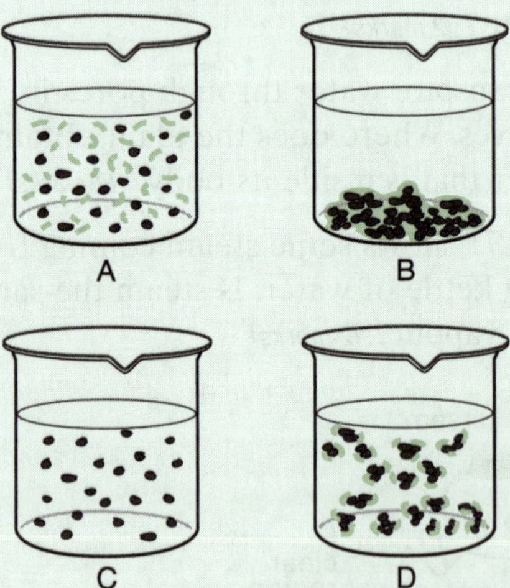

Figure 3.72 Four stages in the flocculation of colloidal particles

36. Table 3.9 provides information on the phosphorus (P) and nitrogen (N) compounds present in greywater from various locations in a typical home.

Table 3.9 Phosphorous and nitrogen compounds in greywater

Compounds	Kitchen	Laundry	Shower/ bath	Wash basin
(P) Phosphate (mg/L)	15.5	10.0	1.5	45.5
(N) Nitrate (mg/L)	0.5	1.5	1.0	0.5
(NH_4^+) Ammonium (mg/L)	4.5	10.5	1.5	0.5

a) Which site in the home is the greatest source of phosphorus pollutants in greywater? *(1 mark)*

b) Which site in the home is the greatest source of total nitrogen pollutants in greywater? *(1 mark)*

c) In one household the total volume of greywater released from the bath and shower was 300 L. Calculate the total mass of phosphate compounds released in the shower/bath water. *(1 mark)*

37. Explain the differences between the ways in which traditional Aborigines managed and cared for waterways to the way water has been used by farmers since European settlement. *(2 marks)*

38. Complete the cloze passage by inserting the correct words. *(5 marks)*

Heat islands are created through urbanisation. As a city grows, trees a)................... cut down to make room for commercial development, roads and suburban growth. Forest growth normally reduces the amount of heat and smog generated b) .... populated areas. Plants and water-retaining soils absorb heat during the day, and then carry the heat away through evaporation. Urban growth transforms the environment, creating a uniquely altered zone c) .... weather. Because urban areas both generate and trap heat, a bubble or urban heat island forms around the city. The temperature can **d)** .... several degrees higher than outlying areas, and this excess heat may produce increased rainfall and thunderstorms. As the heat in a city builds, hot air rises. Colder air rushes into the low pressure zone, creating winds. The warmer, ascending air forms clouds **e)** .... drop water as they continue to rise. These urban heat islands can cause convective clouds to form over the city. Convective clouds typically produce rains that are intense and localised; they may also produce thunder and lightning.

39. Humans have a significant impact on the water cycle. In the following table, match the headings in column 1 to the statements in column 2. *(4 marks)*

Heading	Statement
A urbanisation and population growth	E loss of minerals from the soil
B deforestation	F increased formation of deserts
C irrigation	G flooding due to large increases in storm water
D hydro-electricity production	H reduction in natural water flows and habitat destruction

40. Identify the first stage in the production of drinking water when the water will look clear and colourless. *(1 mark)*

Go to pp. 197–201 to check your answers.

CHAPTER 4

Forces

Overview

In this chapter you will learn about:

- balanced and unbalanced forces
- methods of measuring force
- the effect of friction
- contact forces
- constant motion
- accelerated motion
- field forces
- simple machines
- cars, bikes and safety equipment
- forces and sports.

Glossary

Accelerate—to increase in speed over time

Buoyancy—a contact force that acts on an object when it is partly or completely immersed in a fluid. This force opposes the force of gravity. Buoyancy also refers to the tendency to float in water or another liquid.

Contact forces—forces that involve physical contact between bodies

Decelerate—to decrease in speed over time

Effort—the amount of force required to lift a load

Field—a space in which a body experiences a force

Field forces—force that act at a distance due to the presence of a field

Force meter—a spring balance that measures force in newtons (N)

Fulcrum—a stationary pivot point around which a lever rotates

Lever—a simple machine consisting of a pivoted bar or rod which moves loads when an effort force is applied

Load—the object to be raised by an effort force

Pulley—a rimmed wheel which holds a rope; pulleys are used to raise loads by applying an effort force to one end of the rope

4.1 Balanced and unbalanced forces

Forces cause all movements. When you catch a cricket ball in your hand, your muscles exert a force that stops the ball from moving. A magnet attracts small pieces of iron because a magnetic force causes them to move.

Forces are pulls, pushes or twists that are applied to objects. Forces can cause a:

- change in speed (e.g. pushing harder on bicycle pedals will make the bicycle move faster along a flat road)
- change in direction (e.g. twisting or turning the handle bars of a bicycle allows the bike to change direction)
- change in shape (e.g. squeezing a squash ball will deform it and change its shape until you let go and it returns to its old shape).

How do we measure force?

Forces are measured using force meters. A spring balance is an example of a force meter. Inside a spring balance is a metal spring that stretches when a force is applied. A body hanging vertically from the hook at the end of the spring may cause this force. Gravity is pulling the body down and the spring stretches until there is a balance between the force of gravity downwards and the force applied upwards by the spring. Bathroom scales also measure force by compressing or stretching a spring.

The spring balance measures the size of the force in units called newtons. The symbol for newtons is N. In Figure 4.1, a 1.0-kg mass is hung from the spring's hook and the pointer attached to the spring moves and comes to rest at 9.8 N. This value is called the weight of the 1.0-kg mass. Weight is an example of a force caused by gravity. Often, however, weighing scales are given in kilograms or grams, as most people are not familiar with newtons.

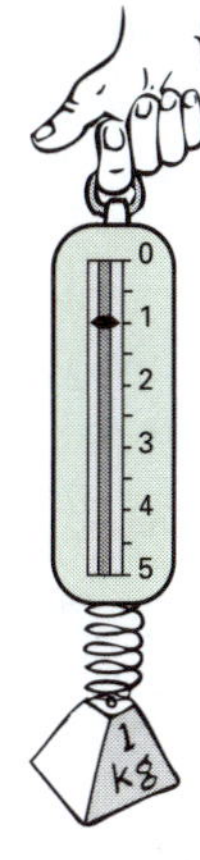

Figure 4.1 Using a spring balance to measure force

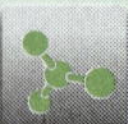

Case study 1: Moving a crate across the floor

When an object is heavy, like a big crate, it takes a lot of muscular strength to push it across the floor. If the man in Figure 4.2 can exert enough force on the crate to overcome the frictional forces between the crate and the floor then the crate will move across the floor. The act of starting a body moving from rest requires a pushing or pulling force that is greater than the opposing forces such as friction.

Figure 4.2 The pushing force must be greater than the frictional force to get the crate to move from rest.

The spring balance or force meter can be used to measure the force applied to a rubber band when it is stretched. Figure 4.3 shows a rubber band looped over a fixed support and a spring balance measuring the force as the band is stretched to its maximum length by a pulling force.

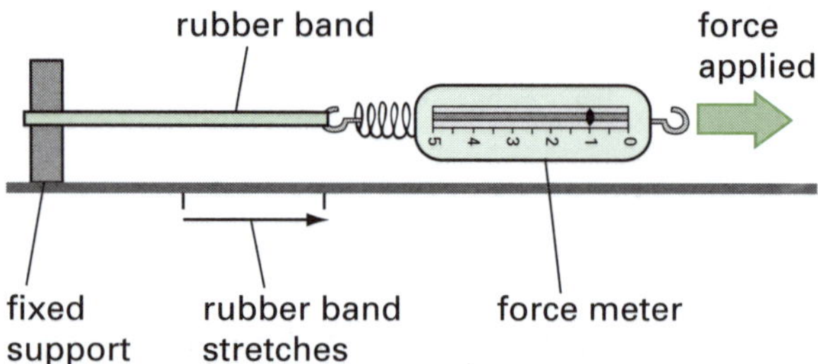

Figure 4.3 A spring balance can be used to measure the pulling force applied to a rubber band in order to stretch it.

Case study 2: Object sliding down a ramp

Consider the experiment in Figure 4.4. A block of wood is placed on a wooden plank, which is slowly raised at one end until the block starts to slide down the ramp at a fairly constant speed. In this experiment, that motion occurs when the plank makes an angle of 30° to the horizontal.

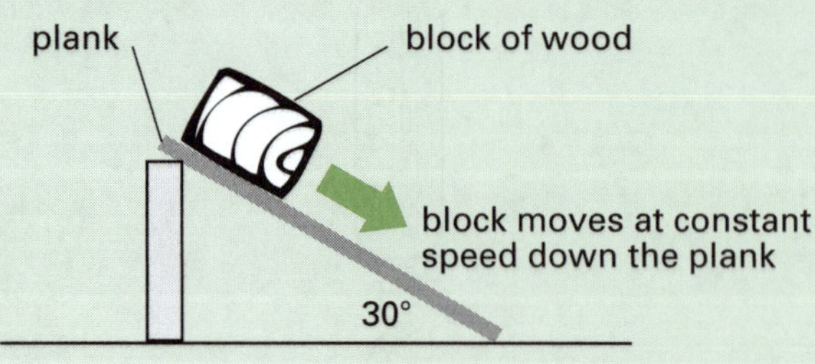

Figure 4.4 A block sliding down a plank

The block of wood does not initially move until the plank is at an angle of 30°. Once movement occurs, the block slides at a fairly constant speed down the ramp.

If the experiment is repeated at an angle greater than 30°, then the block is observed to move at an increasing speed down the plank. It accelerates (speeds up) rather than moving at constant speed.

1. When the wooden block is lying on the plank in the horizontal position, then there is no movement. In this case there are two balanced forces as shown in Figure 4.5. One force is the weight (W) of the block pushing down on the plank. The

plank pushing back on the block with a force equal to the weight force causes the second force. This upward force is called the reaction force (R). These two forces balance each other out and there is no motion of the block.

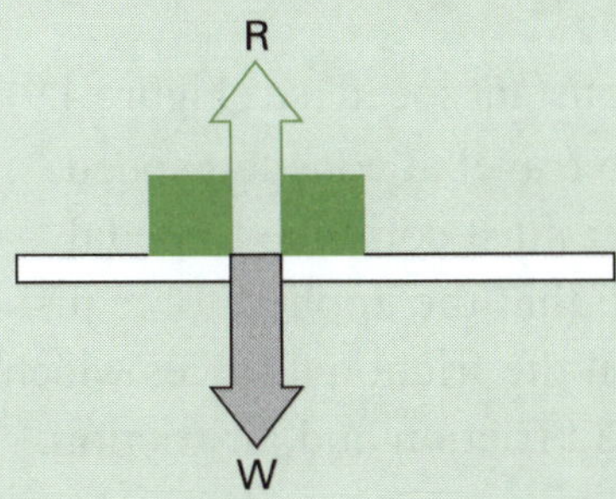

Figure 4.5 Balanced forces when bodies are at rest

2. The plank is now raised at one end and at any angle less than 30° (in this example) the block stays stationary and does not slide. This also means that the forces acting on the block are balanced. When the angle gets to 30°, the block then starts to slide down the slope at a constant speed (see Figure 4.6). When objects move at constant speed, the forces also balance. In this case the force of friction between the block and the plank is balancing the force that causes the object to slide down the plank. The rougher the surface, the greater is the friction.

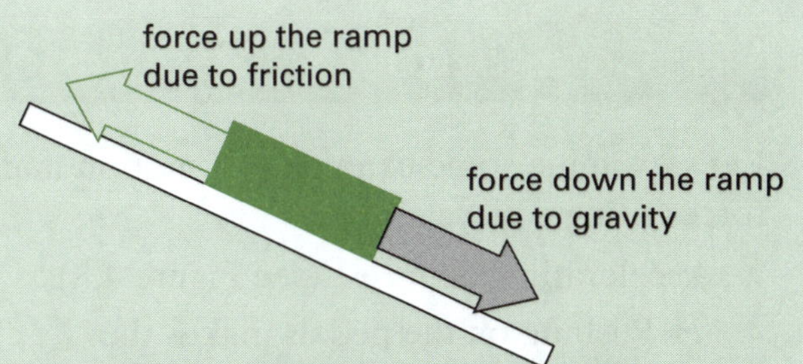

Figure 4.6 Balanced forces when bodies move at constant speed (notice the lengths of the two arrows are the same, indicating balanced forces)

3. When the plank is raised past 30° the block starts to accelerate down the plank as the force down the slope caused by gravity is now greater than the frictional forces up the slope. The forces no longer balance and accelerated motion occurs (see Figure 4.7). If the plank became vertical then the block would experience the full force of gravity and fall straight down with increasing speed.

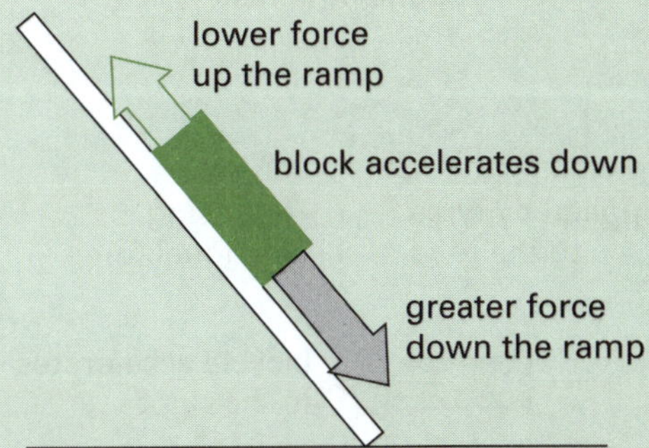

Figure 4.7 Unbalanced forces cause bodies to accelerate (notice the lengths of the two arrows are not the same, indicating unbalanced forces).

In ball sports, such as cricket and tennis, a bat or racquet is used to strike the ball. The force applied by the bat or racquet is a pushing force, and the ball then travels at high speed for a short time but, because of air resistance, it eventually slows down. The air pushes back on the ball and slows its movement. The force of gravity also acts on the ball to bring it back down to the ground.

Contact forces

Friction is an example of a contact force. Tension forces in strings and ropes and friction between a crate moving over the floor are also examples of contact forces. When you kick a football your foot applies a contact force to the ball.

Buoyancy: a contact force

Boats, icebergs and corks float in water. Buoyancy is a contact force that acts upwards on

Case study 3: Changing the motion of a bicycle

Let's examine some examples of motion and forces using a bicycle.

1. Accelerating from rest (see Figure 4.8):
   - Pushing on the pedals makes the pedals and gear wheel rotate.
   - The chain attached to the gear wheels transfers the force to the back wheel and the back wheel rotates.
   - The rotating back tyre pushes against the track as long as there is sufficient friction.
   - The track pushes back against the tyres to propel the bicycle forwards. This pushing back force is called a reaction force.
   - The bicycle's speed increases. It is accelerating.

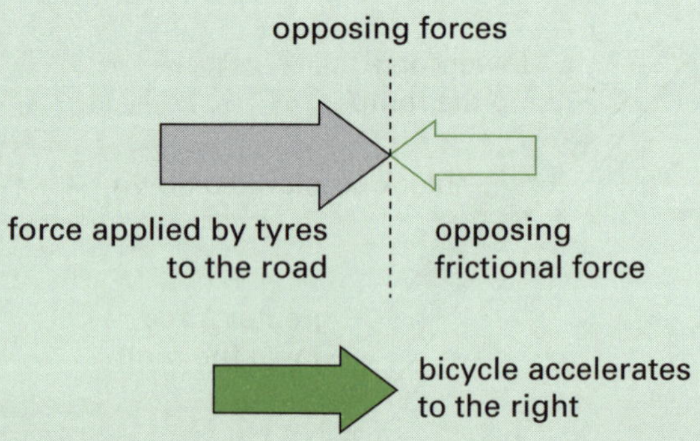

opposing forces

force applied by tyres to the road

opposing frictional force

bicycle accelerates to the right

Figure 4.8 Bicycle accelerates as the applied force is greater than the opposing frictional force

Although friction is necessary to provide the grip between the tyre and the track, frictional forces that oppose the motion can slow the forward motion of the bicycle. If the track is very rough, or the tyres are not properly inflated, you will have to apply a greater force through your muscles to overcome this extra resistance.

Forces come in pairs that oppose one another.

2. Travelling at constant speed (see Figure 4.9):
   - In order to travel at constant speed, a bike rider must continue to pedal at a rate so that the applied force just balances all the frictional forces which include road friction and air friction.

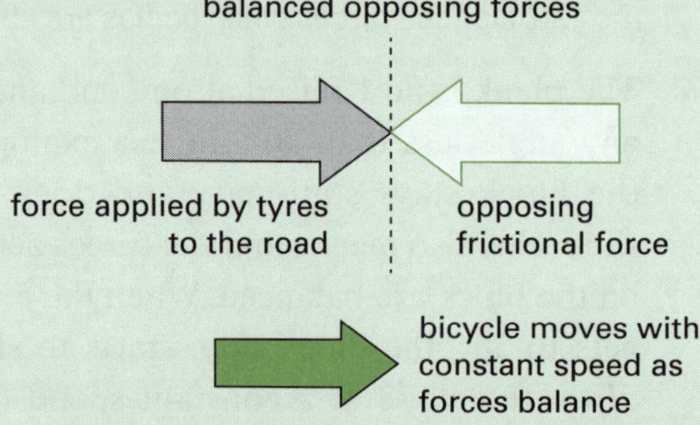

Figure 4.9 Bicycle moves at constant speed when opposing forces balance

You can further minimise friction on your bike by oiling the chain and gears. The oil helps your gears turn more easily, and keeps the gears from wearing out. At high speeds, making the upper part of your body horizontal rather than vertical further minimises air resistance and improves speed.

an object in a fluid that opposes the weight of the object acting down. When the body floats, these opposing forces are balanced (see Figure 4.10).

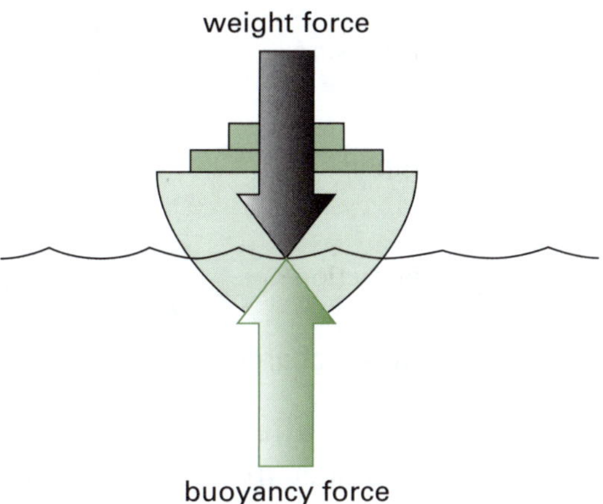

Figure 4.10 Buoyancy

Archimedes (287–212 BC) found that displacing more water could increase the buoyancy force in water. We use this principle today in steel ships. The steel is very heavy but because the ship is quite hollow (i.e. mainly air), it occupies a large volume and can displace large volumes of water. The greater the weight of displaced water, the greater the buoyancy force.

Field forces

Not all forces are contact forces. A second group is called field forces. Magnetic forces, electrostatic forces and gravitational forces are examples of field forces.

What does the term *field* mean? A field is a space in which an object experiences a force even though no physical contact is made. Let's look at some examples.

Magnetic field

If two bar magnets are brought close together, their ends either attract or repel each other. Attraction is an example of a pulling force. Repulsion is an example of a pushing force. The ends of a bar magnet are called the poles and are labelled with the letters 'N' (north-seeking pole) and 'S' (south-seeking pole). They are labelled this way because a freely suspended magnet moves until it points north–south.

Experiments show that:

- like magnetic poles repel
- unlike magnetic poles attract
- iron and steel objects are attracted by magnetic fields.

Thus, two N poles repel and two S poles repel. An N pole attracts an S pole (see Figure 4.11).

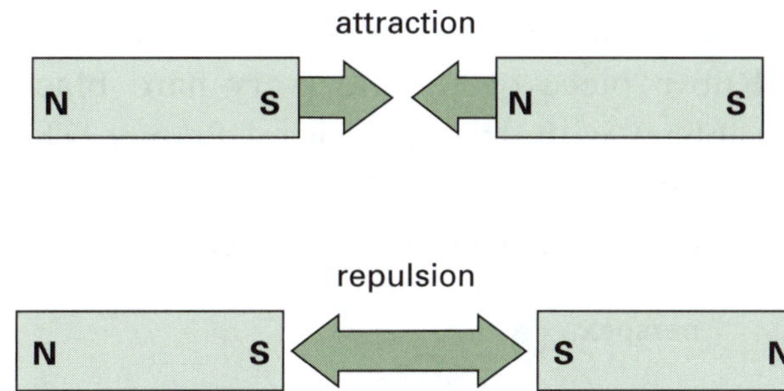

Figure 4.11 Like poles repel; unlike poles attract.

The magnetic field between the two magnets is invisible. We can use fine particles of iron called iron filings to show where the field is. The iron filings form a pattern around the magnets. Where the filings cluster close together, the field is strong and the forces felt by iron objects are strong. This is shown in Figure 4.12.

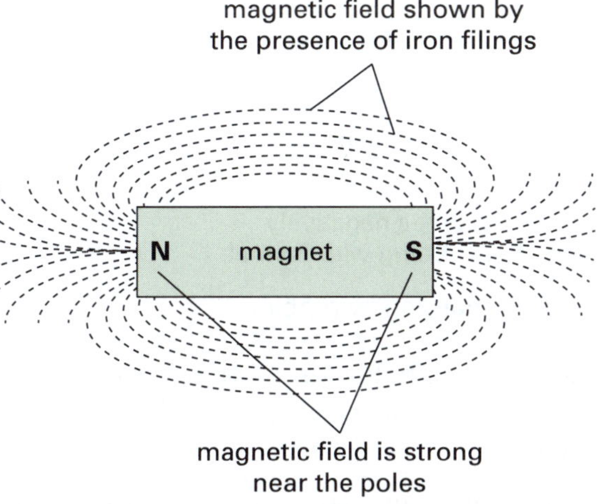

Figure 4.12 The magnetic field around a bar magnet is shown by sprinkling iron filings around the magnet that is lying on a sheet of paper.

Electrostatic field

Have you ever combed your hair on a dry day and had it stand on end? Have you ever rubbed your plastic ruler and picked up small bits of paper with it? These are examples of substances that have developed static (non-moving) charge.

There are two types of electric charge: positive and negative. Rubbing objects with various materials can produce positive and negative charges on them (see Figure 4.13).

- Rub a piece of glass or perspex with silk. The glass or perspex becomes positively charged and the silk becomes negatively charged.
- Rub a piece of ebonite (very hard black rubber) with cotton or wool flannel. The ebonite becomes negatively charged and the flannel becomes positively charged.

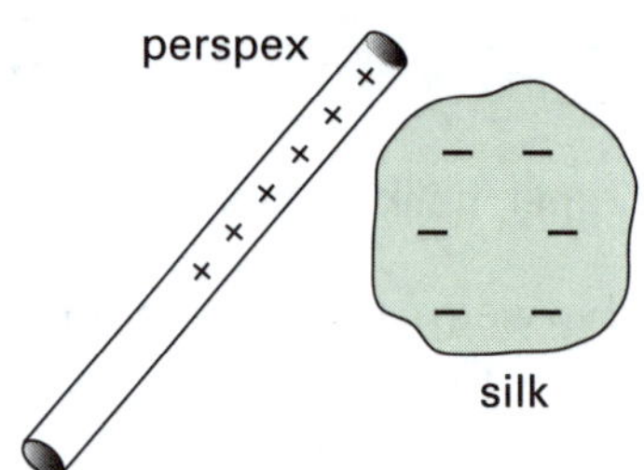

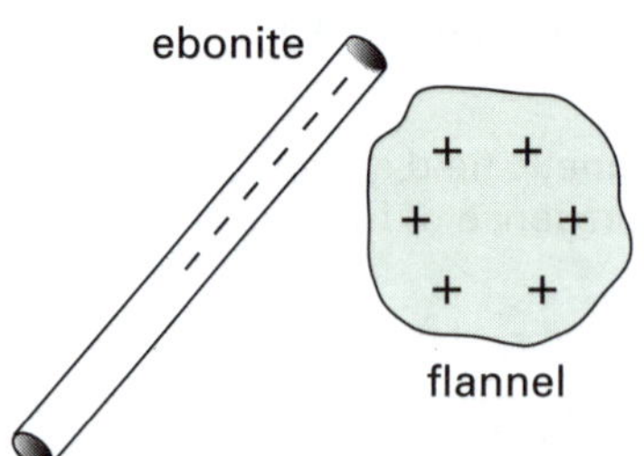

Figure 4.13 Charging by rubbing

An electrostatic field exists around any charged object. Certain objects in that field are attracted or repelled.

- Repulsion occurs when the fields around two positively charged objects interact.
- Repulsion occurs when the fields around two negatively charged objects interact.
- Attraction occurs when the fields around a positively charged object and a negatively charged object interact.

Therefore, as Figure 4.14 shows:

- like charges repel
- unlike charges attract.

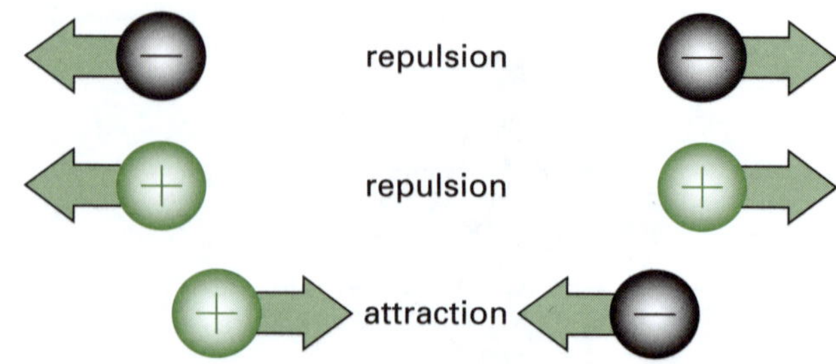

Figure 4.14 Forces between charged objects

Neutral objects are also affected by electric fields. Inside a neutral object are equal numbers of positive and negative charges. If a neutral object is brought near a charged object, then the charges inside the object separate due to the electric field. The object, however, is still neutral. This phenomenon is shown in Figure 4.15.

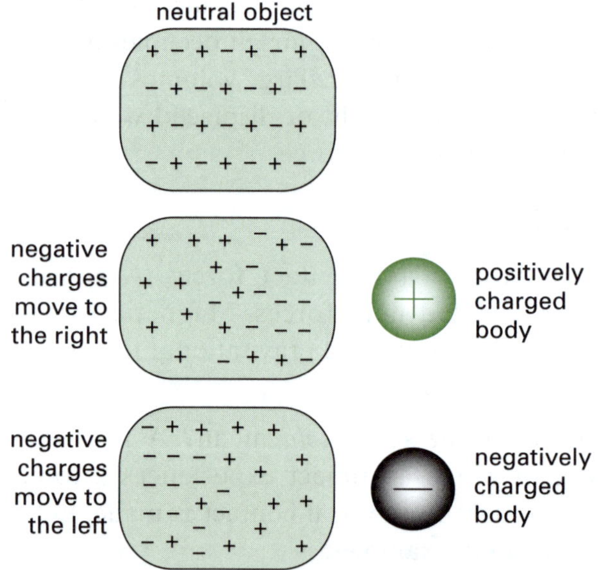

Figure 4.15 Charge separation in neutral objects (notice it is the electrons, which are free to move to one side or the other, that migrate)

Experiment 1

Charged bodies

Consider the following experiment involving charged rods and small pieces of paper.

Aim

- To investigate the behaviour of small pieces of paper near a charged rod

Method

1. A perspex rod is rubbed with silk.
2. The rod is brought close to small, uncharged fragments of paper. Observe what happens.

Results

The paper was initially attracted and stuck to the charged rod. Some pieces then jumped off the rods. Explain why the pieces of paper jumped off the rod at the end of the experiment.

Analysis

Analyse the results. Go to p. 201 to check your answer.

Conclusion

Write a suitable conclusion.

Go to p. 201 to check your answer.

Gravitational field

Large gravitational fields exist around all astronomical bodies such as stars and planets. In fact, gravitational fields exist around any mass no matter how small. A gravitational field is a space in which masses experience a force. This force is always attractive. Gravitational fields are the weakest of the field forces.

In a planet's gravitational field, small bodies appear to fall towards the surface of the planet. We can measure the strength of the gravitational force on a ball by attaching it to a spring balance. The extension of the spring is used to measure the force of attraction between the ball and the Earth. When the spring stops extending and comes to rest, the tension force in the spring exactly balances the weight force of the body.

The strength of the gravitational field (and therefore gravity) becomes less the further you are away from the Earth. In the Apollo missions to the Moon, as the Apollo spacecraft travelled further from Earth their tendency to fall back towards the Earth decreased. At a point about nine-tenths of the way to the Moon, the gravitational attraction of the Moon exactly balanced that of the Earth. Beyond that point, the spacecraft fell towards the Moon.

Test yourself 1

Part A: Knowledge

1. Weight is an example of *(1 mark)*
   - **A** a field force.
   - **B** a contact force.
   - **C** friction.
   - **D** a magnetic force.
2. When a mass experiences an unbalanced force it will *(1 mark)*
   - **A** move backwards.
   - **B** move forwards.
   - **C** move with constant speed.
   - **D** accelerate.
3. When a ship floats in water *(1 mark)*
   - **A** the buoyancy force balances the weight force.
   - **B** the forces are not balanced.
   - **C** magnetic forces balance the buoyancy force.
   - **D** the buoyancy force is greater than the weight force.
4. Which statement is true about magnetic poles and magnetic fields? *(1 mark)*
   - **A** Like poles attract.
   - **B** Unlike poles repel.
   - **C** The magnetic field around the magnet is invisible.
   - **D** The North Pole is stronger than the South Pole.
5. Which statement is true about electric fields and charges? *(1 mark)*
   - **A** Glass becomes positively charged when rubbed with silk.
   - **B** Ebonite becomes positively charged when rubbed with wool flannel.
   - **C** Attraction occurs when two positively charged rods are brought close to each other.

D Repulsion occurs when one negatively charged rod and one positively charged rod are brought close to each other.

6. Complete the following restricted-response questions by inserting the appropriate word. *(1 mark each)*
   a) Force can be measured by a change in the .... it produces in an object.
   b) A .... is a space where an object may experience a force.
   c) Of all the field forces, ... is the weakest.
   d) The Earth has a greater gravitational field than the Moon because the Earth has a greater ....
   e) Objects .... when acted on by an unbalanced force.

7. Use the code letters to match the terms or phrases in each column. *(1 mark for each part)*

Column 1	Column 2
A balanced forces	F unlike magnetic poles
B like charges	G constant velocity
C mass	H unbalanced forces
D deceleration	I gravity
E attract	J repel

Part B: Skills

8. A fish was caught and it was then attached to a spring balance. Its weight when suspended in air was found to be 10 newtons. The fish was now lowered into a tub of water. How will the readings on the spring balance change? Explain. *(2 marks)*

9. A group of children played a game in which they placed a large sheet of plastic on the ground and then ran towards the plastic and dived onto it. They tried to see how far they could slide on the plastic sheet. Suggest a way of increasing the distance they can slide. Explain why your method works. *(2 marks)*

10. Figure 4.16 shows a student's interpretation of the results of an experiment with a neutral piece of cardboard and a positively charged rod which is brought near, but not touching, the cardboard.

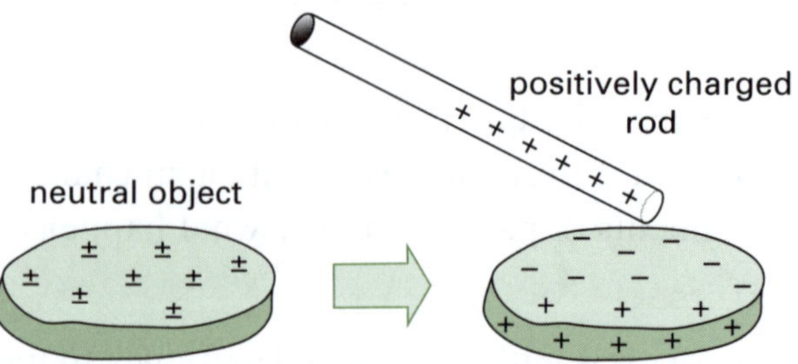

Figure 4.16 An experiment with a neutral piece of cardboard and a positively charged rod

What conclusion did the student make from this experiment? *(2 marks)*

11. A car is travelling at a uniform speed. Which of the graphs shown in Figure 4.17 best shows this type of motion? *(1 mark)*

(a)

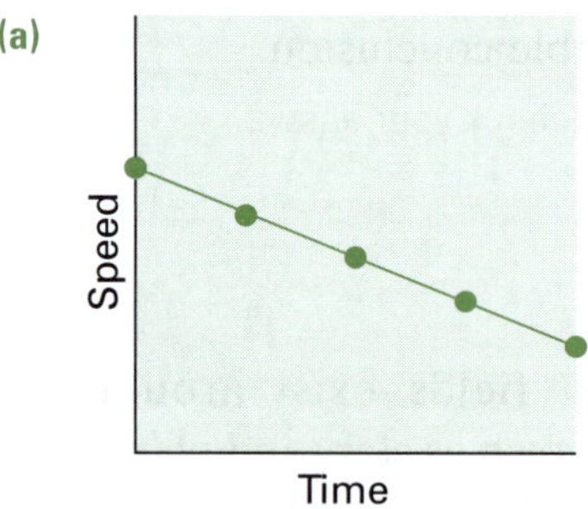

(b)

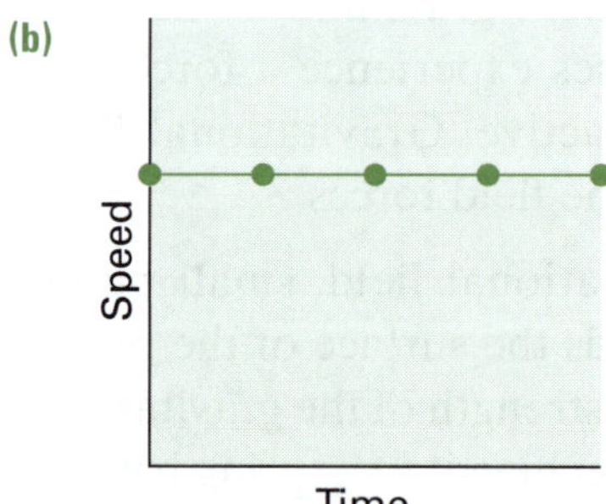

(c)

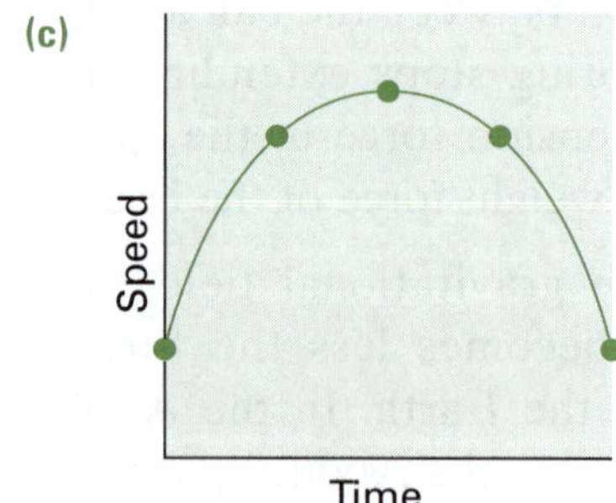

Figure 4.17 Graphs showing speed and time

12. An aeroplane decelerates rapidly as it lands.
   a) Explain what the term decelerate means. *(1 mark)*
   b) Which graph in Figure 4.18 best shows this type of motion? *(1 mark)*

(a)

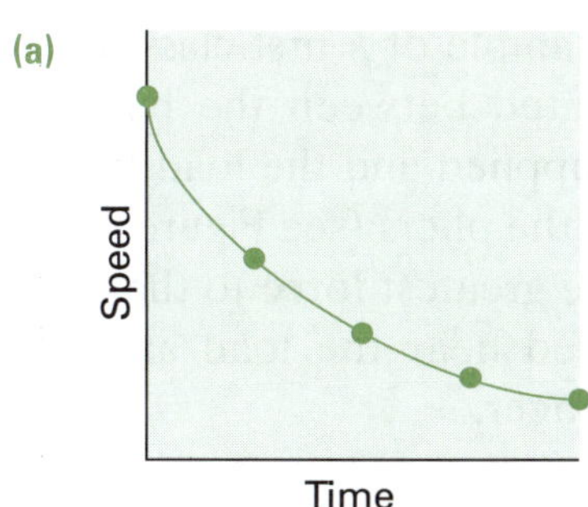

(b)

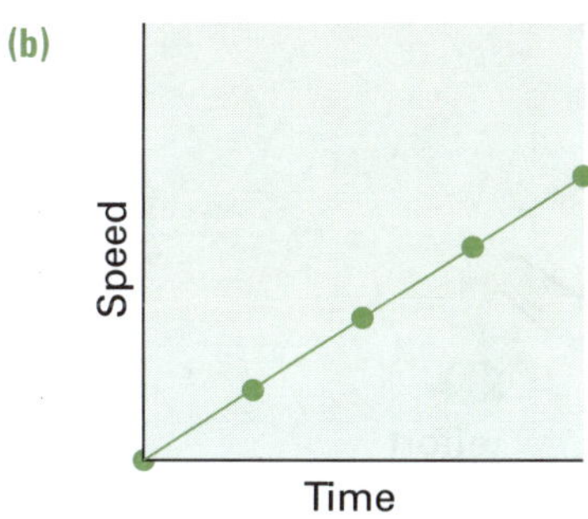

(c)

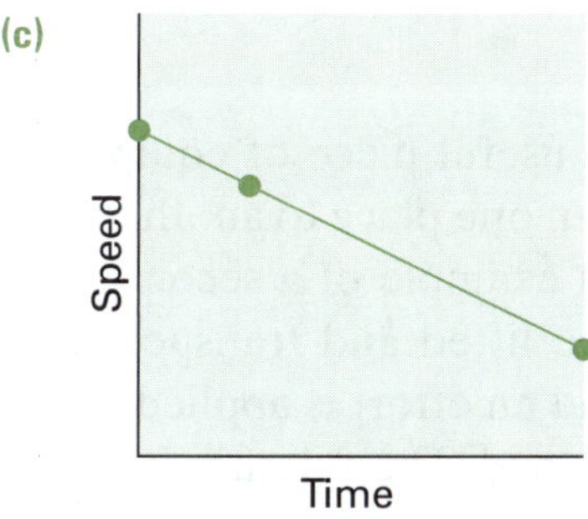

Figure 4.18 Graphs showing speed and time

13. A student set up the apparatus shown in Figure 4.19 using a spring balance, a steel cylinder and three magnets taped together.
   a) The student noticed that the weight of the steel cylinder changed when the magnets were placed on the table beneath. Explain why this happens and state whether the balance reading will be higher or lower than the reading in the absence of the magnets. *(2 marks)*
   b) Identify another variable that will increase the reading on the spring balance without changing the number of magnets used. *(1 mark)*

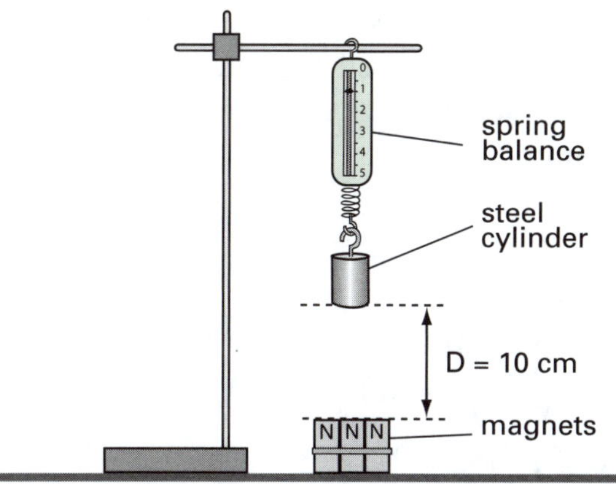

Figure 4.19 Apparatus showing a spring balance, a steel cylinder and three magnets taped together

14. Four steel balls (A = 20 kg; B = 10 kg; C = 5 kg; D = 4 kg) were dropped simultaneously from a height of 25 m above the ground. All four balls hit the ground at the same time. State one conclusion that could be made from such an experiment. *(1 mark)*

Go to p. 202 to check your answers.

4.2 Simple machines

Humans invented the first simple machines tens of thousands of years before the Egyptian pyramids were built. These ancient humans developed stone and bone tools to help them hunt, cut up flesh and sew skins together.

Machines are devices that make work easier. Machines are energy converters. They change energy into a more useful or convenient form. Some machines reduce the effort that needs to be applied to lift a load. Other machines simply change the direction in which forces act.

Levers and pulleys are examples of simple machines.

Levers

The shaduf was an early example of a lever (see Figure 4.20). It was used by ancient Egyptians to draw water from the Nile. The shaduf consisted of a bucket made of animal skins sewn together. A rope lowered this bucket into the river. The rope was attached to one end of a wooden lever that was counterbalanced by a heavy weight.

Once the bucket was full, the counterweight moved down and raised the bucket out of the water. The bucket was then swung over the land to be emptied.

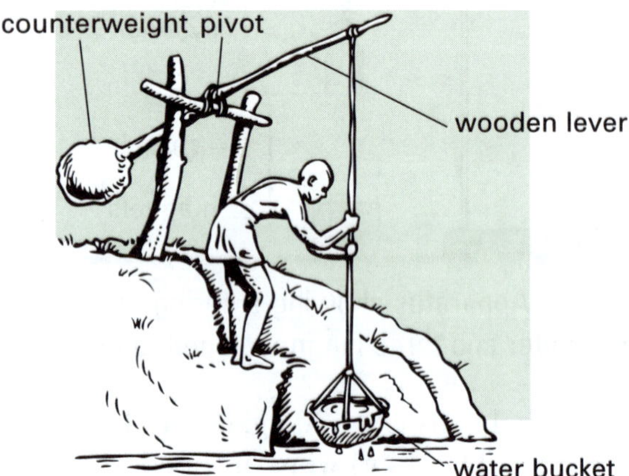

Figure 4.20 The shaduf

Crowbars, spanners, hammers and seesaws are all members of the vast family of machines called levers. They all have a main rigid bar moving around a stationary point called the pivot or fulcrum. The position of the fulcrum (F) in relation to the load (L) and effort (E) produces a wide variety of simple and complex levers.

There are three different classes (or orders) of levers as shown in Figure 4.21.

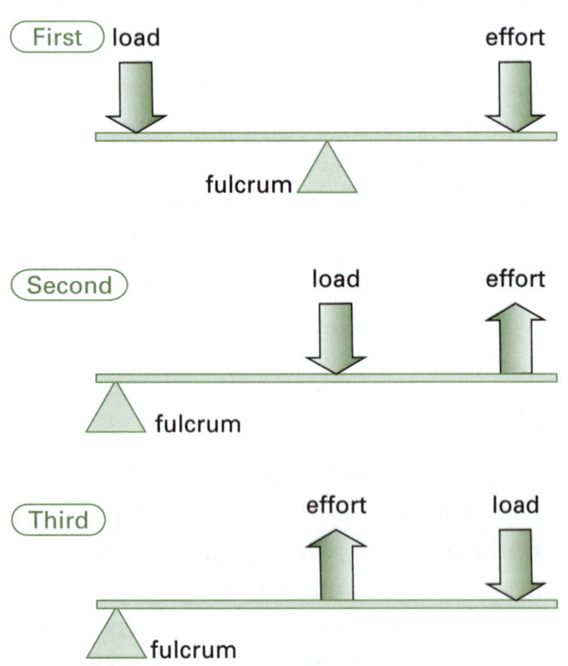

Figure 4.21 Classes of levers

- In a first-class lever the fulcrum is placed between the load and the effort.
- In a second-class lever the load is placed between the fulcrum and the effort.
- In a third-class lever the effort is applied between the load and the fulcrum.

Pliers are a good example of a first-class lever. The fulcrum is located between the handles where the effort is applied and the load that is between the jaws of the pliers (see Figure 4.22). In order to apply the greatest force to the load, the fulcrum is placed near the load and the handles are made longer.

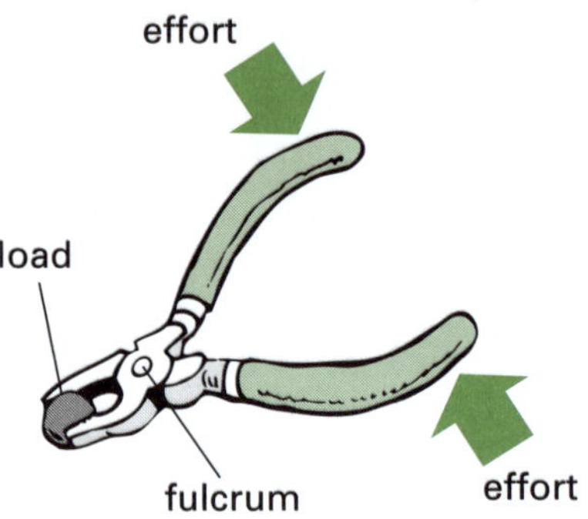

Figure 4.22 Pliers are an example of a first-class lever.

A wheelbarrow is a useful piece of equipment for moving loads from one place to another (see Figure 4.23). It is an example of a second-class lever. The load to be lifted and transported is placed in the tray and an effort is applied at the handles to raise the load. The fulcrum is at the axle of the wheel. The barrow can be now pushed along easily. The effort is reduced if the load is closer to the fulcrum. This can be achieved using longer handles.

Figure 4.23 A wheelbarrow is an example of a second-class lever.

Salad tongs are an example of a third-class lever (see Figure 4.24). The effort force is applied along the rigid bars to clamp the salad item in the jaws at the end. The fulcrum is at the opposite end of the two bars.

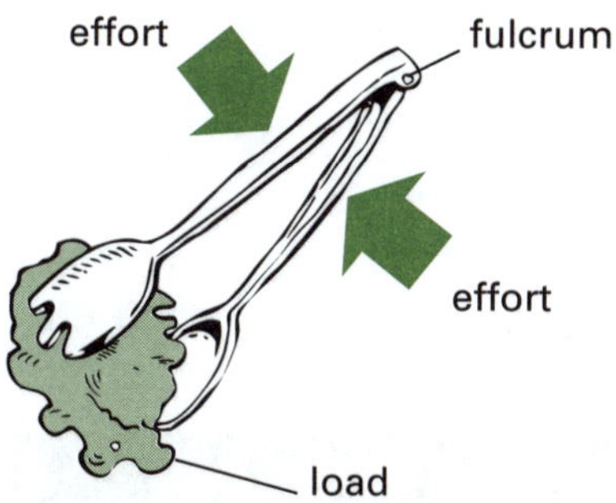

Figure 4.24 Salad tongs are an example of a third-class lever.

Some levers seem to give you increased strength. First- and second-class levers are examples of machines that allow you to lift heavy loads by applying less force. This means that the lever gives you a mechanical advantage.

However, if the load requires a larger effort to move it, then the machine gives you a mechanical disadvantage. Third-class levers are like this. Why would you bother using such a machine? A machine that gives you a mechanical disadvantage must give you an advantage in some other way, such as a change in direction or a change in distance or speed.

Experiment 2

The mechanical advantage of first-class levers

Aim

- To find out where the load and effort should be placed relative to a central fulcrum so that the effort force is kept to a minimum

Method

1. Set up the lever with the fulcrum in the centre of the ruler.
2. Make up a series of masses using the mass carrier and various slotted masses. These are the effort masses (E).
3. Use a 50-g mass carrier and a 50-g slotted mass (100 g in total) as the load (L).
4. Attach the load (L) to a cotton loop 20 cm from the fulcrum. This is D_L in Figure 4.25. Keep this distance constant for the following experiments.
5. On the other side of the fulcrum, investigate what effort mass will balance the lever when attached to a cotton loop placed 10 cm from the fulcrum.
6. Repeat this procedure by changing the effort distance to 20 cm, then to 25 cm and finally to 40 cm from the fulcrum. What effort mass in each case will now balance the lever? Record your results in a table.

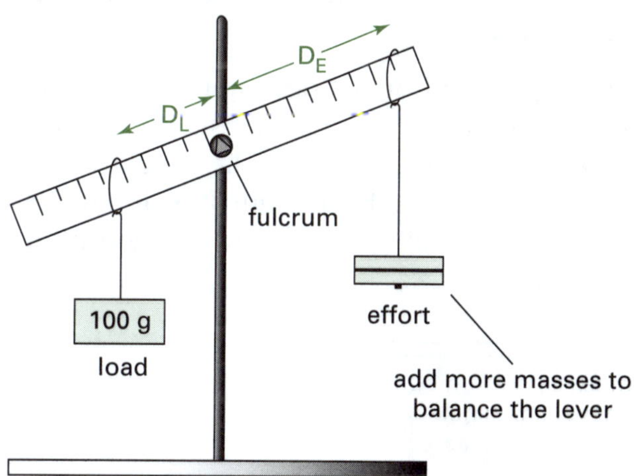

Figure 4.25 First-class lever

Results

Table 4.1 shows some typical results for this experiment.

Table 4.1 Results of experiment with a first-class lever

Load (L) (g)	D_L (load distance) (cm)	Effort (E) (g)	D_E (effort distance) (cm)
100	20	200	10
100	20	100	20
100	20	80	25
100	20	50	40

Analysis

Analyse these tabulated results. Determine which combination of masses produces a first-class lever with the greatest mechanical advantage. Go to p. 201 to check your answer.

Conclusion

Write a suitable conclusion.

Go to p. 201 to check your answer.

Pulleys

A pulley is a wheel with the rim adapted to hold a rope or cable. Pulleys are particularly effective when loads need to be lifted vertically, as it is easier to pull a rope down than to lift the load.

Single fixed pulley

The single fixed pulley does not decrease the size of the effort, but changes the direction of the effort (see Figure 4.26). The weight of the operator, as well as the strength of their muscles, can be used to pull the rope down and therefore raise the load.

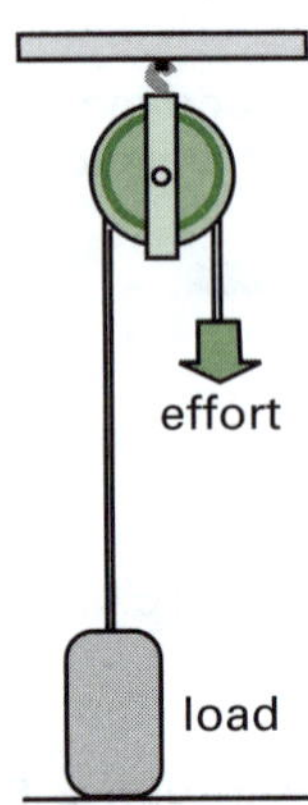

Figure 4.26 Single fixed pulley

Block and tackle

The block and tackle uses two or more pulleys combined in blocks (see Figure 4.27). The tackle refers to the ropes or chains used to operate them. This system has a mechanical advantage as the load can be raised with less effort. However, the effort rope has to be pulled through a long distance for a small movement of the load.

Figure 4.27 Block and tackle

Experiment 3

The mechanical advantage of pulleys

Aim

- To compare the mechanical advantage of using pulleys with different numbers of pulley wheels in each block

Method

1. In the first experiment use a pulley system with one pulley in each block (1×1). In the second experiment use a system with two pulleys in each block (2×2). (See Figure 4.28.)
2. Use the same load in each experiment (e.g. 400 g). Use a spring balance or force meter to measure the effort force required to lift the load.
3. Record the effort force required in each experiment.

Results

Here are some typical results of these experiments:

- a 1×1 pulley system: load = 400 g, effort = 200 g
- a 2×2 pulley system: load = 400 g, effort = 100 g.

Analysis

Analyse the results of this experiment. Which system had the greater mechanical advantage? Go to pp. 201–202 to check your answer.

Conclusion

Write a suitable conclusion.

Go to pp. 201–202 to check your answer.

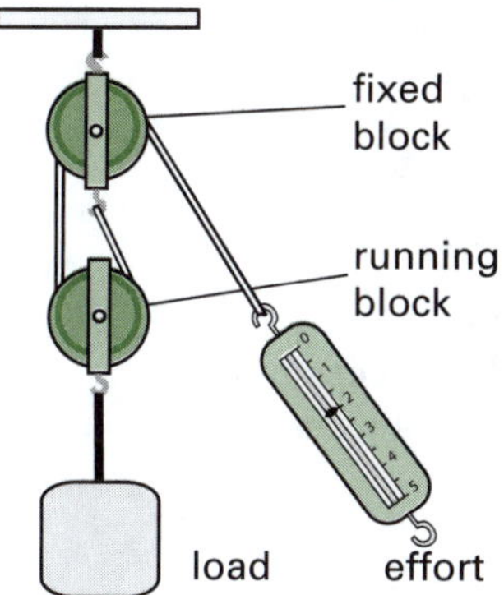

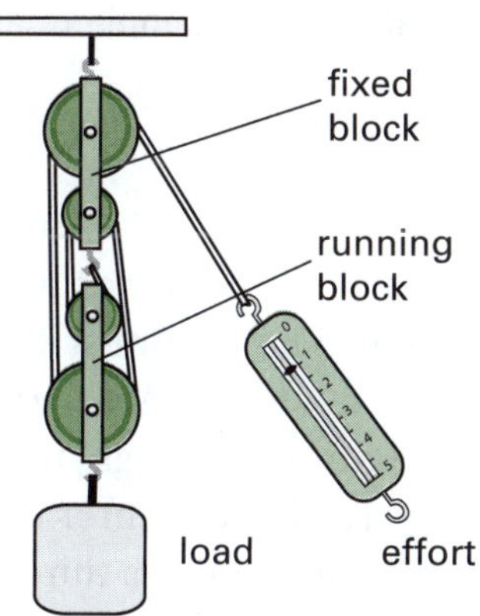

Figure 4.28 Pulley experiment

4.3 Cars, bikes and safety equipment

Safety helmets

In Australia, riders of bicycles or motorcycles are required by law to wear safety helmets. This law has helped to reduce the serious injuries to the head when an accident occurs.

Safety helmets are made from rigid plastic shell casing with an inner layer of expanded polystyrene foam at least 1 cm thick.

- The rigid outer shell protects the rider's head from abrasions and puncture wounds. This rigid shell is usually constructed from fibreglass which has been strengthened with carbon fibre or Kevlar.
- The inner layer of expanded polystyrene absorbs some of the energy of impact and spreads the impact force over a longer time. Less trauma to the head and brain will occur.
- An inner padded, soft lining holds the head snugly.

Car safety equipment

Modern cars are fitted with various safety features that help to reduce injuries to people in an accident. These features are designed to reduce impact energy by increasing the time over which the people and car are brought to rest (see Figure 4.29).

- Seat belts. Without a seatbelt, objects inside the car continue to move forwards in a head-on collision even though the car has come to a sudden stop. This means people and other loose objects collide with the dashboard and front window causing serious damage and often death. The seatbelt reduces sudden movements of the body during a collision by holding it firmly against the back of the seat. Passengers are prevented from being thrown forwards and hitting their heads on the dashboard or other hard objects. Seat belts absorb some of the impact energy because they are designed to stretch a little during the sudden deceleration of the impact.
- Crumple zones. Crumple zones help to slow down the time of the collision so the vehicle and its occupants come to rest more slowly. Crumple zones are designed to deform and absorb a lot of the impact energy. These zones are steel frames that deform during the impact. They protect the rigid parts of the car such as the engine block and passenger compartment.
- Airbags. Airbags inflate rapidly during a collision to prevent drivers and passengers from decelerating too quickly. The people are brought to rest more slowly and so impact forces cause less damage. It is particularly

important to prevent the front seat driver and passenger from hitting the steering wheel or the dashboard.

Figure 4.29 Force arrows show how a seatbelt and airbag can prevent the sudden forward movement of the driver during a collision.

4.4 Forces and sports

The human body consists of a number of simple levers. The bones of our skeleton provide the rigid bars and out muscles provide the effort force to raise various loads.

Let us look at some examples that are related to sport.

Weightlifting: barbell curls

Figure 4.30 shows the bones and muscles of your arm. In sports such as weightlifting you can improve muscular strength by raising heavy weights. A barbell curl is an example of such an exercise. Your biceps muscle contracts and applies an effort force (E) to the bone of your forearm. At the same time, your triceps muscle relaxes. The effort force from the biceps muscle is applied quite close to your elbow joint which acts as the fulcrum. The heavy weight, as well as your forearm, is the load (L) to be raised. This is a third-class lever.

When you straighten your arm again, your triceps muscle contracts and the biceps relaxes. In this case the arm acts as a first-class lever.

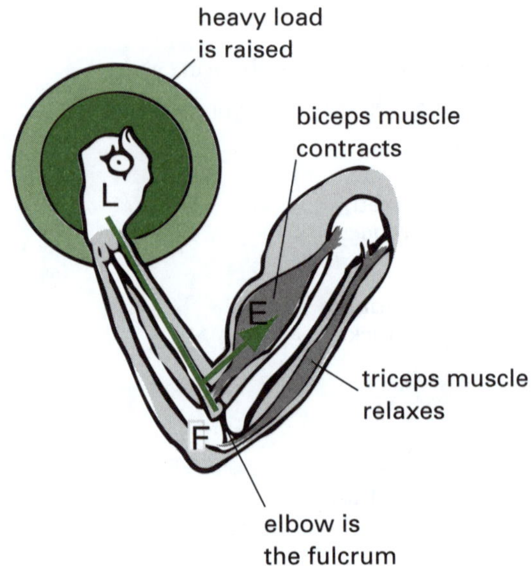

Figure 4.30 Raising your forearm is an example of a third-class lever.

Football: kicking

When a footballer kicks the ball, the lower leg pivots around the knee joint. The leg must be straightened when the ball makes contact with the boot. In order to straighten the legs, the front muscle in the lower leg (called the anterior tibialis) contracts. This lever action is also a third-class lever and so the speed of the ball increases even though there is a mechanical disadvantage. The effort force applied comes from the muscles in the lower leg, which are attached just below the knee (see Figure 4.31).

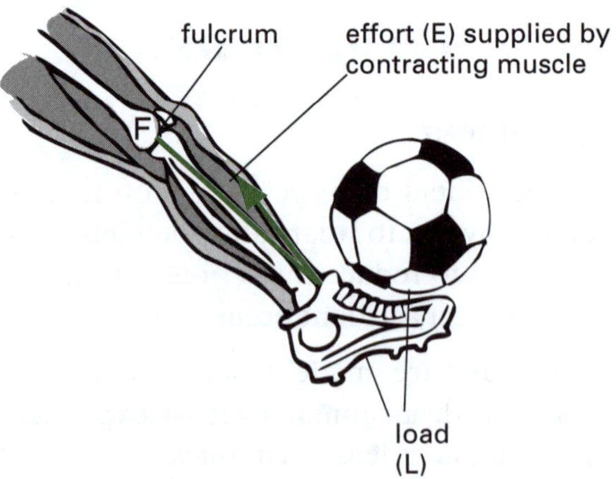

Figure 4.31 Kicking a football

Sprinting: moving hip, knee and ankle joints

The joints of the body are also important in moving the body during sport. The joints are hinges around which the bones can move when they are subjected to load or effort forces. Between the hip and the upper leg bone (femur) is a joint called a ball-and-socket joint. This type of joint allows leg rotation in many directions. Your shoulder joint is also a ball-and-socket joint. Between the femur (upper leg bone) and the tibia (lower leg bone) is the knee joint which acts as a hinge allowing the bones of the leg to bend and straighten (see Figure 4.32).

At the ankle is a type of pivot joint that allows up and down movement of the foot and some sideways motion.

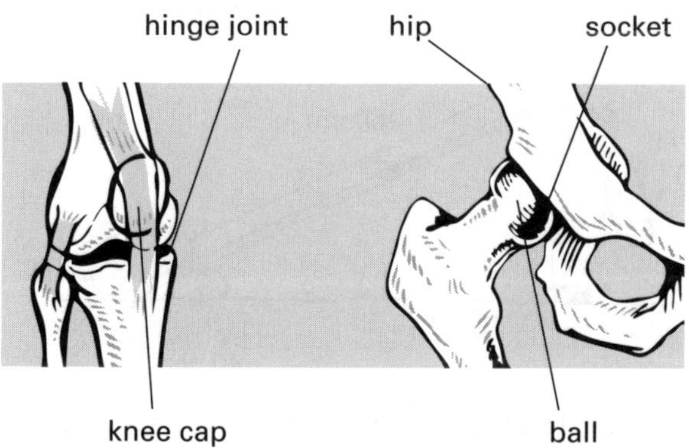

Figure 4.32 Joints in the knee and hip

In a sprinting race the athletes use all these joints to run down the track. The muscles of each part of the leg and foot apply forces to move the legs and feet as well as move the joints.

Figure 4.33 shows the location of the hamstring muscles in the legs. There are three muscles which form this group. They are involved in bending the leg at the knee and also straightening the leg at the hip. They provide power and stride length during sprinting. If the muscles are not kept flexible by training exercises, insufficient power will result. But not all muscles connect to move bones. For example, muscles around the eye allow you to move it around in various directions.

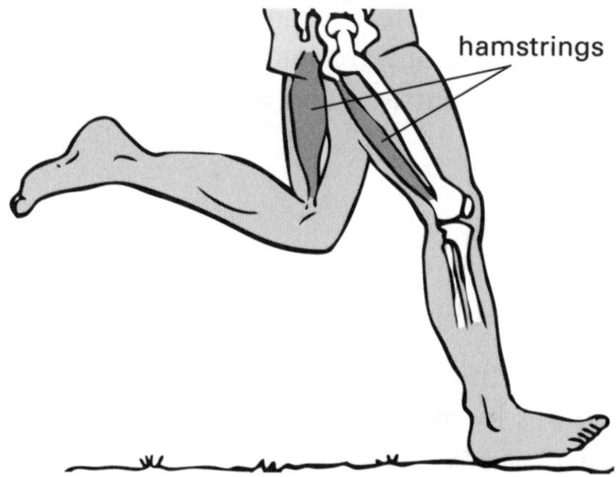

Figure 4.33 Hamstring muscles

Test yourself 2

Part A: Knowledge

1. Which of the following is *not* a way machines help humans? *(1 mark)*
   A by increasing the force applied
   B by producing a change of speed
   C by producing a change of direction
   D by making the effort used always less than the load
2. Which one of the following levers has the load between the effort and the fulcrum? *(1 mark)*
   A scissors
   B salad tongs
   C a fishing rod
   D a nut cracker
3. Levers are found in which of the following? *(1 mark)*
   A screws
   B pliers
   C a block and tackle
   D a ramp
4. A bicycle helmet is made of a number of layers. The function of the inner layer of expanded polystyrene is to *(1 mark)*
   A hold the head snugly
   B protect from abrasions
   C protect from puncture wounds
   D absorb some impact energy and spread the force over a longer time

5. When a weightlifter strengthens his arms by performing upward barbell curls, the muscle that contracts and applies an effort force is the
   A hamstring.
   B triceps.
   C biceps.
   D anterior tibialis.

6. Complete the following restricted-response questions by inserting the appropriate word. *(1 mark for each)*
   a) The ankle is a type of .... joint that allows up and down movement of the foot and some sideways motion.
   b) When you straighten your arm, your .... muscle contracts.
   c) Car seatbelts reduce sudden movements of the body during a collision by holding the body .... against the back of the seat.
   d) In a block and tackle system of pulleys, the effort rope has to be pulled through a .... distance for a small movement of the load.
   e) Pliers are a good example of ....-class levers.

7. Use the code letters to match the terms or phrases in each column. *(1 mark for each part)*

Column 1	Column 2
A airbags	F first-class lever
B kicking football	G mechanical disadvantage
C hedge cutters	H leg bending
D third-class lever	I third-class lever
E hamstrings	J slower deceleration

Part B: Skills

8. Figure 4.34 is a simplified model of a person standing on tiptoes. This is an example of a lever action. Name the class of lever and identify the labels A, B and C. *(4 marks)*

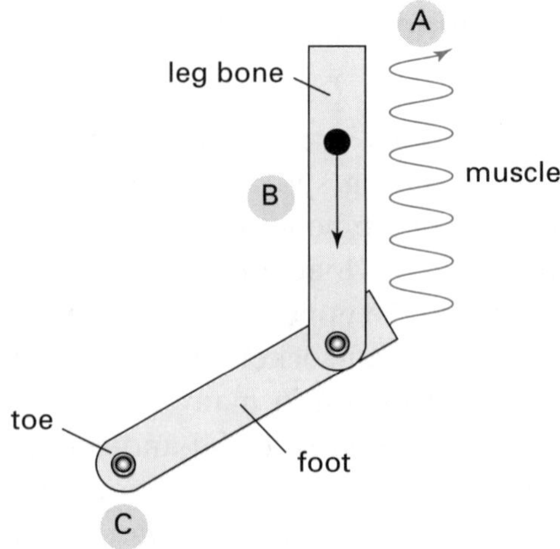

Figure 4.34 A simplified model of a person standing on tiptoes

9. A student uses a strong, rigid plank of wood to lift a heavy rock as shown in Figure 4.35.

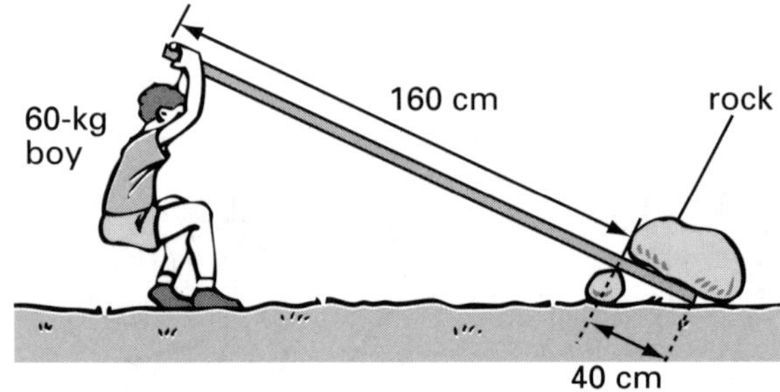

Figure 4.35 Using a strong, rigid plank of wood to lift a heavy rock

The aim is to lift the rock until the plank is horizontal.

   a) Use the following formula to calculate the mass of the rock that the student can lift by using his own body weight. *(2 marks)*

$$E \times D_E = L \times D_L$$

where E = effort
L = load
D_E = length of the effort arm
D_L = length of load arm

   b) Identify the class of lever. *(1 mark)*

10. Levers are simple machines. A children's seesaw is a lever. In each of the following examples, children of different weights wish to play on the seesaw which is 6 m long. Draw a diagram to show where each child should sit on the see-saw so they can balance properly and enjoy the game.

a) Nerida (45 kg) and Melanie (45 kg) *(1 mark)*

b) Wills (30 kg) and Alec (60 kg) *(1 mark)*

11. In most doors, the handle is placed as far away from the hinges as possible. In some doors, the handle is placed near the centre of the door, but never closer to the hinges than the centre. Explain why this is so. *(2 marks)*

12. A worker using a shovel to toss soil from a pile into a garden bed holds one hand at the top of the handle and lifts the weight with the other hand placed closer to the soil.

a) What class of levers is the shovel an example of? *(1 mark)*

b) Does this lever have a mechanical advantage? Explain. *(2 marks)*

13. Figure 4.36 shows a single movable pulley which is to be used to life a load off the floor. Explain whether or not this system has a mechanical advantage. *(2 marks)*

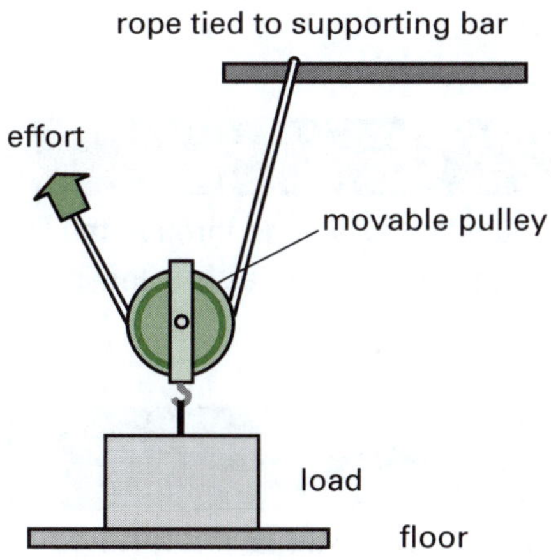

rope tied to supporting bar
effort
movable pulley
load
floor

Figure 4.36 A single movable pulley which is to be used to life a load off the floor

14. Figure 4.37 compares the bone lengths in the foot of a lion and a cheetah. Cheetahs can sprint at very high speed whereas lions cannot. Suggest an hypothesis that could explain this difference in speed. *(2 marks)*

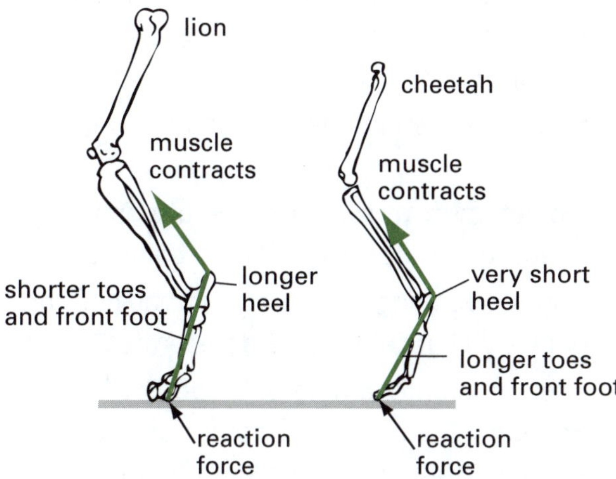

Figure 4.37 Comparison of the bone lengths in the foot of a lion and the foot of a cheetah

Go to pp. 202–203 to check your answers.

Summary

1. All movements are caused by forces. Forces are pulls, pushes or twists.
2. Forces are measured using force meters. A spring balance is an example of a force meter.
3. Forces come in pairs that oppose one another.
4. Objects stay at rest or move at constant speed when opposing forces balance.
5. When opposing forces do not balance, an object will accelerate or decelerate.
6. Friction is an example of a contact force. Friction is caused by contact between surfaces that are not smooth. Frictional forces oppose action forces.
7. A field is a space in which an object experiences a force even though no physical contact is made. Fields include gravitational, magnetic and electrostatic fields.
8. Like magnetic poles repel. Unlike magnetic poles attract.
9. Like charges repel. Unlike charges attract.
10. Machines are devices that make work easier. Machines are energy converters. Levers have a main rigid bar moving around a stationary point called the pivot or fulcrum.
11. There are three classes of lever. First-class levers have the fulcrum between the effort and the load. Second-class levers have the load between the effort and the fulcrum. Third-class levers have the effort between the load and the fulcrum.
12. A pulley is a wheel with the rim adapted to hold a rope or cable. Pulleys are particularly effective when loads need to be lifted vertically, as it is easier to pull a rope down than to lift the load.
13. The block and tackle uses two or more pulleys combined in blocks. They are used to lift very large loads with a small effort.
14. Modern cars are fitted with various safety features that help to reduce injuries to people in an accident. These features are designed to reduce impact energy by increasing the time over which the people and car are brought to rest.
15. The human body consists of a number of simple levers. The bones of our skeleton provide the rigid bars and our muscles provide the effort force to raise various loads.

Syllabus checklist

Are you able to answer every syllabus question in this chapter? Tick each question as you go through the list if you are able to answer it. If you cannot answer it, turn to the appropriate page in the guide as is listed in the column to find the answer.

For a complete understanding of this topic		Page no.	✓
1	Can I recall the types of forces that cause motion?	111	
2	Can I state an example of a force meter?	111	
3	Can I explain why objects stay at rest or in constant motion when acted upon by balanced forces?	111–113	
4	Can I explain Newton's idea that forces come in opposing pairs?	114	

For a complete understanding of this topic		Page no.	✓
5	Can I explain why objects accelerate or decelerate when acted upon by unbalanced forces?	113	
6	Can I recall the type of contact forces that oppose motion?	113–115	
7	Can I recall the definition of a field force and can I recall examples of field forces?	115–116	
8	Can I recall the rules of attraction and repulsion for like and unlike magnetic poles?	115	
9	Can I recall the rules of attraction and repulsion for like and unlike charges?	116	
10	Can I define the word machine and explain why machines are examples of force converters?	119	

For a complete understanding of this topic		Page no.	✓
11	Can I recall the positions of the load, effort and fulcrum in the three classes of levers?	120–121	
12	Can I describe the structure of a pulley system?	122	
13	Can I explain why a block and tackle system of pulleys can be used to lift very heavy loads?	122	
14	Can I recall some of the safety features in modern cars and describe how they reduce impact energy?	123–124	
15	Can I describe examples of lever systems in the human body?	124–125	

Chapter test

Go to p. v for *Tips for tests and examinations* 80 MIN

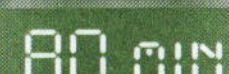

Part A: Multiple-choice questions

(1 mark for each)

1. The force acting downwards on a body because of gravity is called
   - **A** friction.
   - **B** weight.
   - **C** mass.
   - **D** newtons.
2. Which of these forces can be described as a contact force?
   - **A** gravitational force
   - **B** magnetic force
   - **C** electrostatic force
   - **D** frictional force
3. Which of these statements is *not* true about electrostatic forces?
   - **A** Charged objects attract uncharged objects.
   - **B** Objects that have the same charge repel each other.
   - **C** Objects that have opposite charges attract each other.
   - **D** Uncharged objects attract uncharged objects.
4. Figure 4.38 shows a fish swimming at a depth of 5 m underwater.

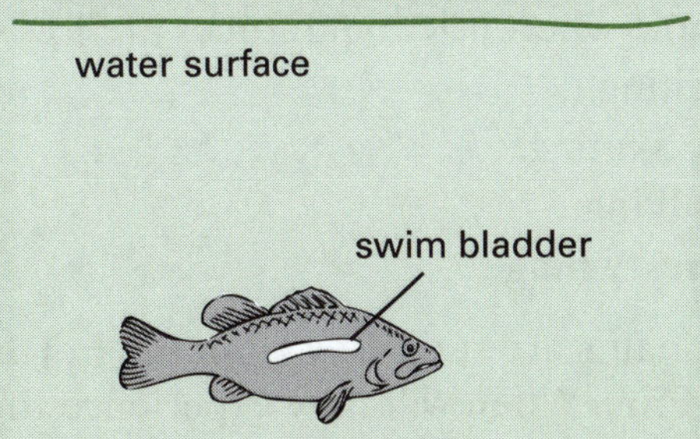

Figure 4.38 A fish swimming underwater

Identify the force that prevents the fish sinking.
   - **A** gravity
   - **B** magnetic force
   - **C** buoyancy
   - **D** friction

5. Four objects of different masses are put in a container of water. Their stationary positions in the water are shown in Figure 4.39.

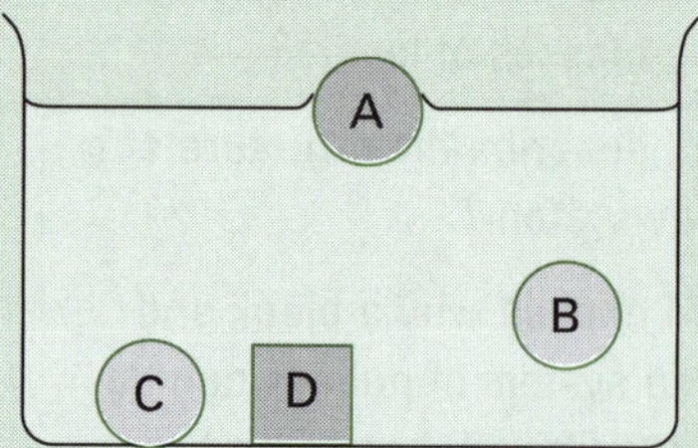

Figure 4.39 The stationary positions of four objects of different masses in a container of water

In which position are the forces balanced?

A position B only
B positions C and D
C positions A, B and D
D positions A, B, C and D

6. Which one of the following statements is true?

A A fishing rod is an example of a second-class lever.
B In a pair of tweezers, the effort is applied between the fulcrum and the load.
C A pair of scissors is a second-class lever.
D A single movable pulley allows a load to be lifted by applying a force downwards.

7. Frictional forces oppose motion. One example where frictional forces are low is

A writing.
B ice skating.
C walking.
D a rusty hinge.

8. Astronauts are travelling from the Earth to the Moon. When will they experience the greatest forces on their bodies?

A On the Moon's surface
B On the Earth's surface
C As the rocket lifts off the launch pad on Earth
D As the lunar lander takes off from the Moon's surface

9. When the astronauts arrive at the Moon, they need to detach the lander from the command module and descend onto the surface. How does the lander avoid crashing into the Moon?

A Retrorockets are fired to oppose the force of gravity.
B A parachute is used so the craft slowly descends.
C The craft glides in at a low angle.
D Gases are released so the craft descends like a balloon.

10. Which of the following objects will be attracted into a magnetic field?

A plastic spoon
B stainless-steel pin
C copper wire
D glass cup

Part B: Short-answer questions

11. A glass of water is filled to overflowing. A square of thin cardboard is slid over the rim of the glass so that no air is trapped. When the glass is carefully turned upside down and the supporting hand withdrawn, the water and card do not fall down (see Figure 4.40).

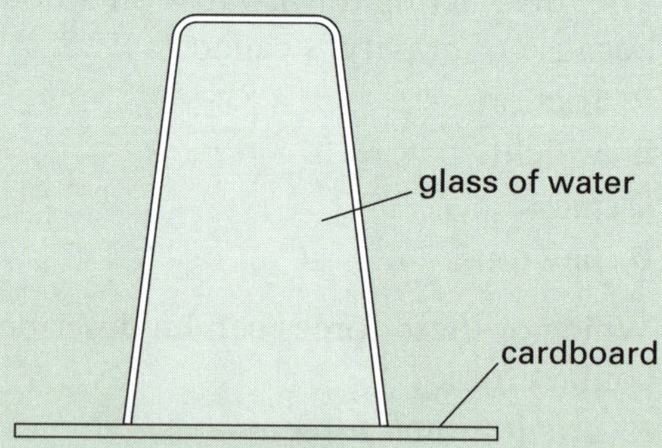

Figure 4.40 A glass of water resting upside down on a square of cardboard

a) Are the forces balanced or unbalanced in this example? *(1 mark)*
b) Copy the diagram of the glass of water and use arrows to draw in the opposing forces. *(1 mark)*
c) What causes the force that opposes the weight force of the water? *(1 mark)*

12. A tennis player plays a lob shot as shown in Figure 4.41. Copy this drawing and label all the forces acting on the ball at each of the stages A, B and C during its flight. *(6 marks)*

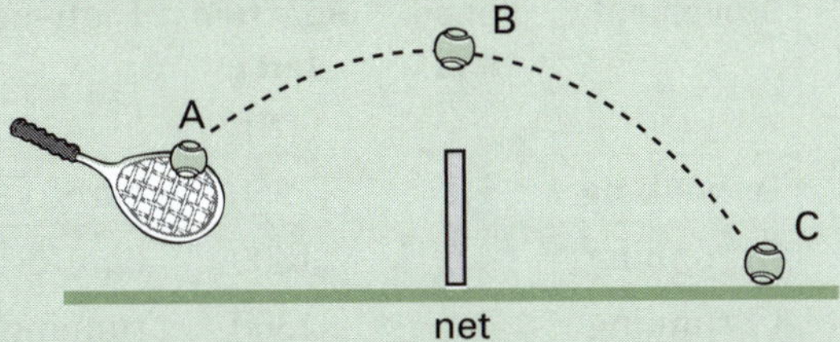

Figure 4.41 Forces acting on a tennis ball

13. As Figure 4.42 shows, steel ball bearings are used to separate the wheel casing and the axle of castors on skateboard wheels. As the wheel moves, the ball bearings roll over the surface of the axle and the inside of the wheel surface. How is this a useful way of reducing friction? *(1 mark)*

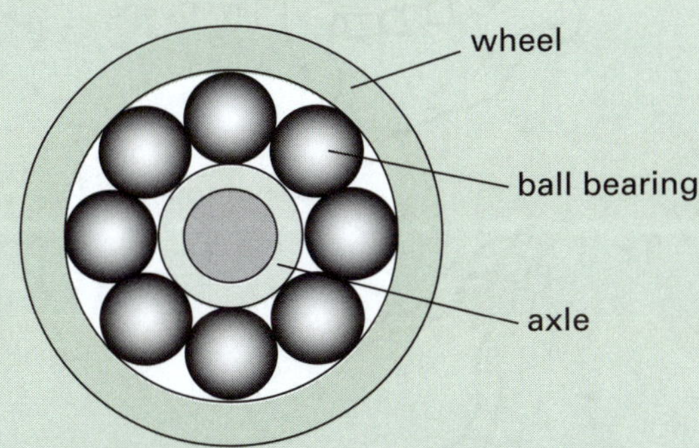

Figure 4.42 Inside of a skateboard wheel

14. Air resistance (or drag) can slow down moving objects such as racing cyclists. Information collected on drag is shown in Table 4.2.

Table 4.2 Amount of drag at different speeds

Speed of object (m/s)	Drag (N)
0.0	0
4.5	5
9.0	15
13.5	30

a) Plot this data as a line graph. Plot the speed along the horizontal axis. Give your graph a title. *(5 marks)*

b) How does the air resistance change as the speed of the object increases? *(1 mark)*

c) What must designers try to do to reduce drag for racing cyclists? *(2 marks)*

15. The first column of Figure 4.43 represents the information you are given about a closed box with one corner marked with X, and a magnet. In the second column, the magnet is moved and there is a resulting movement in the closed box.

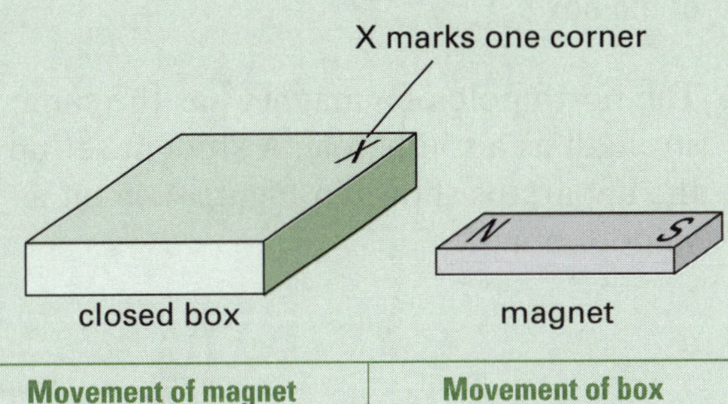

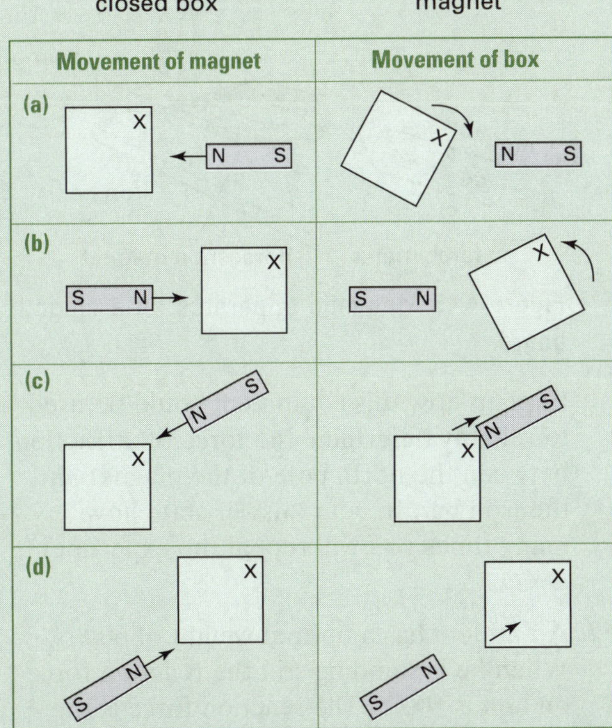

Figure 4.43 Information about a closed box marked with an X, a magnet and the resulting movements

a) What do you deduce is inside the closed box? *(1 mark)*

b) Which of the diagrams in Figure 4.44 most accurately describes the position of the contents of the closed box? *(1 mark)*

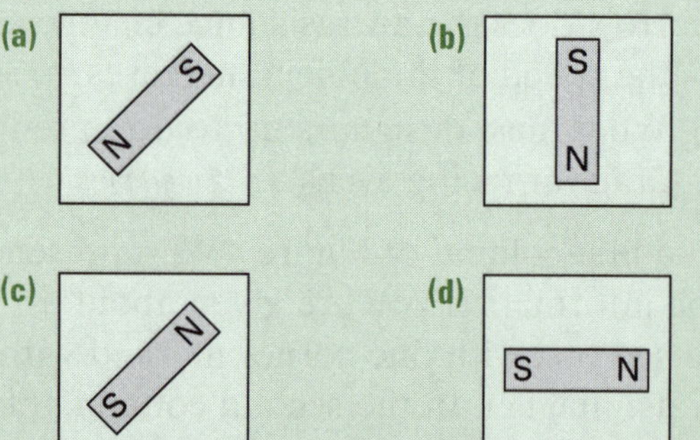

Figure 4.44 Possible positions of the contents of the box

16. The north pole of a magnet has the same strength as a south pole. A student set up the apparatus shown in Figure 4.45 on a smooth board.

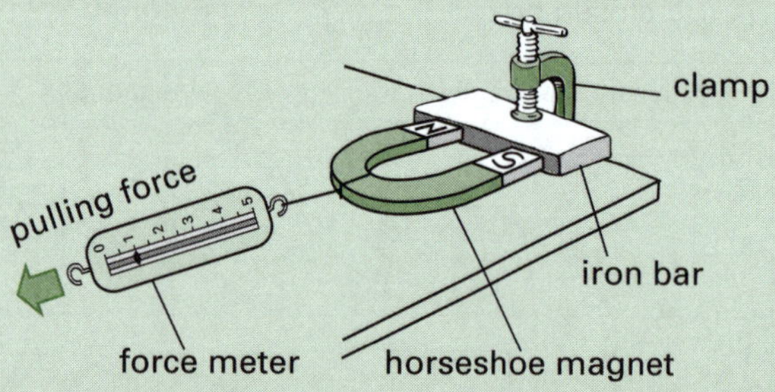

Figure 4.45 Magnetic apparatus on a smooth board

Explain how this equipment could be used to reliably determine the force of attraction between the north pole of the magnet and the iron bar. In your answer state how many times you will repeat the experiment. *(5 marks)*

17. An athlete has a normal weight of 600 N. When he is standing still the reaction force on him is 600 N. The reaction force is the force that the track applies to the feet of the athlete.
Table 4.3 shows the information that was collected on the reaction force produced by this athlete while walking or running on a track.

a) What evidence is there that running shoes can act as shock absorbers and reduce the reaction force on the body? *(1 mark)*

b) What evidence is there in the data that increased speed will cause more damage to the joints? *(1 mark)*

Table 4.3 Reaction force produced by an athlete

Movement	Speed (m/s)	Reaction force (N)	Footwear
A: walking	1.2	350	barefoot
B: running	4.6	1800	barefoot
C: running	4.6	1300	running shoes

18. Classify the levers shown in Figure 4.46 as first-class, second-class or third-class levers. *(3 marks)*

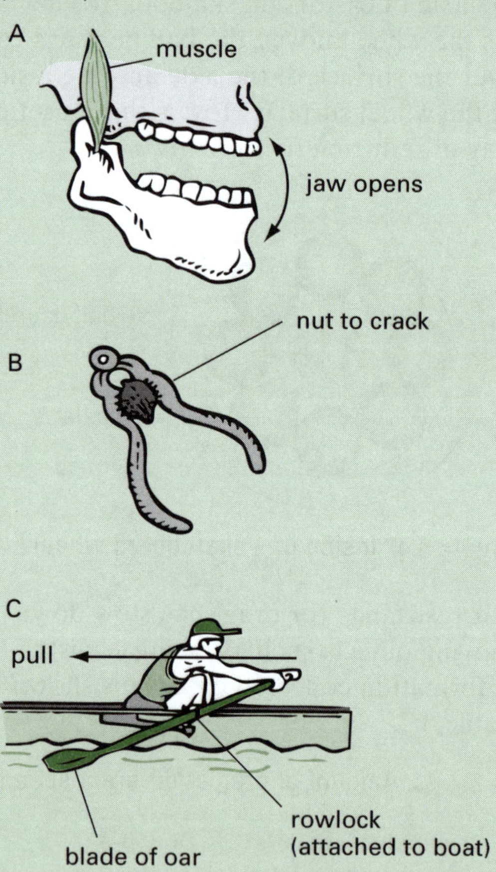

Figure 4.46 Different types of levers

19. Two pulley systems were used to investigate the force required to lift a 300-g mass off the floor (see Figure 4.47). Assume the pulley wheels have very low friction. The spring balances measure in grams.

a) What is the reading on spring balance (1) when the mass has been lifted off the floor? *(1 mark)*

b) What is the reading on spring balance (2) when the mass has been lifted off the floor? *(1 mark)*

c) Which pulley system has a mechanical advantage? *(1 mark)*

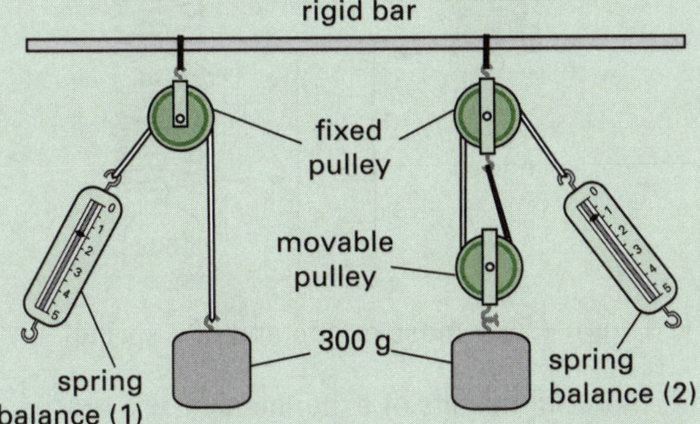

Figure 4.47 Two pulley systems used to lift a 300-g mass

20. a) How is the weight force of a body measured? *(1 mark)*

b) Three bodies A, B and C have masses of 3 kg, 10 kg and 20 kg respectively. Which body has the greatest weight? Explain. *(2 marks)*

21. If a 10-kg mass is dropped from a great height above the Earth's surface, describe the motion of the body in the time before the body contacts the surface. *(2 marks)*

22. a) How are magnetic field forces similar to gravitational field forces? Identify three similarities. *(3 mark)*

b) How are magnetic field forces different from gravitational field forces? Identify two differences. *(2 mark)*

23. a) The Latin word for mass is *massa* which means 'composed of lumps'. Explain how this definition is related to the scientific meaning of mass. *(1 mark)*

b) The Earth would be described as a *massive* object. What is meant by this term? *(1 mark)*

c) The Latin word for weight is *gravitas*. What scientific term is derived from this Latin word? Relate this term to the word weight. *(1 mark)*

24. Complete the following restricted-response questions by inserting the appropriate word. *(1 mark for each)*

a) Strong .... fields surround massive objects.

b) The size of the gravitational force acting on a small body can be measured by the .... of the spring on a force meter to which the body is attached.

25. Figure 4.48 shows three cases of forces acting on a body. The length of each force arrow is proportional to its size. Describe what will happen to the body in each case. *(3 marks)*

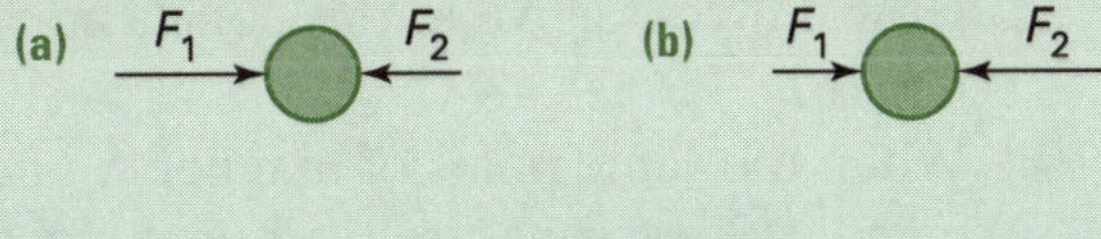

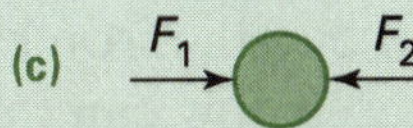

Figure 4.48 Three cases of forces acting on a body

26. Consider Figure 4.49, which shows a mass suspended at the end of a spring. Compare the weight force of the body and the tensional force in the spring. *(1 mark)*

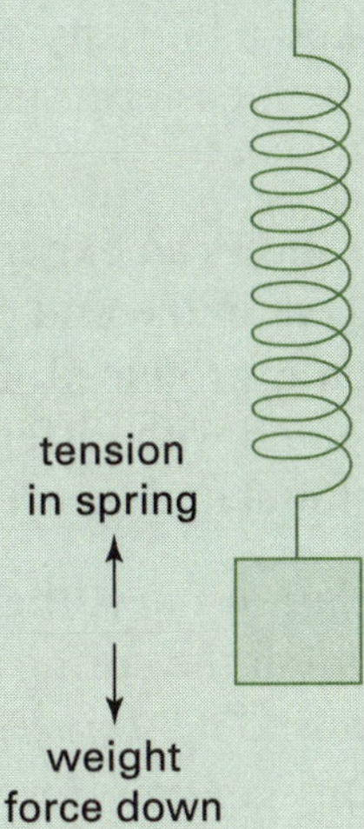

Figure 4.49 A mass suspended at the end of a spring

27. Scientists measure friction by calculating a number called the coefficient of friction. As Table 4.4 shows, the smaller the number, the easier it is to slide one object over the other. Skiers regularly wax their skis. Explain why they do this. *(1 mark)*

Table 4.4 The coefficient of friction

Surfaces	Coefficient of friction
steel on steel	0.6
waxed skis on snow	0.05
rubber on concrete	about 1

28. When the north pole of magnet A is brought close to a pole of magnet B, a repelling force is felt. When a pole of magnet C is brought close to the same pole of magnet B, an attractive force is felt.

a) Is the pole of magnet C north or south? How did you arrive at this conclusion? *(2 marks)*

b) When the same poles of magnet A and magnet B are separately brought close to an iron bar, the iron bar is attracted in each case. How can this be? *(1 mark)*

c) When the same poles of magnet A and magnet B are separately brought close to a copper bar, there is no effect. How can this be? *(1 mark)*

29. Since a force cannot be seen, how do we know it is there? *(1 mark)*

30. Johannes noticed that the chain on his bicycle looked a bit rusty so he oiled it. Now he finds it easier to pedal. Explain. *(1 mark)*

31. Figure 4.50 shows an experiment where a mass is placed on the end of a spring and an extension is produced. The experiment is then repeated with different masses and the results tabulated (see Table 4.5).

a) What relationship is there between the extension on the spring and the mass which caused it? *(1 mark)*

b) How many centimetres would the spring extend if a mass of 800 g was attached? *(1 mark)*

c) What mass would be needed to stretch the spring 36 cm? *(1 mark)*

d) Could the spring be stretched indefinitely using increasing masses? Explain. *(2 marks)*

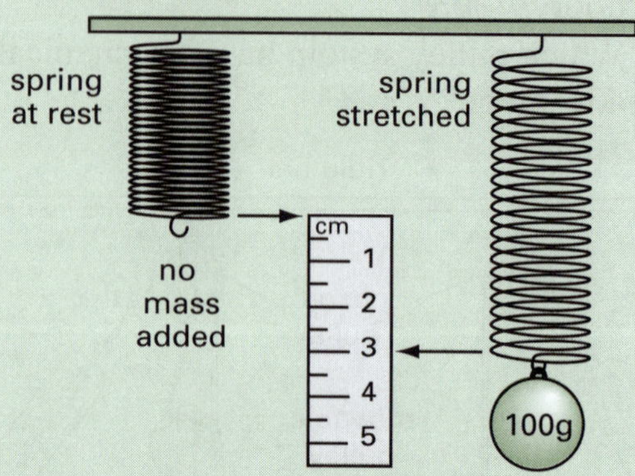

Figure 4.50 A mass on the end of a spring

Table 4.5 Results of experiment

Attached mass (g)	Distance spring is stretched (cm)
100	3
200	6
300	9
400	12
500	15
1000	30

32. Figure 4.51 shows an experiment with two glass rods that have been rubbed with silk.

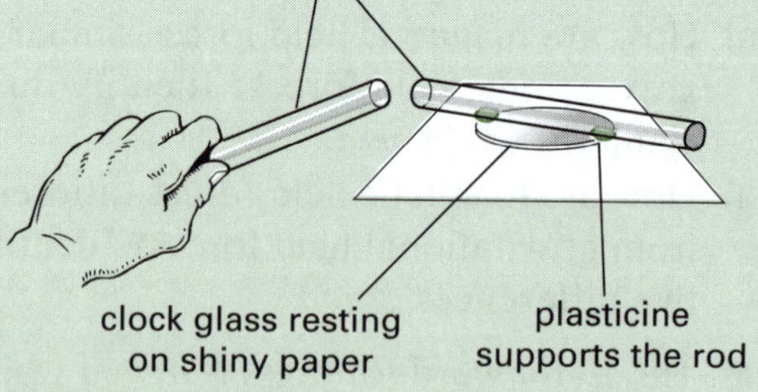

Figure 4.51 Two glass rods that have been rubbed with silk

a) What charge does each rod now carry? *(1 mark)*

b) Explain what will be observed when the second rod is brought near the rod mounted on the clock glass? *(2 marks)*

33. There are many terms that mean essentially the same thing. Deceleration, for example, can be described as retardation, slowing down, decreasing speed or negative acceleration. For each of the following terms, give another word or phrase that has a similar meaning. *(1 mark for each)*

a) acceleration

b) time = 0

c) acceleration = 0

d) speed = 0

34. The following sentences a) to h) explain what happens during a collision, but they are out of order. Rewrite them in the correct sequence. *(8 marks)*

a) The second, and more important, collision is the human collision, where an occupant hits some part of the car.

b) But the passenger compartment comes to a more gradual stop as some of the impact is absorbed in the crushing of, say, the engine bay or boot.

c) They are still moving forwards at their original speed when they slam into the steering wheel, windscreen or some other part of the car.

d) The net result is that the passenger compartment often remains relatively undamaged.

e) In any road crash, there are really two collisions.

f) The part of the vehicle that receives the first impact of the collision stops abruptly.

g) In a crash, unrestrained occupants keep moving inside the passenger compartment during the time it takes the car to stop.

h) The first is the car's collision where the car, hitting something, buckles and bends, and then comes to a stop.

35. a) Copy and complete the following by selecting one of the words or phrases in parentheses. *(6 marks)*

The other day, Mum and I went shopping. In the supermarket, I pulled out an empty trolley from its bay. It was (easy/hard) to push. I noticed I could give it a (small/medium/large) push and it (would/would not) speed up quickly. At one point, I gave the trolley one (small/big) push and it went right down the aisle. Mum told me not to do that as I might hurt someone. As Mum loaded the trolley, I found pushing it became (easier/harder). Towards the end, pushing it hard (did/did not) cause it to speed up as quickly as it did before.

b) Write one sentence that summarises the information in the text above. *(1 mark)*

36. Figure 4.52 compares a car colliding with a concrete barrier versus a wire-rope Brifen safety fence. Both these barriers are placed in the middle of roads to avoid head-on collisions. Explain why the Brifen fence is a better option than the concrete fence. *(2 marks)*

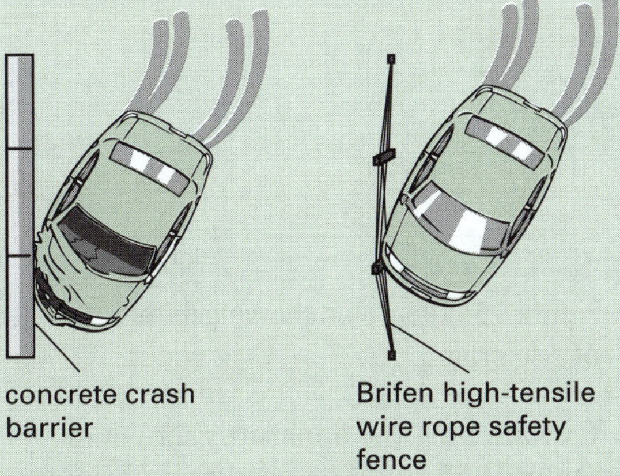

Figure 4.52 A car colliding with a concrete barrier compared to a wire-rope Brifen safety fence

37. Cars usually have head supports fitted. They are not there to rest your head on while driving. Suggest a safety reason for head supports. Refer to Figure 4.53 in your answer. *(1 mark)*

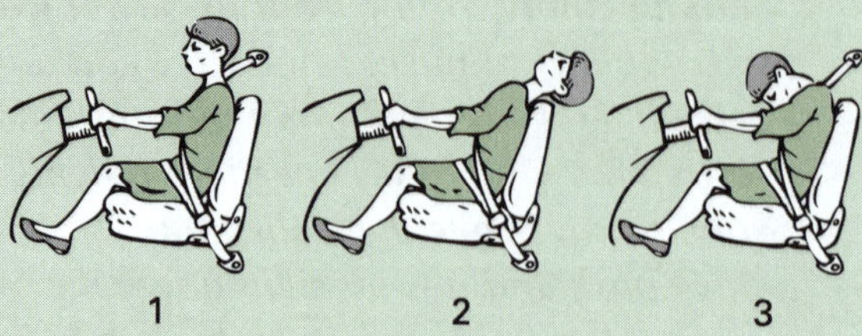

Figure 4.53 A car accident without a head support fitted

38. Look at Figure 4.54 and identify the better position of the fulcrum to allow the load to be raised with the least effort. *(1 mark)*

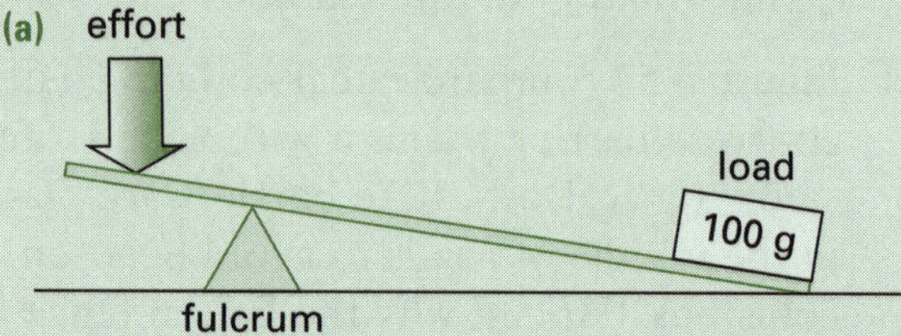

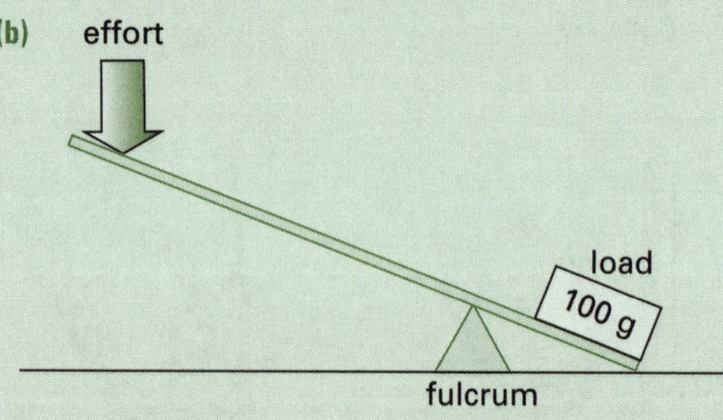

Figure 4.54 Diagrams showing different positions of a fulcrum

39. Explain how the apparatus shown in Figure 4.55 could be used to determine the strength of the electromagnet at each setting of the power supply. *(3 marks)*

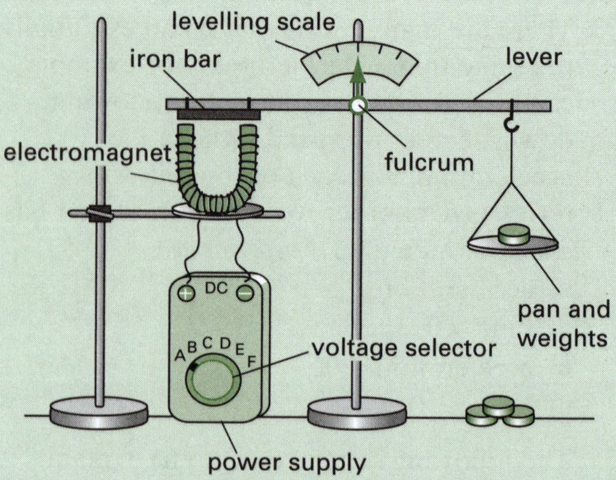

Figure 4.55 Apparatus to determine the various strengths of an electromagnet

40. Archimedes (287–212 BC) was a great scientist and mathematician. One important area of his research was the lever. He once stated, 'Give me a place to stand and a lever long enough and I will lift the world'. If Archimedes was able to stand on Mars and had a lever long enough to move the Earth, suggest where he might place his pivot or fulcrum? *(1 mark)*

Go to pp. 203–206 to check your answers.

CHAPTER 5

Investigations and problem solving

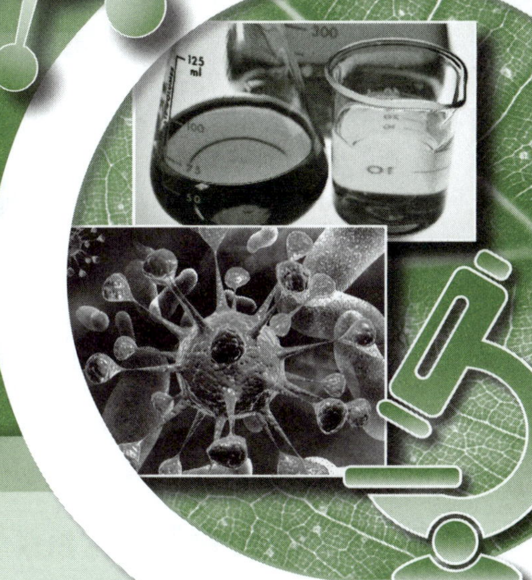

Overview

In this chapter you will learn about:

- the advantages and responsibilities of working individually or collaboratively
- conducting background research using secondary sources
- investigating a scientific problem
- fair testing and variables in first-hand investigations
- safety and hazards when undertaking first-hand investigations
- choosing appropriate equipment and increasing accuracy and reliability
- collecting, recording, processing and presenting data in scientific reports
- conclusions, discussions and evaluations from your research
- analysing sample practical reports.

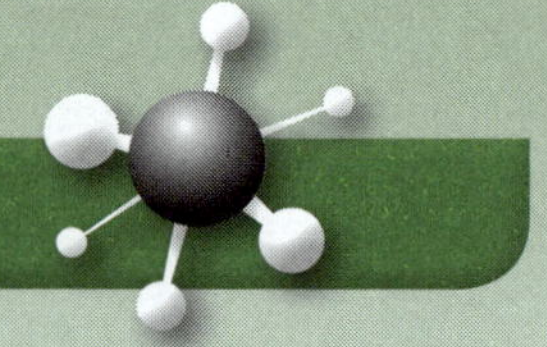

Glossary

constants—factors that are kept the same during an investigation

dependent variable—a measurement that changes due to the independent variable

first-hand investigation—an investigation where original material is collected by the investigator

generalisations—statements that are true in the majority of cases

hypothesis—a broad, general explanation of a collection of observations that predicts a result and suggests a way of testing it

independent variable—a factor which you change on purpose

inference—a possible and reasonable explanation of an observation

parallax error—an error in reading an instrument due to the observer's eye and the pointer not being in a line perpendicular to the plane of the scale

second-hand investigation—an investigation where the original material sourced from the work of other people is used

5.1 Working as an individual and working collaboratively

Students need to be able to work individually on some occasions as well as collaboratively in teams on others. There are advantages to both ways of working. There are also responsibilities that must be considered.

Working individually requires students to:

- appreciate the importance of taking personal responsibility in planning and performing a task
- work to realistic timelines and goals
- persevere with an activity to achieve a reasonable end point
- accept personal responsibility for a safe working environment for themselves and others in the laboratory, and
- evaluate the effectiveness of their work in completing the task.

Working collaboratively in teams requires students to:

- identify the different roles of each team member
- negotiate and allocate responsibility to each team member
- accept and perform the roles decided by the team
- collaborate with others in setting and working to agreed timelines and goals
- accept personal responsibility for maintaining a safe working environment for all team members
- work cooperatively to monitor progress of the investigation, and
- evaluate the process used by the team and the effectiveness of the team in completing the task.

5.2 Conducting background research using secondary sources

The following points should be noted when doing background research and gathering information from secondary sources.

- Use a wide range of resources to gather secondary information:
  - library books
  - internet
  - technical magazines
  - CDs/DVDs.

  Ensure you reference sources from which you have collected information.
- Use key words to locate information or to find the meanings of words and terms. You could look in:

- the index of a book
- the table of contents at the front of the book
- a glossary to help you understand difficult technical words.

- Collate and summarise the collected information. You must not plagiarise (or copy) the work collected. You must read and make notes into a scaffold and then use these note summaries to construct your own text.
- Extract data. Information from secondary sources may contain various types of graphs (e.g. histograms, sector graphs, line graphs, divided bar graphs), flow charts and diagrams. You can use this type of information to extract data.
- Is information relevant or irrelevant? You must decide this as you gather second-hand data. For example, if you are gathering information on water movement in flowering plants then any data on non-flowering plants would be irrelevant.
- Not all the gathered data will be reliable. You need to check its reliability by comparing it with other sources. Science articles that appear in respected journals and that have been peer reviewed are more likely to be reliable than an article that is self-published.
  - Articles written by creation scientists are not reliable as they do not follow the scientific method.
  - Many articles from companies, rather than educational or research websites, may be biased.
  - Scientific product claims may be written by commercial companies. In advertising their product, various companies use scientific terms and data to back up the claim that their product is superior. For example, 'Scientific tests show that Sorbo paper towel is more water absorbent than all other brands'. These statements cannot be accepted as reliable without independent testing.
- Organise your data. Your collected data needs to be organised for easy analysis. Tables, graphs, scaffolds and diagrams are commonly used to organise information. Computer databases and spreadsheets are becoming increasingly important as ways of organising collected data (see Figure 5.1).

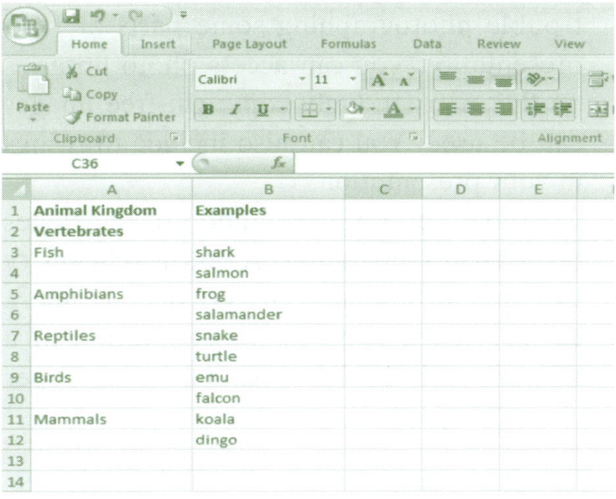

	A	B	C	D	E
1	Animal Kingdom	Examples			
2	Vertebrates				
3	Fish	shark			
4		salmon			
5	Amphibians	frog			
6		salamander			
7	Reptiles	snake			
8		turtle			
9	Birds	emu			
10		falcon			
11	Mammals	koala			
12		dingo			
13					
14					

Figure 5.1 Using a spreadsheet to organise information

5.3 Investigating a problem: observing

The first step of a scientific investigation takes place when you make an observation. This might be an observation of an event such as a pine cone falling from a tree. It might be an observation of some feature of an animal such as the presence of webbing between the toes of a bird. These observations may then lead you to think about the cause or reason behind the observation.

Observations can be made using various technologies. Digital cameras and video cameras are ideal to make a recording of your observations. The digital files can be published or sent via the internet through social media applications to other students for review and comment.

The following are examples of experiments using digital technology.

1. Observing the metamorphosis of a caterpillar into a butterfly (see Figure 5.2). Digital photos can be taken each day of the change in the caterpillar as the metamorphosis occurs. These photos can be combined to produce a digital presentation. If you have trouble photographing some parts of your presentation, you can access photo libraries on the internet. Only copyright-free images can be used. You can also include drawings in your presentation.
2. Creating a photo presentation of the stages involved in a separation process in a chemistry experiment. Students can use a digital camera to record the steps in the separation of a mixture such as plant pigments using paper chromatography.

Experiment 1

Boiling water

Observation is a powerful and important process in science. While this is a simple practical, a large number of observations can be made.

Aim

- To boil a beaker of water and make observations about what is occurring

Method

Set up the apparatus shown in Figure 5.3 and allow the water to reach boiling. If you want to do this experiment at home, use a pyrex saucepan.

Figure 5.3 Boiling water in a beaker

Analysis

Make as many observations as you can over the next half-hour.

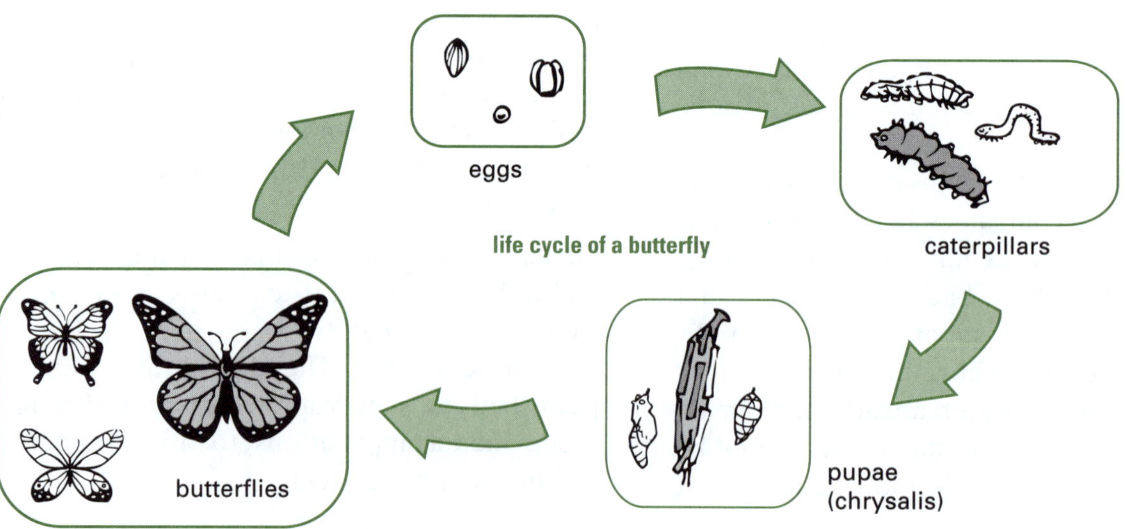

Figure 5.2 Using digital technology to create a presentation

When complete, turn off the Bunsen burner and allow the apparatus to cool down before discarding the water and putting the equipment away.

Analyse the results. Go to p. 206 to check your answer.

Conclusion

Write your own conclusion.

Go to p. 206 to check your answer.

5.4 First-hand investigations: fair testing and variables

Some problems will be investigated by conducting research in the laboratory or in the field. When you actually research and come up with original data, this is usually called a first-hand investigation. Some problems will be investigated by second-hand data research. This involves using information (data) developed by other persons and working with it. Table 5.1 shows some of the steps of the typical scientific method.

Table 5.1 Steps and examples of the typical scientific method

Some of the steps of the typical scientific method	Simple example
Observations—observed or measured results	'I have seen quite a few bird feathers in the garden lately and so far they are either big or small feathers.'
Inferences—following an observation, a scientist may make an inference; this is a reasonable explanation drawn from a small sampling of data. An inference may ultimately turn out to be incorrect.	'Maybe large birds have big feathers and small birds have small feathers.'
Predictions—the next step is to test the inference by making a prediction and performing an experiment to test the prediction. These experiments may support the inference or they may not. A new explanation may need to be sought.	'I'll have to check this out. So I'll examine a number of different-sized birds I haven't seen before, but I think the big birds will have the bigger feathers.'
Generalisations—as more supporting data is collected, the scientist may see a pattern emerging. At this point he or she may be able to make a generalisation about the topic under investigation. Generalisations are statements that are true in the great majority of cases.	'I can now conclude: large birds have big feathers and small birds have small feathers.'

Making a hypothesis

The next step in the scientific method is to generate a hypothesis. What is a hypothesis?

- It is a general explanation of the collected observations or a logical solution to a problem being investigated.
- A hypothesis is *not* a wild guess. You might say that it is an educated guess.
- Hypotheses are temporary statements that lead towards the development of scientific theories.

Hypotheses are only made after looking at all the available evidence and information that has been collected. This is why background reading of identified data sources is so important. The work of others may help you to produce a good hypothesis. Hypotheses are the result of a scientist making inferences, making predictions and developing generalisations.

The hypothesis must be able to:

- predict the results of future investigations about the problem, and
- suggest ways in which it can be tested to determine whether or not it is true.

Hypotheses are not always correct. If the data collected during the investigation does not support the hypothesis, then a new hypothesis must be developed.

Fair testing and variables

Many factors can influence the results of an experiment. These factors are called variables. In order to investigate a problem experimentally you must keep all, but one, variable constant so that only the variable being investigated can affect the result. This is known as fair testing.

- Experimental results are only reliable if the tests are fair.
- During an investigation, the experimenter tests what happens when he or she alters one variable. The effect of a change in this independent variable on another variable (the dependent variable) is investigated.
- The independent variable is the one that the investigator systematically allows to change. It is the change you allow to happen when conducting an experiment.
- The dependent variable is affected by the change in the independent variable.
- The controlled variables are the other factors that can alter the result if they are not held constant.

Experiment 2

Investigating a hypothesis

A hypothesis is a logical proposal about how something works. Often you can write a hypothesis as: 'If [I do this], then [this] will happen.' You should replace what is in the square brackets [ ] with appropriate information from your own experiment (e.g. 'increasing the temperature of a given beaker of water increases the amount of sugar that dissolves in it').

Your hypothesis should now be something that you can actually test, a *testable hypothesis*. That is, you need to be able to measure both 'what you do' and 'what will happen'.

Aim

- To investigate the hypothesis that an ice cube will melt faster in salt water compared to one in fresh water

Method

1. Set up two identical beakers as shown in Figure 5.4.
2. In one beaker place 200 mL of tap water, and in the second beaker place 200 mL of salt water. (Dissolving 3 to 4 teaspoons of salt in the water can make the water salty.)
3. Ensure the temperature of the water is the same in both beakers.
4. Obtain two identical ice cubes and place one in the first beaker and one in the second beaker.
5. Time how long it takes for each of the ice cubes to completely melt.

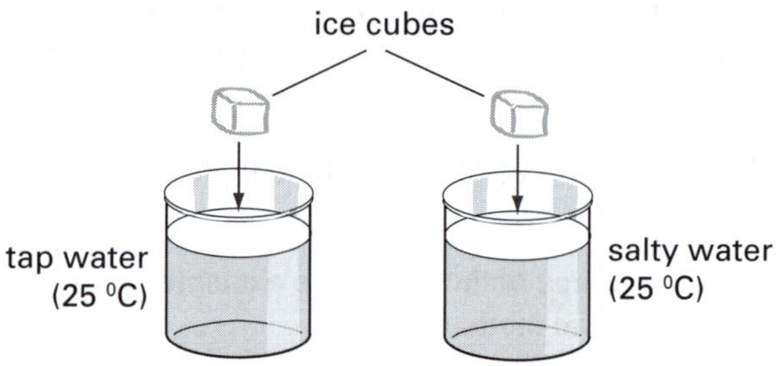

Figure 5.4 Setting up the experiment

Analysis

1. Why is it important that:
   a) the beakers are the same?
   b) the quantity of water used is the same?
   c) the temperature of the water remains the same?
   d) the ice cubes are the same size?
2. Which is the independent variable?
3. Which is the dependent variable?

Analyse the results. Go to pp. 206–207 to check your answer.

Conclusion

Write your own conclusion.

Go to pp. 206–207 to check your answer.

Further examples

Example 1: Buoyancy

- **Hypothesis:** the buoyancy force acting on a floating body increases as liquids become more dense
- **Independent variable:** density of the liquid
- **Dependent variable:** buoyancy force
- **Variables to be kept constant:** mass of floating body; size of floating body; shape of floating body; temperature of liquids tested; type of force meter

Example 2: Levers

- **Hypothesis:** in a simple first-class lever, less effort force is required to lift a load the further the effort force is from the fulcrum
- **Independent variable:** distance between the effort force and the fulcrum
- **Dependent variable:** size of the effort force
- **Other variables to be controlled:** length of lever; mass of load; type of force meter; type of fulcrum; material of the lever

Control

Measurements may only be reliable in some experiments if a control is used. This is particularly true in biological or biochemical systems where many factors can alter a result. A control is an experiment that is performed as a comparison with those in which the independent variable is allowed to change.

Example 3: Fertilisers and plants

A gardener wishes to test which brand of fertiliser leads to the greatest plant growth in potted geraniums. She will conduct the test and compare the results with a plant that has not had any fertiliser added to the soil. This experiment is shown in Figure 5.5.

- **Control:** plant with no fertiliser
- **Independent variable:** brand of fertiliser
- **Dependent variable:** height of plant
- **Controlled variables:** same type of potting mix; same amount of potting mix; same volume of water added; same time to add water each day; same position in the yard; same mass of fertiliser

Figure 5.5 A controlled experiment growing geraniums with and without different brands of fertiliser

5.5 Safety in a first-hand investigation

Experiments should always be done in a safe manner to avoid injuries. You must always listen to and follow the teacher's instructions. Your teacher will inform you of the safety issues for each experiment.

Safety signs can be found in the laboratory. Chemicals have safety signs on their labels. Read the label carefully before using the substance. Figure 5.6 shows an example of a safety sign.

Figure 5.6 Poisonous substances have this safety sign on their label.

Here are some examples of safety issues:

1. Using a Bunsen burner. When using a Bunsen burner to heat a beaker of water supported on a tripod and gauze, ensure that a blue flame is used (the inlet hole is open) and that the Bunsen burner is not placed under the gauze until all the apparatus has been assembled. After the heating is completed the Bunsen should either be turned off at the tap or placed to the back of the bench with the hole closed to produce a yellow safety flame. These yellow flames are easy to see and so you are less likely to accidentally burn yourself. Do not touch the hot equipment until it has cooled down. Make sure your safety glasses are on.
2. Evaporation of salt solutions. When you evaporate a saltwater solution in an evaporating basin, the apparatus gets very hot. Ensure you do not burn yourself. When the evaporation is almost complete the salt crystals may jump out of the basin. Make sure you have your safety glasses on and that your arms are covered to avoid burns from hot salt. Reduce the chance of hot salt jumping out of the basin by reducing or removing the heat source before all the water has evaporated. Allow the apparatus to cool down before disassembling it.
3. Distilling a solution. When assembling a distillation apparatus, ensure that several retort stands and clamps are used to support the flask as well as the condenser. The apparatus must be prevented from falling down. Ensure water is flowing from the tap through the jacket of the condenser both before heating commences and after heating has finished. Should some chemicals spill onto your skin, wash them off with plenty of water. Allow the apparatus to cool down before disassembling it.

Do not twist, bend or kink the gas hose. This can interrupt the flow of gas and the Bunsen burner can go out. If the gas is not then promptly turned off, gas will begin to flow into the room. Also do not allow the gas hose to inadvertently come into contact with hot equipment. The heat can melt the rubber and possibly cause a fire in the laboratory.

5.6 Choosing equipment and increasing accuracy and reliability

The selected equipment needs to be assembled to produce an apparatus that will perform the required task. Assembling pieces of equipment and using the apparatus requires good manipulation skills. Figure 5.7 provides some examples of common laboratory equipment and how they can be used.

Accuracy

Making accurate measurements is important when conducting an investigation. This involves choosing the most appropriate measuring devices available. Reducing errors and choosing equipment that can measure in a more precise way can improve accuracy.

One important error is called parallax error. This is an error that can be avoided. For example, when measuring the volume of a liquid on a scale, your eye must be placed directly over the scale. If you look at the scale from an angle the reading will be incorrect, as Figure 5.8 shows.

Reliability

Reliability can be improved by making repeated measurements over a number of trials. The number of trials that are performed depends on the type of experiment. For example:

1. Testing a new drug for migraine headaches. In this case the drug will need to be tested on thousands of volunteers to determine whether there are adverse side effects to the users and whether the drug helps those who suffer from migraines compared to those who do not.
2. Testing a sample of sea water for its salt content. In this case a minimum of five repeats

Equipment	What it is used for	Equipment	What it is used for
test tube 	• observing reactions when chemicals are mixed • heating solids, liquids and solutions over a Bunsen burner, using a wooden test-tube holder to hold the test tube	evaporating basin	• evaporation of solutions to dryness or partially to allow crystallisation
beaker 	• temporary storage of liquids and solutions • observing reactions involving larger quantities of liquids than in test tubes • heating solutions	watch-glass	• natural evaporation of small quantities of solutions • weighing small quantities of solids
conical flask	• observing reactions in solutions • often used when the contents need to be kept well mixed by swirling • temporary storage of liquids and solutions • boiling solutions	Petri dish	• temporary storage of specimens • growing cultures of microorganisms • weighing small quantities of solids
tripod	• support for gauze and equipment during experiments	measuring cylinder 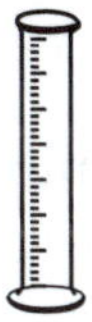	• measuring quantities of liquids or solutions
gauze	• support for beakers, conical flasks or evaporating basins during heating experiments	filter funnel 	• support for filter paper during filtration
mortar and pestle	• grinding materials to fine powders or to extract materials into liquids	Bunsen burner 	• heating materials and equipment during experiments

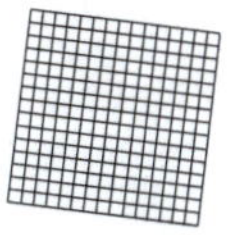

Figure 5.7 Apparatus and its uses

is all that is needed. The five repeats should give answers that cluster closely. An average of the five results can then be calculated.

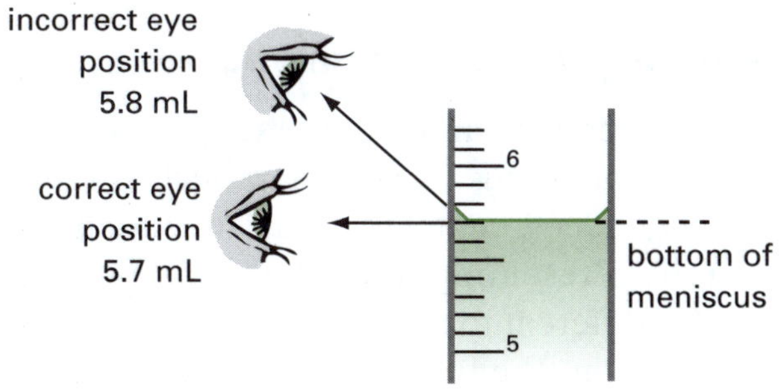

Figure 5.8 Avoiding parallax error in reading a scale

Using alternative technologies in experiments

Figure 5.9 shows two methods that can be used to record motion. In (a), a stopwatch is used to record the time for a vehicle to move certain distances along a marked track. In (b), a motion sensor beam is used to monitor the movement of a person walking towards the sensor. The computer program calculates the position of the person with time and uses a graphing application to plot this data on a graph.

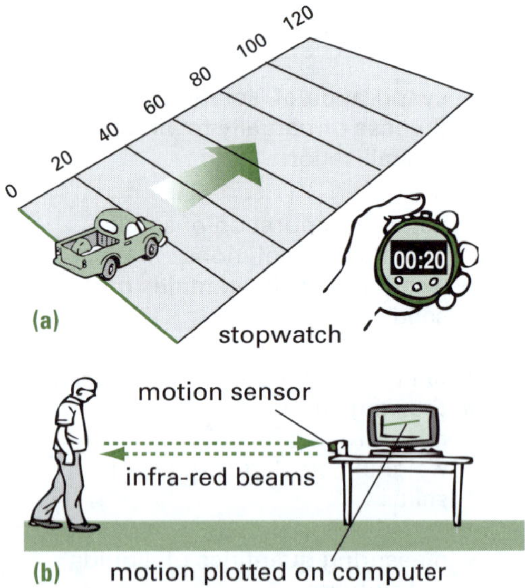

Figure 5.9 Alternative methods to measure motion

The motion sensor in (b) produces a more accurate result than the method used in (a). Greater errors occur in experiment (a) because it relies on the hand–eye coordination of the experimenter. There is a delay between observing the vehicle cross a line on the track and the depression of the stop button on the stopwatch.

Experiment 3

Using a data logger with a motion sensor

The details of this experiment will depend on the data logging system you use.

Aim

- To use a data logger to plot distance–time graphs

Method

1. Set up the sensor of a data logger to record movement.
2. Produce live position–time graphs of yourself as you walk towards and away from the sensor.
3. Use this to predict the distance–time graph.
4. Now obtain an almost frictionless ramp and place the sensor at the top of the ramp as shown in Figure 5.10.
5. Record the position–time graph of a trolley rolling on the ramp, when the ramp makes a small angle to the horizontal. Predict and check the distance–time graph.
6. Repeat the previous step, but with the ramp steeper so the trolley accelerates. Record the distance–time graph.

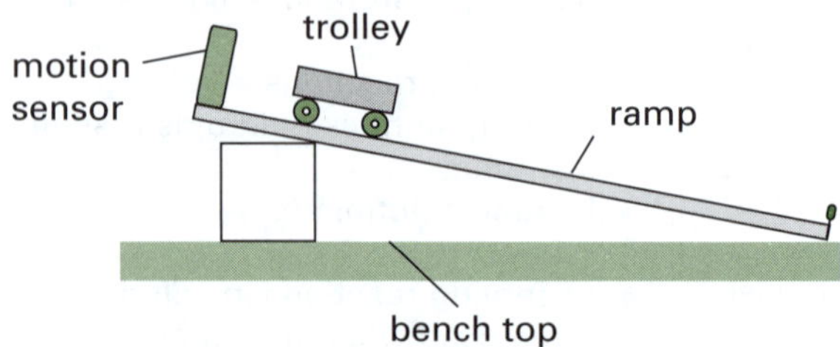

Figure 5.10 Setting up the motion sensor on a ramp

Analysis

Obtain a printout of your graphs and paste them into your book. Explain what each of your graphs shows. Go to p. 207 to check your answer.

Conclusion

Write a suitable conclusion.

Go to p. 207 to check your answer.

Using technology can enhance the results from an experiment. While using a stopwatch has its merits, the individual's reaction time means it is not as accurate. Additionally, you need to measure distances and plot the graphs manually, or insert the data into a computer spreadsheet. Motion sensors linked to data loggers are more accurate with outputs provided almost instantaneously.

5.7 Scientific reports

Recording and presenting data

Scientific reports can be presented in a variety of ways including a written report, photos, sketches, videos, computer slide show, debates or orally. Presenting a written experimental report is different to the way you would present an exposition. The following are some different methods you can use to present information.

- Procedure. This text type is used to describe how an experiment is to be performed. Present tense is used. The steps of the procedure are usually numbered or presented as sequential dot-points.
- Experimental record. This extends the procedure to include results and conclusions. Results are written in past tense, while the conclusion can be written in the present tense, past tense or both.
- Procedural recount. This text type is used to record the procedure of the experiment that has been performed. Past tense is used in a recount.
- Report. This type of text is used to present information on a particular topic (e.g. the different types of frogs in Australia) that has been investigated using first-hand and/or second-hand data. Factual and descriptive information is presented.
- Discussion. This text type identifies issues and presents points for and against. In the end you decide on a viewpoint and justify it.

At the end of your report, you must acknowledge all sources of information in your bibliography. Use an accepted method of citing references. This will include the author, name of the article/book, chapter or page reference, publisher and date of publication. Also, for sources on the internet you should state as much of this information as you can (not just the web address).

Processing collected data

Tables, graphs and drawings are common methods of recording and presenting data in a scientific report.

Tables

Scientists often record their information in tables and as graphs. Table 5.2 shows the gas pressure in a container of gas as a reaction occurs. The table has the following features.

- There are two columns. (Other tables may have more columns for data.)
- Each column has a clear heading.
- The units (seconds and kilopascals) are shown in the column headings and are not repeated with each measurement.
- The table is enclosed by lines.

Table 5.2 Gas pressure versus time

Time (s)	Gas pressure (kPa)
0	63
10	71
15	75
25	82
30	84
45	89
50	91
60	93
75	94
90	95
100	96
105	97
110	98
120	99

Graphs

The information in a table can usually also be shown as a line graph. Here are the steps you would take to draw Table 5.2 as a graph.

1. Draw your axes. Time is usually shown on the horizontal axis, so gas pressure would be on the vertical axis. Since there are not any gas pressure values below 60, start the vertical scale at that point.
2. Mark the points on your graph with a small cross (×) or a large dot (•) in pencil. A pencil allows you to easily erase points if you make a mistake.
3. Join the crosses/dots with a smooth curve or straight line. If the line is straight, use your ruler. If the plotted points on a line graph seem not to line up, draw a line of best fit. This is a smooth line, straight or curved, that passes through as many points as possible. The points it does not pass through should be scattered equally on both sides of the line. The graph shows a line of best fit.

Gas pressure versus time
Gas pressure (kPa)
100
95
90
85
80
75
70
65
60
0
10
20
30
40
50
60
70
80
90
100
Time (s)

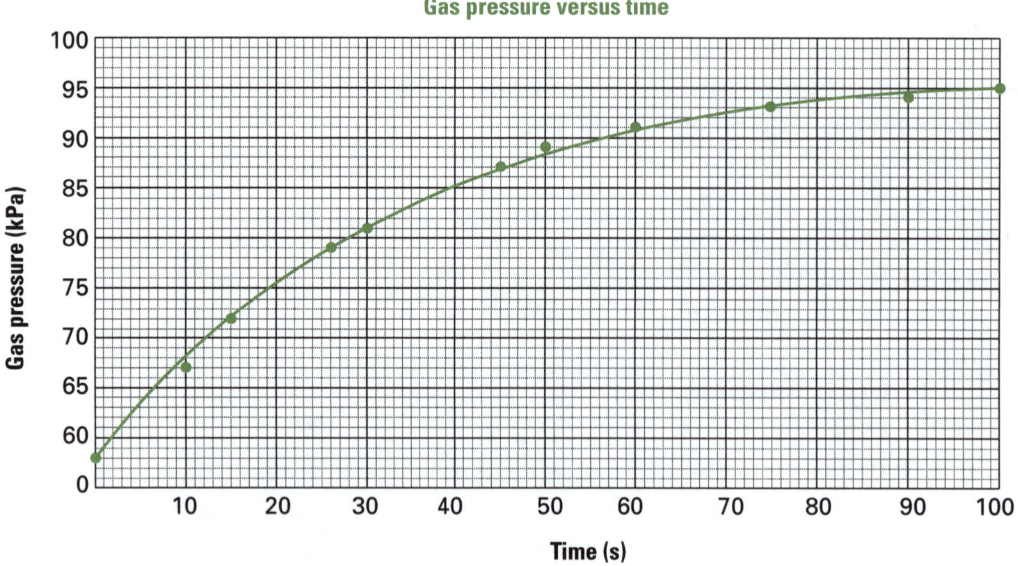

Figure 5.11 Gas pressure versus time

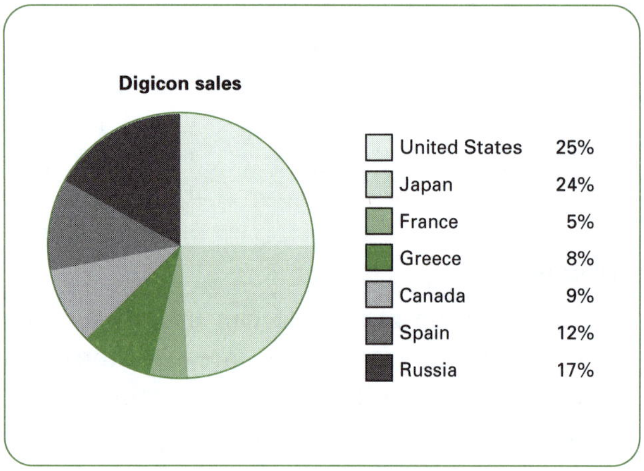

Digicon sales
United States 25%
Japan 24%
France 5%
Greece 8%
Canada 9%
Spain 12%
Russia 17%

Sector (pie) graph

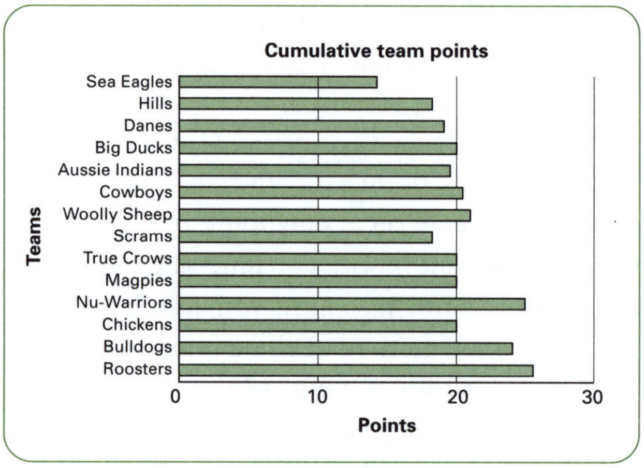

Cumulative team points
Teams
Sea Eagles
Hills
Danes
Big Ducks
Aussie Indians
Cowboys
Woolly Sheep
Scrams
True Crows
Magpies
Nu-Warriors
Chickens
Bulldogs
Roosters
0
10
20
30
Points

Bar graph

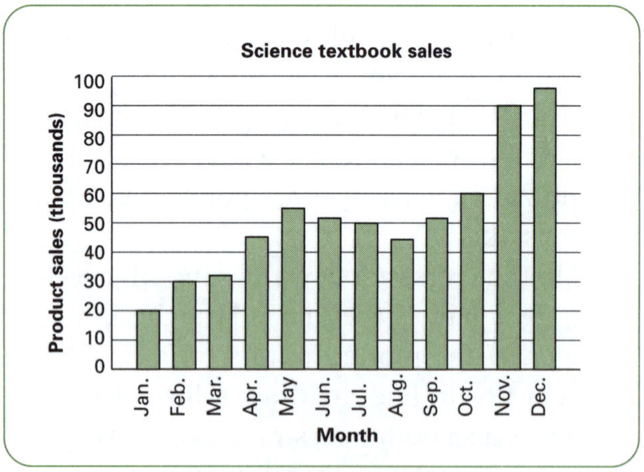

Science textbook sales
Product sales (thousands)
100
90
80
70
60
50
40
30
20
10
0
Jan.
Feb.
Mar.
Apr.
May
Jun.
Jul.
Aug.
Sep.
Oct.
Nov.
Dec.
Month

Column graph

Car crashes: the effect of alcohol
Risk of crash
35
30
25
20
15
10
5
0
0.01
0.02
0.03
0.04
0.05
0.06
0.07
0.08
0.09
0.10
0.11
0.12
0.13
0.14
0.15
0.16
Blood alcohol concentration

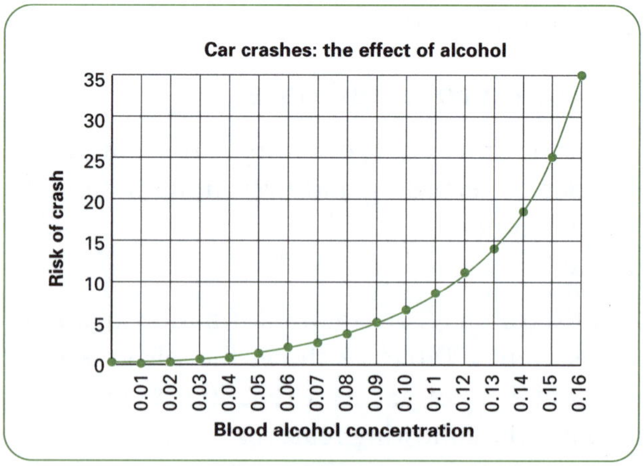

Line graph

Figure 5.12 Types of graphs

Remember to always give your graph a title. Figure 5.11 shows the completed graph of the data in Table 5.2. Other types of graphs are commonly used in science to display and analyse data. Some of these other types are shown in Figure 5.12.

Drawing diagrams

Scientific drawings summarise a lot of important information about an experiment. Here are some important rules to follow.

- Use a pencil, rubber and ruler. A pencil and rubber allow you to make corrections easily.
- Your diagram should be at least one-third of a page (10 to 12 cm) in size.
- Draw the apparatus using a two-dimensional perspective.
- Remember to use your ruler wherever straight lines are needed.
- Use printing to label the parts of your diagram.
- Connect the labels to the diagram using ruled straight lines. No arrow heads are needed at the end of lines that have labels.

Figure 5.13 shows an example of a basic scientific drawing showing the apparatus used in an experiment.

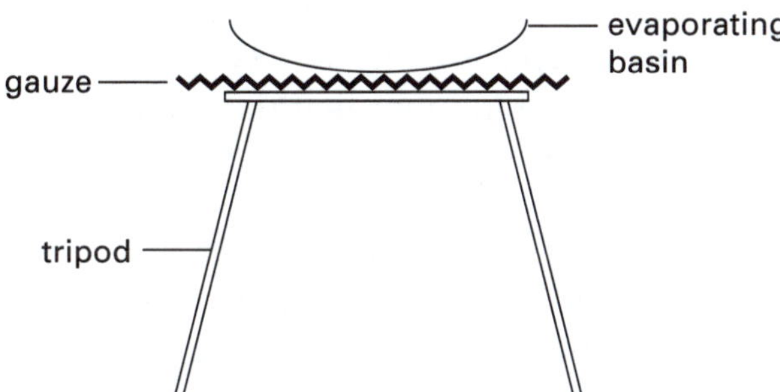

Figure 5.13 Scientific drawing

Test yourself 1

Part A: Knowledge

1. An inference is *(1 mark)*
   A a statement that is true in the great majority of cases.
   B a statement that predicts what will happen in the next set of experiments.
   C a theory.
   D a reasonable explanation of an observation.

2. The independent variable in an investigation is *(1 mark)*
   A the one that is allowed to change systematically.
   B affected by the changes made during the experiment.
   C the control.
   D controlled so that its value does not change.

3. Which of the following statements is not a scientific hypothesis? *(1 mark)*
   A Wine is a preferable drink to beer.
   B Cricketers wear light-coloured clothes rather than dark clothes when playing in the summer.
   C Sugar dissolves more readily when the water is warmer.
   D Roosters crow when they see the first light of day.

4. A prediction is a *(1 mark)*
   A new observation.
   B statement that suggests the result of a future experiment.
   C theory that accounts for all the observations.
   D logical explanation for an observation.

5. Which of the following is *not* a factor in designing an experimental procedure? *(1 mark)*
   A identifying the variables that need to be kept constant
   B determining which will be the dependent variable and which will be the independent variable
   C ensuring that all measurements are reliable and accurate
   D collecting only those results that will support your hypothesis

6. Complete the following restricted-response questions using the appropriate word. *(1 mark for each)*
   a) Further experiments may .... an inference or they may not.
   b) Generalisations are statements that are .... in the majority of cases.

c) Hypotheses are .... statements that lead eventually to the development of scientific theories.
d) The factors that influence the results of experiments are called ....
e) The dependent variable is affected by a change in the .... variable.

7. Use the code letters to match the terms or phrases in each column. *(5 marks)*

Column 1	Column 2
A control	F predictive explanation
B reliability	G allows comparison
C parallax	H repetition
D hypothesis	I error
E inference	J reasonable explanation

Part B: Skills

8. Suggest a logical inference that could be made for each of the following observations. *(1 mark for each)*
   a) Arthur noticed that moss only grew in the garden on the southern side of his house.
   b) Eight boys suddenly turned up at sickbay after lunch at school last Tuesday. They all had stomach cramps.
   c) The label on the brown bottle of hydrogen peroxide said to store it in the refrigerator after opening.
   d) Kerosene floats on water.
   e) All six children of a family have pale skin and blue eyes.

9. The data in Table 5.3 shows the time it takes for two chemicals to completely react together at different temperatures. Use it to predict the time for the reaction between the two chemicals at 50 °C. *(1 mark)*

Table 5.3 Time taken for two chemicals to completely react together at different temperatures

Temperature (°C)	Time (s)
10	400
20	200
30	100
40	50
50	

10. Billy observed that magnesium metal dissolved in dilute hydrochloric acid and a colourless gas was released. He noticed that on some days the reaction seemed faster than on other days. He decided to investigate the reaction between the acid and magnesium. He made the following hypothesis which he then tested:

    Magnesium dissolves faster in dilute hydrochloric acid as the temperature of the acid increases.

    a) Name the independent variable in Billy's experiment. *(1 mark)*
    b) Name the dependent variable. *(1 mark)*
    c) Name the variables that need to be controlled. *(6 marks)*

11. A student wanted to compare the water-holding capacity of a given quantity of various soils: clay, sand and loam.
    a) List at least four pieces of equipment the student may use. *(4 marks)*
    b) What are the three variables the student must keep constant? *(3 marks)*
    c) Why is it important that the student start off with dry soil in each case? *(3 marks)*
    d) Describe a procedure to allow a student to experimentally determine how much water the soil can hold. *(6 marks)*

12. Suggest two advantages of working in teams. *(2 marks)*

13. Make a list of the variables which could affect the following measurements.
    a) the time for a weather balloon to reach its target height *(2 marks)*
    b) the stopping distance of a car when the brakes are applied *(2 marks)*
    c) the distance that light can penetrate into a pool of water *(2 marks)*

14. Mary investigated how quickly a blue solution was decoloured by a mild bleaching agent. She conducted her investigation at different temperatures. The results of her experiment are shown in Table 5.4.

Table 5.4 Time taken for a blue solution to be decoloured by a bleaching agent

Temperature (°C)	Time to decolourise (s)
20	80
30	40
40	19
50	11
60	5

a) What variables did Mary need to control in her experiments? *(3 marks)*

b) Construct a line graph of Mary's data. Plot temperature on the horizontal axis. *(5 marks)*

c) How did the speed of the reaction change as the temperature increased? *(1 mark)*

d) Do Mary's results support the following hypothesis?

The rate of a reaction doubles for each 10-degree rise in temperature.

Explain. *(2 marks)*

Go to pp. 207–208 to check your answers.

5.8 Conclusions, discussions and evaluations

Once you have processed your results, you will need to draw valid conclusions. The conclusions must relate directly to the experimental results. You cannot make a conclusion about information that you have not gathered or measured. Your conclusion must be brief and answer the aim of the experiment.

As part of the scientific method, you will then provide plausible explanations of the phenomena being investigated. These possible explanations will be written in the discussion section of the report. There is also an opportunity at this point of justifying any inferences you have made about the collected data. In the discussion you may make some predictions that could lead to further experimentation.

As more experiments are completed, you may be able to evaluate how successful or otherwise your experiments have been. If you are working in a group, discuss what other members of your group think.

Case study 1: Solubility of sugar in water

Hypothesis

- Sugar dissolves more rapidly in water as the temperature increases.

Aim

- To measure the time that it takes to dissolve a fixed mass of sugar in water at various temperatures

Equipment and chemicals

- beaker (250 mL)
- water bath
- magnetic heater and stirrer
- measuring cylinder (100 mL)
- electronic balance
- sugar

Method

1. Weigh a 5-g sample of table sugar onto a glossy square of paper.
2. Use a measuring cylinder to measure 100 mL of tap water and empty the water into a 250 mL beaker.
3. Place the beaker in a water bath held at a fixed temperature using an electrical heater/magnetic stirrer as shown in Figure 5.14.
4. Measure this temperature with a thermometer.

(cont.)

Case study 1: Solubility of sugar in water (cont.)

5. Add the sugar to the water at zero time. Set the magnetic stirrer to a fixed (slow) stirring rate.
6. Record the time when all the sugar has dissolved.
7. Repeat the procedure five times at the selected temperature. Take an average of the readings.
8. Repeat the procedure five times at selected higher temperatures.

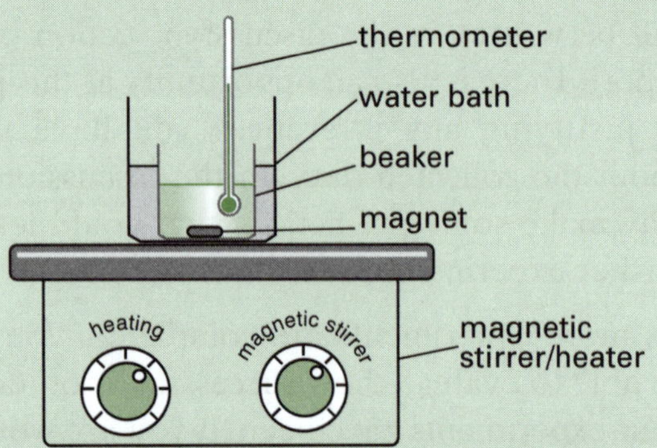

Figure 5.14 Experimental apparatus to determine the time for sugar to dissolve in water at various temperatures

- The variables to be kept constant are the mass of sugar, grain size of sugar, volume of water, rate of stirring and size of the beaker.
- The independent variable is the temperature of the water.
- The dependent variable is the time for the sugar to dissolve.
- The accuracy of the measurements depends on the equipment selected and how carefully the experiment is conducted. The accuracy in determining the mass of the crystals can be improved by using an electronic balance that weighs to 2 or 3 decimal places. The accuracy of measuring volume can be improved by using more accurate measuring vessels. The accuracy of the temperature measurement can be improved by using a temperature probe connected to a data logger.
- The measurements are reliable if repeated experiments (five or more) lead to consistent results. Sometimes in a large set of repeated measurements, an individual measurement may differ widely from the other measurements. You may be justified in ignoring this measurement before calculating an average. Measurements cannot be reliable if they are not performed with high precision and accuracy.

Typical results

Table 5.5 shows the final average times for a fixed mass of sugar to dissolve at five temperatures.

Table 5.5 Final average times for a fixed mass of sugar to dissolve at different temperatures

Temperature (°C)	Average time for sugar to dissolve (s)
20	150
30	70
40	33
50	17
60	8

Graphical results

Figure 5.15 shows the graphical results for this experiment. The following points should be noted.

- Axes have titles and units.
- The independent variable is labelled across the horizontal axis, while the dependent variable is labelled along the vertical axis.
- The graph has a title.
- The scales on each axis allow the data to fill most of the grid space.
- A curved line of best fit is drawn through or near the data points.

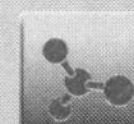

Case study 1: Solubility of sugar in water (cont.)

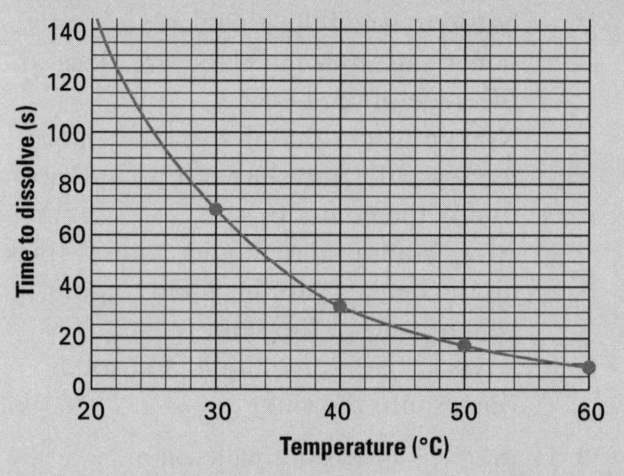

Figure 5.15 Sugar solubility versus temperature

Conclusion

As the water temperature increases, the time for the sugar to dissolve decreases. Therefore, sugar dissolves faster as the temperature of the water increases.

Test yourself 2

Part A: Knowledge

1. A science class conducted a survey of the plant species found in the native bush area near the school. They found several trees that they didn't recognise. Which of the following would be most helpful in identifying these trees? *(1 mark)*
   - **A** Fossil records from a local university.
   - **B** A book of native plants in the council area.
   - **C** An encyclopaedia.
   - **D** Local newspapers and periodicals.
2. A Year 7 science class decides to measure the height of each student and then to calculate the average height of students in the class. Which of the following is a possible source of error in this exercise? *(1 mark)*
   - **A** how many boys and girls are in the class
   - **B** the total number of students in the class
   - **C** the ages of students in the class
   - **D** how accurately these measurements can be made
3. Which of the following cannot be answered because the quantity cannot actually be measured? *(1 mark)*
   - **A** how fast a kangaroo can move
   - **B** the weight a giraffe puts on as it is growing
   - **C** how happy a dog is when given his favourite food
   - **D** how much a crocodile eats in a week
4. A student experimented by measuring plant growth in six different soil samples. She planted several seeds of one particular plant species in identical containers, left them in full sunlight and watered them regularly with the same volume of water. She monitored the growth of each plant over the next month. What is the independent variable in this experiment? *(1 mark)*
   - **A** soil
   - **B** seeds
   - **C** container
   - **D** light
5. Vivienne observed the Ulysses butterfly during a field trip. Which one of the following is an observation she made about this butterfly? *(1 mark)*

A The butterfly has a scientific name of *Papilio ulysses.*

B The butterfly is closely related to the Orchard butterfly (*Papilio aegeus*).

C The butterfly enjoys flying from bush to bush.

D The butterfly has a wingspan between 100 and 130 mm.

6. Complete the following restricted-response questions using the appropriate word. *(1 mark for each part)*

a) A .... is a duplicate setup of an experiment with everything identical except for the variable being tested.

b) .... are things that are kept the same for all trials.

c) In a .... test you change only one factor at a time while keeping all other conditions the same.

d) The singular of the word 'data' is ....

e) A first-hand .... is where original material is collected by the investigator.

7. Use code letters to match the terms or phrases in each column. *(1 mark for each)*

Column 1	Column 2
A line of best fit	F variable
B reliability	G report
C graph	H visual representation
D dependent	I repeated measurements
E scientific	J show the trend between two sets of data

Part B: Skills

8. Read the following statements. Each one describes a student doing something wrong in a school laboratory. Write down what is wrong, and the correct way of performing the task. *(1 mark for each)*

a) Anita uses her hand to scoop out a small amount of solid chemical from a jar.

b) Gianni looks down the barrel of a test tube being heated to see what is going on inside it.

c) Charles leaves the beaker boiling on the tripod to go and talk to his friends.

d) Frances sticks her nose directly over a bottle to smell the chemical.

e) James sets up some glassware close to the edge of the bench.

f) Nihal puts her thumb over the end of the test tube and shakes it up and down to mix the contents.

g) After boiling some liquid, Andrew picks up the beaker with his hand to pour the contents down the sink.

h) Alyssa pours some liquid and solid wastes into the sink.

9. Figure 5.16 shows the scale on a thermometer at two different times of day. Determine the temperature at each time and calculate the rise in temperature. *(3 marks)*

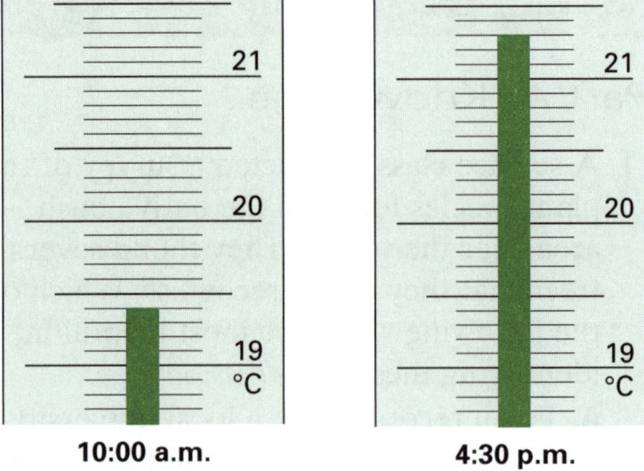

Figure 5.16 The scale on a thermometer at two different times of day

10. Figure 5.17 shows a student reading the scale on the side of a burette.

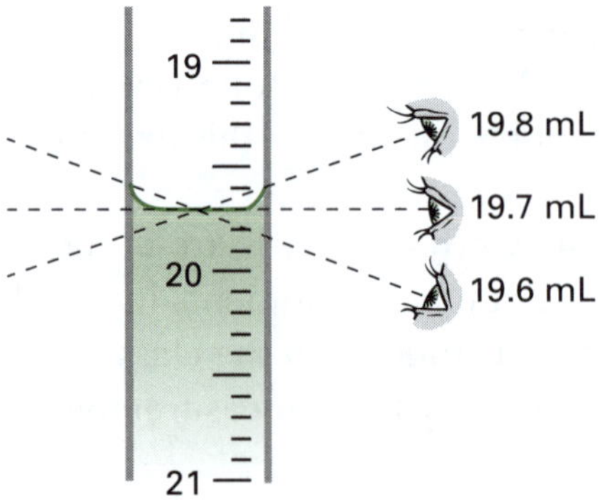

Figure 5.17 A student reading the scale on the side of a burette

a) Which is the correct reading for this volume? *(1 mark)*
b) What error do the other two readings show? *(1 mark)*
c) How can this error be avoided? *(1 mark)*

11. a) Define collaborative research. *(1 mark)*
b) Suggest two reasons why scientists often collaborate in research. *(2 marks)*
c) Give three advantages of collaborative research. *(3 marks)*

12. a) Define the term reliability. *(1 mark)*
b) What is validity? *(1 mark)*
c) The word valid comes from the Latin *validus*, meaning 'strong'. Is this an appropriate derivation for this word? *(1 mark)*
d) Consider the bullseye diagrams in Figure 5.18.

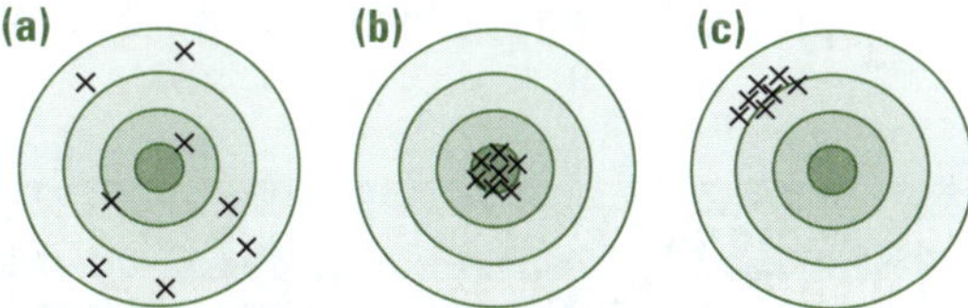

Figure 5.18 Bullseye diagrams

Assuming the diagrams represent validity and reliability, state whether the results of each are valid and/or reliable. *(3 marks)*

13. What is the difference between an inference and a prediction? Give an example of each. *(4 marks)*

14. a) Explain the term accuracy. *(1 mark)*
b) Which of the measuring vessels shown in Figure 5.19 will more accurately measure 200 mL? *(1 mark)*

Figure 5.19 Measuring vessels

c) A typical medicine cup is wide at the top and narrow at the bottom, as shown in Figure 5.20.

Figure 5.20 A typical medicine cup

Use the information in this question to explain a possible reason for the shape of the medicine cup. *(4 marks)*

Go to pp. 208–210 to check your answers.

Summary

1. The scientific method involves observing, inferring, producing, experimenting, formulating hypotheses, testing and modifying hypotheses, making conclusions, generalising, and developing theories and laws.
2. Fair testing involves controlling the variables. The independent variable is allowed to change and the dependent variable is affected by this change. Other factors are the constant variables.
3. Experiments should be carefully designed to ensure that they are valid and safe.
4. There are advantages and disadvantages in working individually or in teams.
5. Data can be gathered by first-hand experimentation and/or by second-hand data research.
6. Repeated measurements improve reliability.
7. Data can be presented in a variety of ways including graphs and tables.

Syllabus checklist

Are you able to answer every syllabus question in this chapter? Tick each question as you go through the list if you are able to answer it. If you cannot answer it, turn to the appropriate page in the guide as is listed in the column to find the answer.

	For a complete understanding of this topic	Page no.	✓
1	Do I understand the advantages and responsibilities of working individually or collaboratively?	138	
2	Can I conduct background research using secondary sources?	138–139	
3	Can I investigate a scientific problem?	139–140	
4	Do I understand safety and hazards when undertaking first-hand investigations?	143–144	
5	Can I choose appropriate equipment and increase accuracy and reliability?	144–145	
6	Can I collect, record, process and present data in scientific reports?	146–149	
7	Can I draw conclusions, discussions and evaluations from my research?	151	
8	Can I analyse experimental data and write a report?	146–153	

Chapter test

Go to p. v for *Tips for tests and examinations* 80 min

Part A: Multiple-choice questions

(1 mark for each)

1. In a scientific report, the correct sequence of recording is

A Aim, Method, Results, Conclusion.
B Aim, Method, Conclusion, Results.
C Conclusion, Results, Method, Aim.
D Method, Aim, Results, Conclusion.

2. Geraniums were studied to determine if humidity affected the plant's flowers. Which of these is the dependent variable in this study?

A how much watering the plant received
B the degree of humidity around the plant
C how many flowers are on each plant
D the size of the geranium cuttings

3. Which of the following methods would provide the most accurate information for collecting data about attitudes concerning global warming?

A listen to political speeches
B conduct a survey among various people
C read newspapers and periodicals
D ask scientists at a university

4. Consider these facts:

- 1869: DNA discovered inside cell nuclei
- 1940s: Scientists establish that chromosomes are made of DNA and protein
- 1950s: Scientists demonstrate that DNA is responsible for heredity
- 1953: Three-dimensional model of DNA is built

These discoveries best illustrate the importance of

A better ways of conducting research.
B working independently from other scientists.
C repeating experiments.
D collaborative research.

5. Mark consistently measured the mass of solids precipitated from a reaction. How would this practice impact on his work?

A The mass would vary between readings.
B The scales have an inherent fault.
C His measurements would not be precise.
D His measurements would be accurate.

6. If a student needs to do research on solar energy, which of these sources is the least reliable?

A local or national newspapers
B science programs on pay TV channels
C the internet
D science journals

7. Geoff hypothesises that strong wing muscles are an inherited trait in birds. He collects data on a number of birds, and it shows that birds that live outdoors have stronger wing muscles than birds that live in cages. What should Geoff conclude?

A Strong wing muscles may not be due to inheritance alone.
B Birds with strong wing muscles need to fly around more.
C Birds with strong wing muscles will not survive in cages.
D Colourful birds have weak wing muscles.

8. Which of these best shows the importance of using control groups?

A Control groups make calculations a lot easier.
B Control groups minimise the need for repetitions in experiments.
C They allow a comparison to be made between things receiving treatment and those not.
D They are traditionally used in scientific experiments.

9. Demeter devised the hypothesis that more eucalypt seeds germinate after a forest fire. The most valid and reliable test she could conduct of this hypothesis would include an experimental group of eucalypt seeds recovered from a fire area and eucalypt seeds that were
   A obtained from a region without a fire.
   B placed in a fire-prone area.
   C fire tolerant.
   D exposed to fire.

10. Two plant species found in a dry region of central Australia exhibit very different abilities to survive. Species P grows extremely slowly and has few leaves, but is very abundant. Species Q grows quickly with many leaves but is rare. Which hypothesis most likely supports this information?
    A Both plants require plenty of water to survive.
    B Reduced growth rate may give species P an advantage over species Q.
    C Species P has a more extensive root system than species Q.
    D Species Q is the preferred food source for animals living in arid environments.

Part B: Short-answer questions

11. Three students are reading a thermometer (see Figure 5.21).

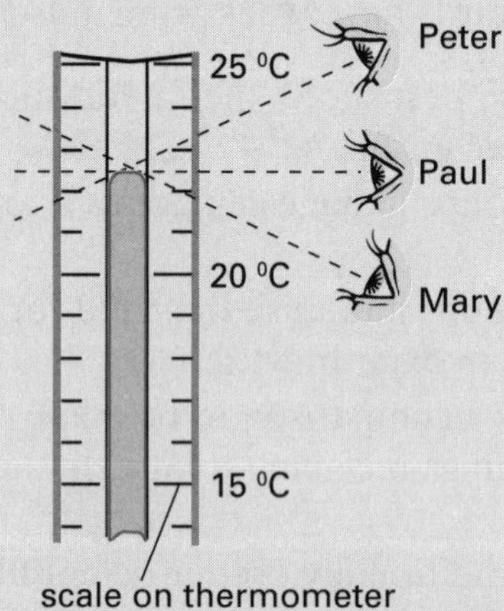

Figure 5.21 Three students reading a thermometer

   a) What is the scale on this thermometer? *(1 mark)*
   b) What value does each person read? *(3 marks)*
   c) Who reads the correct value? *(1 mark)*
   d) What directions should this person give the other two about correctly reading a thermometer? *(2 marks)*

12. Read each of the following statements and decide whether each is an example of an observation, inference, prediction or generalisation.
    a) The crack in the glass window was caused by a sudden change in temperature. *(1 mark)*
    b) Most compounds of copper are blue. *(1 mark)*
    c) The flight feathers of the bird were a different colour to the tail feathers. *(1 mark)*
    d) If a more concentrated acid is added to zinc, the zinc will dissolve faster. *(1 mark)*

13. Read each of the following statements and decide whether or not each statement is a hypothesis. Justify your answer.
    a) Snails live in moist, dark places rather than dry, lit places. *(2 marks)*
    b) Neanderthal people were exclusively cave dwellers but the ancestors of modern humans were not. *(2 marks)*
    c) Jazz music is more pleasant to the ear than rock and roll. *(2 marks)*

14. Figure 5.22 shows an experiment where a student grew a plant and measured its height over a period of time.

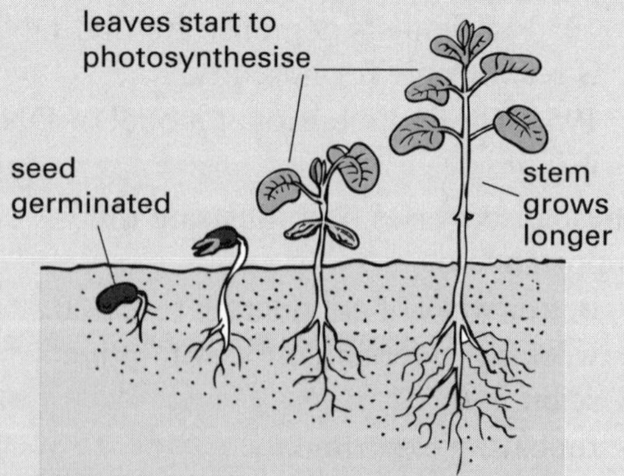

Figure 5.22 A plant grown and measured over a period of time

Which is the dependent variable and which is the independent variable? *(2 marks)*

15. Figure 5.23 shows two students (A and B) measuring the length of a glass prism using a millimetre ruler. Which student will obtain a more accurate reading?

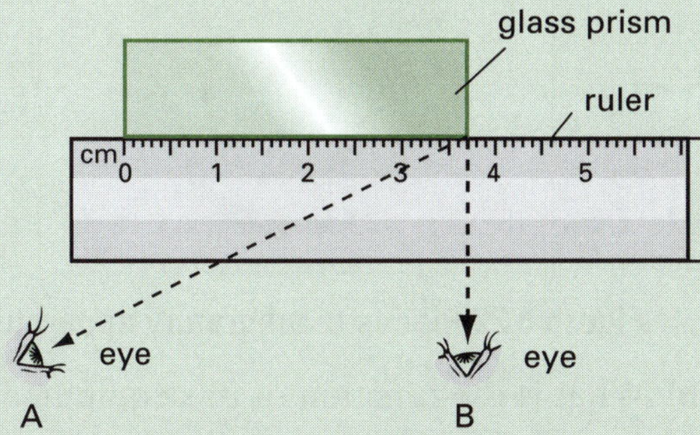

Figure 5.23 Two students measuring the length of a glass prism

16. Consider the scale on an instrument that measures volts (see Figure 5.24).

Figure 5.24 Instrument that measures volts

Three students use the instrument to measure the voltage across a device.

- Patrice: I measured the voltage and got 55 volts.
- Nikolas: I measured the voltage and got 54.955 volts.
- James: I measured the voltage and got 54 or 55 volts.

Which of the three students produced the most accurate measurement? Explain. *(4 marks)*

17. Figure 5.25 shows a measuring cylinder and a beaker. They are both used to measure the volumes of liquids. Which is more accurate? Explain. *(2 marks)*

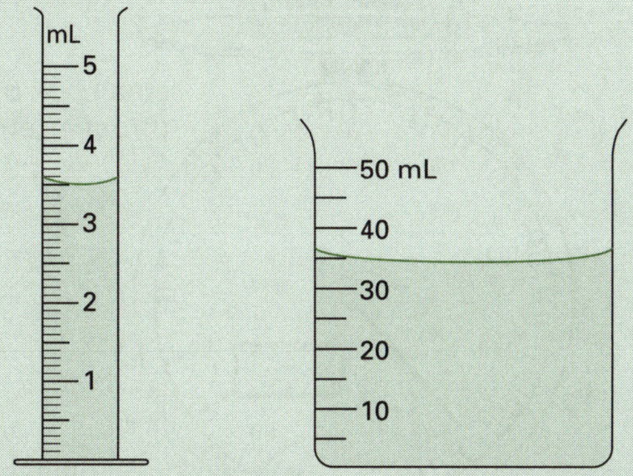

Figure 5.25 A measuring cylinder and a beaker

18. Figure 5.26 shows the scale of a 100-mL measuring cylinder. What is the volume of liquid in the cylinder? *(1 mark)*

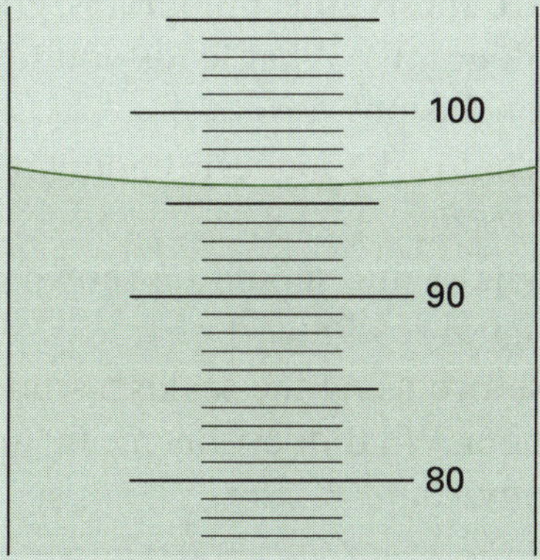

Figure 5.26 The scale of a 100-mL measuring cylinder

19. Draw scientific diagrams of the following pieces of laboratory equipment.

a) conical flask *(1 mark)*

b) filter funnel *(1 mark)*

c) retort stand with attached bosshead and clamp *(2 marks)*

20. Consider the electronic balance pan shown in Figure 5.27. The scale reading can also be displayed as a digital readout.

Figure 5.27 Weighing scales

a) The units have been left off the scale. Are these units milligrams, grams or kilograms? What leads you to that conclusion? *(2 marks)*
b) What is the highest reading it can make? *(1 mark)*
c) What value should be shown on the digital readout? *(1 mark)*
d) Before using the scales a student zeroed them. What does this mean, and why is it important? *(2 marks)*

21. Name the pieces of laboratory equipment shown in Figure 5.28 (not drawn to scale). *(12 marks)*

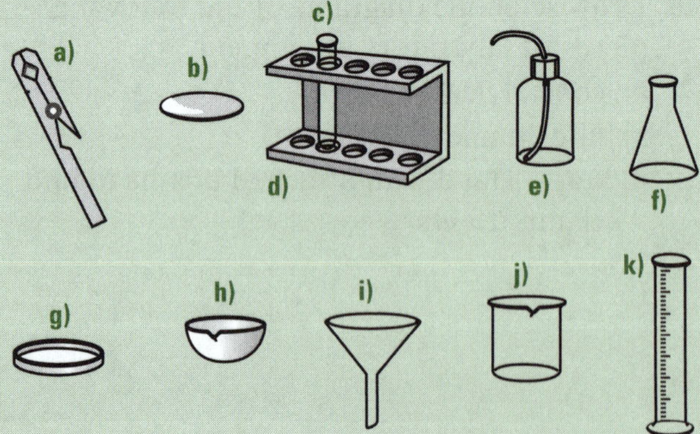

Figure 5.28 Pieces of laboratory equipment

22. a) What pieces of laboratory apparatus are shown in Figure 5.29? (Name each piece separately.) *(2 marks)*

Figure 5.29 Pieces of laboratory apparatus

b) What is the function of this equipment? *(1 mark)*
c) The word *mortar* comes from the Latin *mortarium*, meaning 'receptacle for pounding' while *pestle* derives from the Latin *pistillum*, meaning 'pounder'. Are these suitable derivations for these words? Explain. *(2 marks)*
d) Laboratory mortar and pestles are usually made from porcelain. List at least six features of a good mortar and pestle. *(6 marks)*

23. The following description refers to a piece of laboratory equipment.

It is a simple container for stirring, mixing and heating liquids commonly used in many laboratories. It is generally cylindrical in shape, with a flat bottom and a lip for pouring.

a) Name this piece of equipment. *(1 mark)*
b) This piece of equipment is often graduated with line markings on the side indicating the volume contained. For example, if it can hold up to 250 mL, it might be marked with lines to indicate 50, 100, 150, 200 and 250 mL of volume. What is the purpose of these marks? *(2 marks)*
c) This equipment is commonly made from borosilicate glass. What is an important feature of this glass for such a piece of equipment? *(1 mark)*

24. The following description and Figure 5.30 show one way to transfer a solid into a test tube.

Remove a spoonful of salt from its reagent bottle with a spatula. Place it at the centre of a square piece of glazed paper. Roll the paper into a cylinder and slide it into a test tube that is lying flat on the table. Then lift the tube vertically and tap the paper gently. The solid should slide down into the test tube.

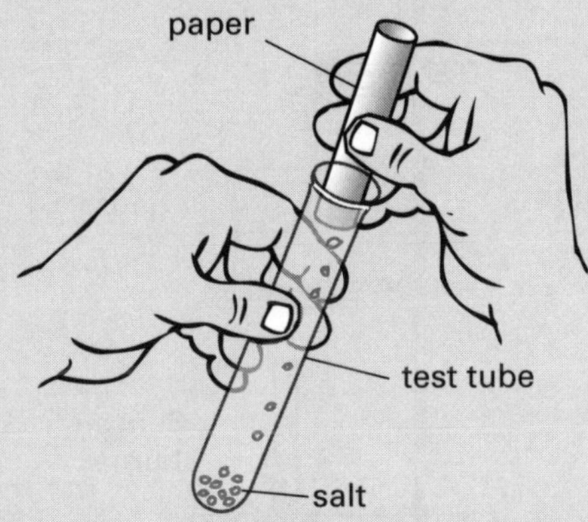

Figure 5.30 Transferring a solid into a test tube

a) Why isn't the salt transferred directly to the test tube with the spatula? *(1 mark)*

b) Why do the instructions recommend glazed paper? *(1 mark)*

c) Why is the cylindrical paper inserted horizontally? *(1 mark)*

25. A burette is more accurate than a measuring cylinder when measuring volumes of liquids. The burette is filled with a liquid, the tap opened and a known amount released (see Figure 5.31).

a) Suggest a reason for this. *(1 mark)*

b) You should never pour a liquid directly from its reagent bottle into the burette. First pour the liquid into a small beaker (say, 50 mL) that is easy to handle. Then pour the liquid from the small beaker into the burette. What is the reason for this? *(1 mark)*

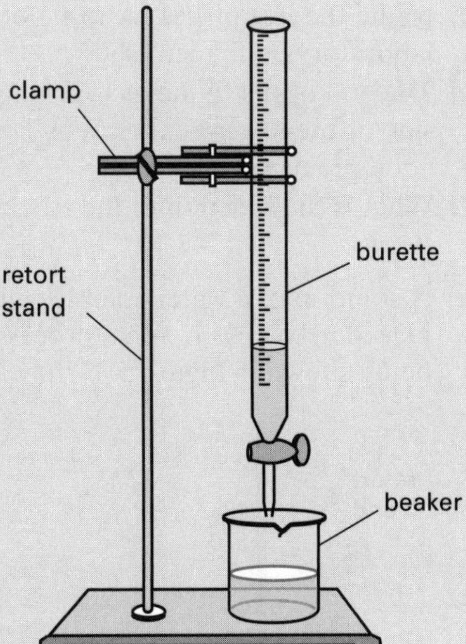

Figure 5.31 The burette tap allows water to be released into the beaker.

26. Safety goggles, gloves and an apron should be worn whenever you measure or use chemicals. Tie long hair at the back of your head and away from the front of your face. Wash your hands thoroughly at the end of experiments. Suggest reasons for these safety tips. *(5 marks)*

27. a) What process is being shown in Figure 5.32? *(1 mark)*

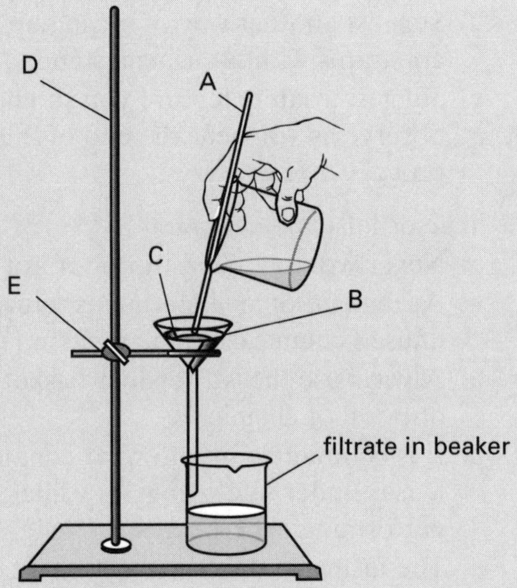

Figure 5.32 Process involving laboratory equipment

b) Name the five pieces (A to E) of laboratory equipment shown. *(5 marks)*

c) The end of the funnel is touching the side of the lower beaker. Why is this good practice? *(1 mark)*

d) What is the function of the stirring rod? *(1 mark)*

28. a) A solution of a water-soluble salt is placed in the basin. What process is being shown in Figure 5.33? *(1 mark)*

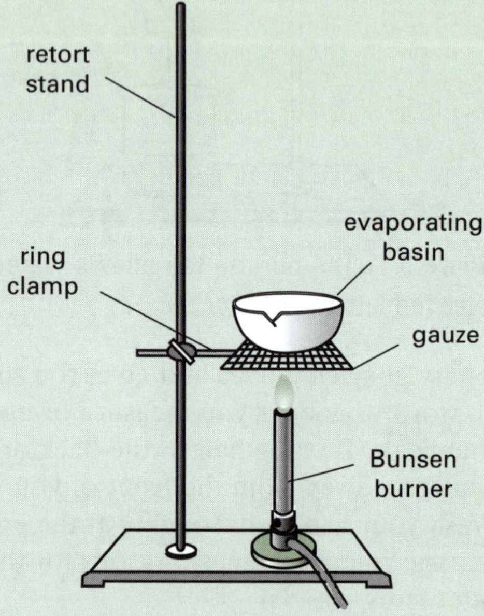

Figure 5.33 Process involving a water-soluble salt in a basin

b) Suggest another way of setting up this apparatus without using a stand. *(1 mark)*

c) Suggest a safety feature you should observe as you near the end of this process. *(2 marks)*

29. True or false *(1 mark for each)*

a) Never work alone in the laboratory.

b) At the end of an experiment return all unused chemicals to the reagent jar.

c) Always use the wastepaper basket for disposal of chemicals.

d) It is completely safe to wear contact lenses under safety goggles while performing experiments.

e) The fastest and safest way to heat liquid in a test tube is to concentrate the flame on the bottom of the test tube.

f) Never lay the stopper of the reagent bottle on the benchtop.

g) A mortar and pestle should be used to grind only one substance at a time.

30. The apparatus in Figure 5.34 is set up to measure the temperature of water as it is being heated.

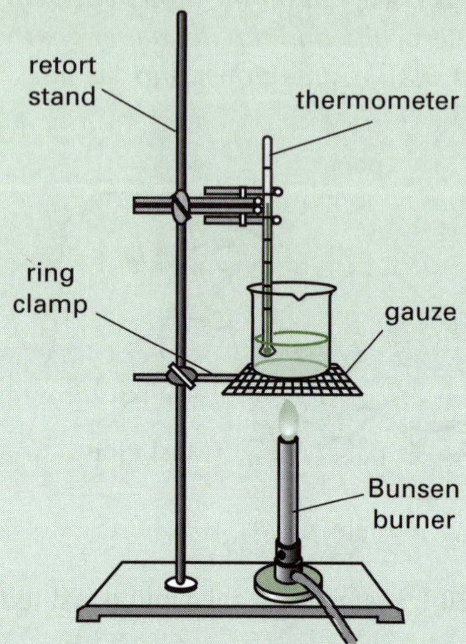

Figure 5.34 Apparatus to measure the temperature of water as it is being heated

Explain why the thermometer is not just allowed to sit in the beaker with the bulb on the bottom of the beaker. *(2 marks)*

31. Figure 5.35 shows the parts of a Bunsen burner.

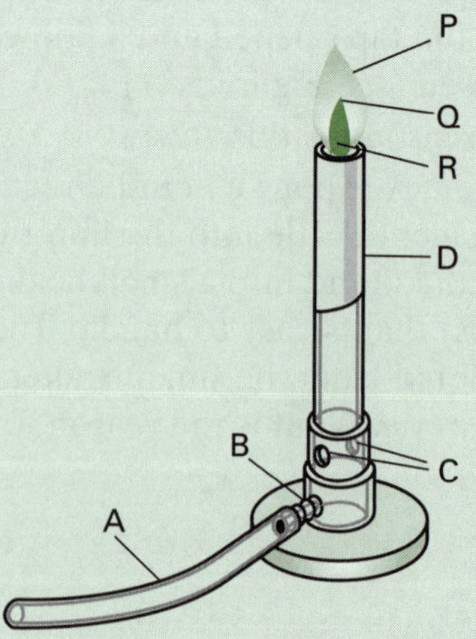

Figure 5.35 The parts of a Bunsen burner

a) Label the missing parts A, B, C and D with the words 'barrel', 'rubber hose', 'air inlet', and 'gas inlet'. *(4 marks)*

b) Which part of the flame (P, Q or R) is the hottest ? *(1 mark)*

c) What is a safety flame? When should it be used? *(2 marks)*

32. Test tube holders are often made from wood (see Figure 5.36). Suggest an important reason for this. *(1 mark)*

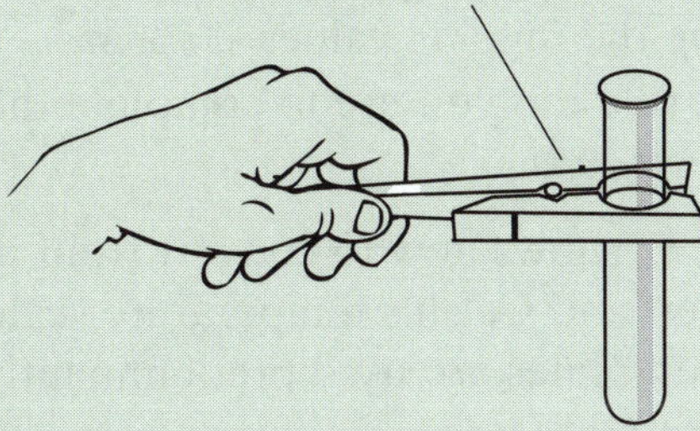

Figure 5.36 Wooden test tube holder

33. In 1876 Samuel Plimsoll introduced a law into British Parliament that prevented ships from sailing if overloaded or not seaworthy. The Plimsoll line, painted on the side of all ships, is shown in Figure 5.37. It shows the level that water should reach when the ship is properly loaded.

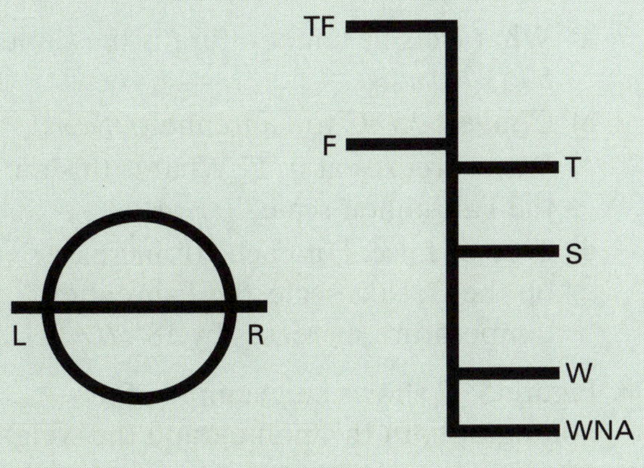

Figure 5.37 The Plimsoll line

a) Explain how this is a safety feature. *(1 mark)*

b) Using the Plimsoll line, which is denser?

i) Salt water in summer, or salt water in winter? *(1 mark)*

ii) Tropical fresh water or tropical salt water? *(1 mark)*

34. Figure 5.38 shows the distance from home a bicyclist travelled. The zero distance on the graph at zero time represents his home. During his journey he travels away from home, rests for a while and then travels part of the way back to home. He then moves further away from home to reach his final destination.

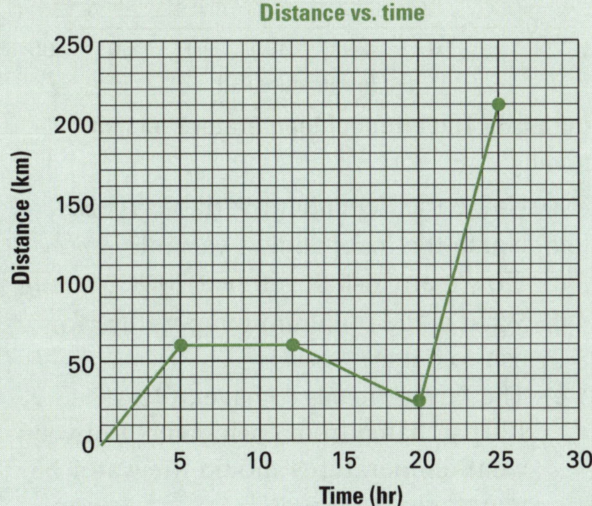

Figure 5.38 The distance from home a bicyclist has travelled from home, expressed as distance versus time

a) How far did the cyclist travel in the first 5 hours? *(1 mark)*

b) Between what times did the cyclist stop to rest? *(1 mark)*

c) How far did the cyclist travel between 12 hours and 20 hours? (1 mark)

d) Between which times did the fastest speed occur? *(1 mark)*

e) What total distance did the cyclist travel? *(2 marks)*

35. A tropical fish farmer kept records about water temperature and the number of fish that hatch from equal-sized batches of fertilised eggs. He drew the graph shown in Figure 5.39.

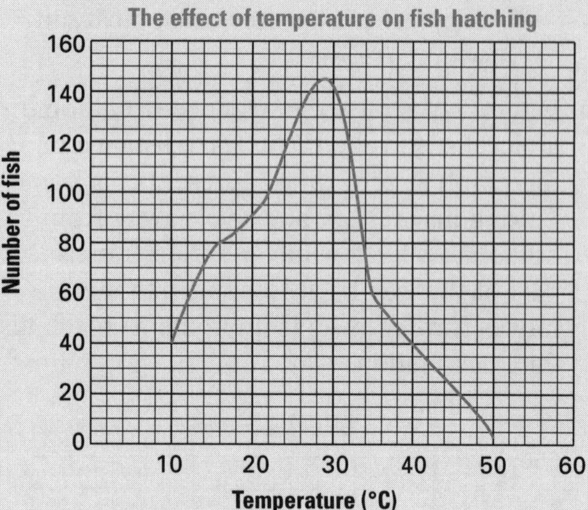

Figure 5.39 The effect of temperature on fish hatching

a) Name the dependent variable. *(1 mark)*
b) Name the independent variable. *(1 mark)*
c) How many fish develop at 20 °C? *(1 mark)*
d) What is the optimum temperature for fish to hatch? *(1 mark)*
e) The farmer needs to have at least 120 fish hatch from each batch. Between what temperatures should the water be maintained? *(2 marks)*

36. Hookworms (Figure 5.40) live attached to the intestinal wall of humans and suck blood from their host. Table 5.6 shows the number of hookworms in the intestine and the volume of blood lost each day.

Table 5.6 Number of hookworms and related volume of blood lost

Number of hookworms	Volume of blood lost (mL/day)
15	7.5
22	11
28	
	18
40	
	28
63	

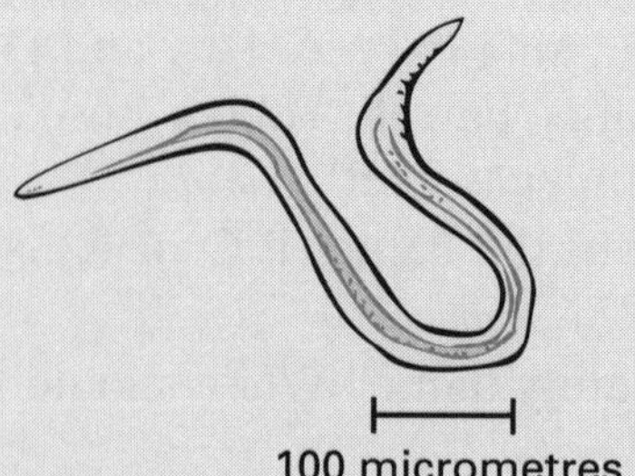

Figure 5.40 Hookworm

a) Name the dependent variable. *(1 mark)*
b) Name the independent variable. *(1 mark)*
c) The table is incomplete. Complete it by finding the missing values. *(5 marks)*
d) What rule did you use to complete the table? *(1 mark)*

37. Figure 5.41 shows how to convert from the Fahrenheit and Celsius temperature scales. Fahrenheit scales are used predominantly in the United States.

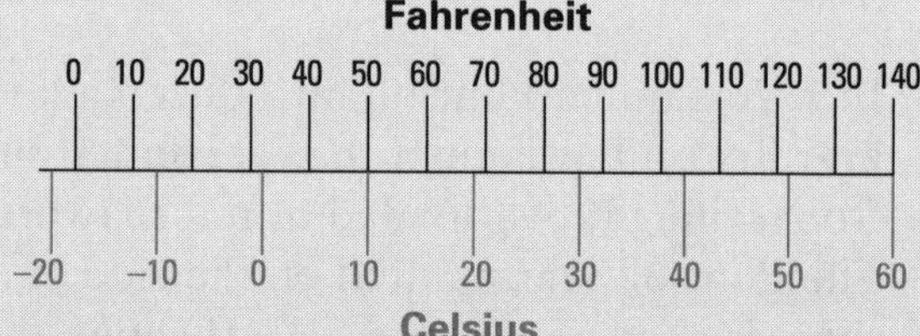

Figure 5.41 Fahrenheit and Celsius temperature scales

a) What Celsius temperature is the same as 50 °F? *(1 mark)*
b) Convert 35 °C to Fahrenheit. *(1 mark)*
c) Water freezes at 0 °C. What is this on the Fahrenheit scale? *(1 mark)*
d) True or false: For each 10° increase on the Celsius scale, the Fahrenheit temperature increases by 18°. *(1 mark)*

38. Figure 5.42 shows an example of a nomograph. In this nomograph the weight of a horse can be estimated using a scale measuring its condition, and another scale measuring its shoulder height (measured in hands).
For example, a horse rated at a condition of 2 and being 16.0 hands high has a weight of about 450 kg.

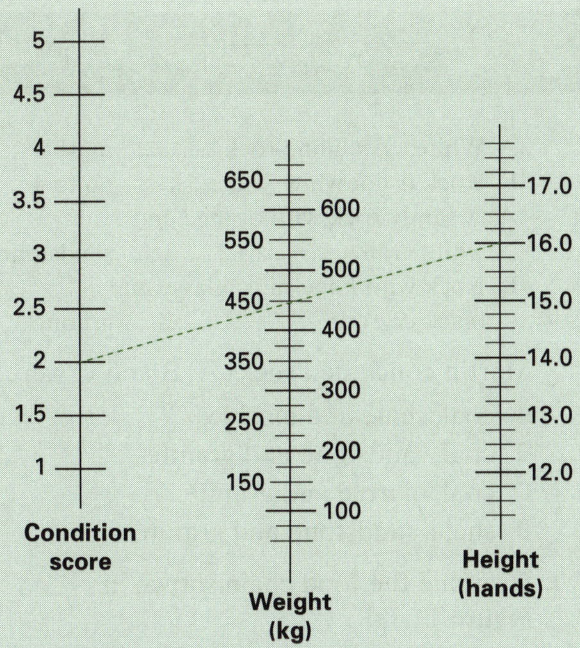

Figure 5.42 Horse rating scales

a) A horse has a condition of 3 and is 14.0 hands high. What is its weight? *(1 mark)*

b) Another horse has a condition of 4 and weighs 300 kg. What is its height? *(1 mark)*

c) A horse is 12.0 hands high and has a weight of 225 kg. What should its condition be? *(1 mark)*

d) Would this chart be suitable to use with other farm animals? Explain. *(2 marks)*

39. The sector (pie) graph in Figure 5.43 shows the main atmospheric gases on Earth.

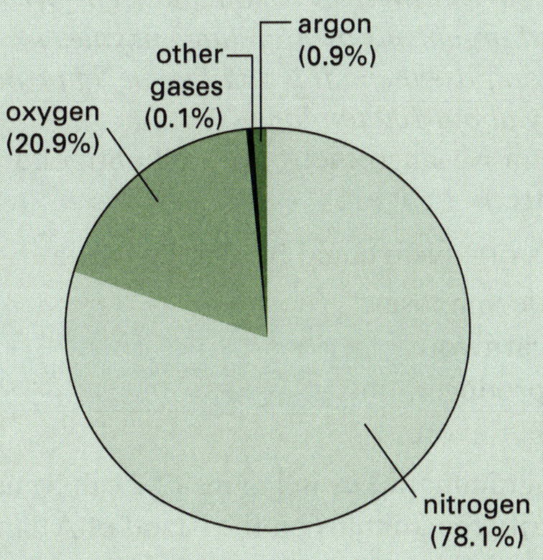

Figure 5.43 Main atmospheric gases on Earth

a) What is the angle at the centre of the circle for nitrogen? *(1 mark)*

b) If the sector angle for oxygen is 75.6°, what percentage of the Earth's atmosphere does oxygen occupy? *(1 mark)*

c) What percentage of the atmosphere is made up of argon and other trace gases? *(1 mark)*

40. Tyre pressure in cars is measured in kilopascals (kPa). An older unit, still in use, is pounds force per square inch (lbf/in^2). (Note: 'lb' is the abbreviation for pounds.)

Table 5.7 gives some values to convert from one set of units to the other.

Table 5.7 Table to convert lbf/in^2 to kPa

lbf/in^2	kPa
25	172.36
26	179.26
27	186.16
28	193.06
29	199.96
30	206.86
31	213.76
32	220.66
33	227.56
34	234.46
35	241.36

Use this table to draw a graph showing the conversion from lbf/in^2 to kPa. Place lbf/in^2 on the horizontal axis, and kPa on the vertical axis. *(6 marks)*

Go to pp. 210–214 to check your answers.

Course test 1

Go to p. v for *Tips for tests and examinations*

This test covers material from all the chapters in this book. It consists of the following.

- Section A: 25 multiple-choice questions *(1 mark for each)*
- Section B: 15 restricted-response questions *(1 mark for each)*
- Section C: 11 knowledge, skill and processing data questions *(total 60 marks)*

Total marks for this test: 100

Time: 2 hours

Part A: Multiple-choice questions

(25 marks—1 mark for each question)

1. Which of the following characteristics allows us to group birds together?

A presence of feathers
B ability to fly
C possession of wings
D lack of teeth

2. Martin was given three rocks and asked to describe them in his workbook. He recorded his results as shown in Table T1.1.

Table T1.1 Rocks

Rock	Description
A	Dark, tiny crystals, shiny, hard
B	Orange and white grains, brown, speckled, dull, grains easily fall off on rubbing
C	Many coloured crystals, some glassy crystals, large and small crystals, shiny

Martin used the following key to identify these rocks.

1A Rock dark in colour .... go to 2
1B Rock not dark .... go to 3
2A Rock is very black and powdery; burns easily .... coal
2B Rock is grey or black with fine crystals .... shale
3A White, crystalline rock .... marble
3B Rock is not white .... go to 4
4A A sandy rock with orange and white grains .... sandstone
4B Rock with large pink, glassy and black crystals .... granite

Martin concluded rocks A, B and C were

A coal, shale and marble.
B coal, sandstone and granite.
C coal, marble and granite.
D shale, sandstone and granite.

3. Examine the food chain shown in Figure T1.1.

Figure T1.1 Food chain

The second-order consumer is

A seagrass.
B sea snails.
C crabs.
D egret.

4. Read the following description of an organism.

This organism lives in soil and feeds off dead organisms by excreting enzymes to decompose them. It is rod-like in shape but only about 100 μm long.
(1 μm = 1 micrometre = 1 millionth of a metre.)

This organism could be classified as a

A decomposer.
B carnivore.
C producer.
D herbivore.

The diagrammatic key in Figure T1.2 can be used to identify the animals on the island of Atlantis. Use this information to answer Questions 5 and 6.

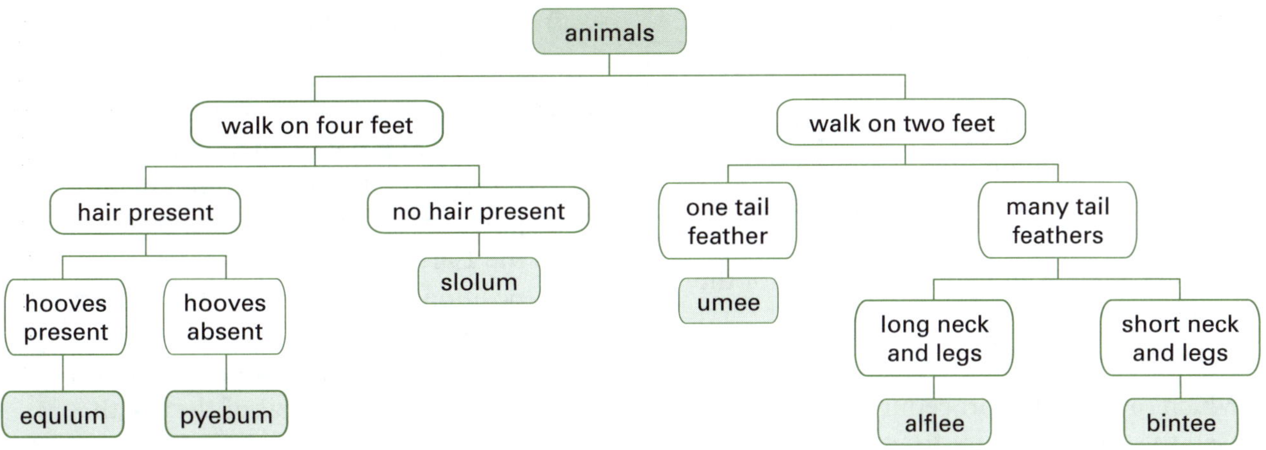

Figure T1.2 Diagrammatic key

5. Read the description of the following animal and determine the identity of the animal using the key.

   The animal is brown with white spots on its head. Its body is long and covered with fine hairs which are very long over its neck and knees. The animal can jump over hurdles. Its four legs are long and farmers often nail metal shoes to its hooves when the animal is tamed for domestic use.

   The animal is

   A pyebum.
   B alflee.
   C slolum.
   D equlum.

6. Use the key to identify the animal shown in Figure T1.3.

Figure T1.3 Animal to classify

   The animal is

   A slolum.
   B alflee.
   C bintee.
   D umee.

7. Consider the food web on a rocky shore, as shown in Figure T1.4.

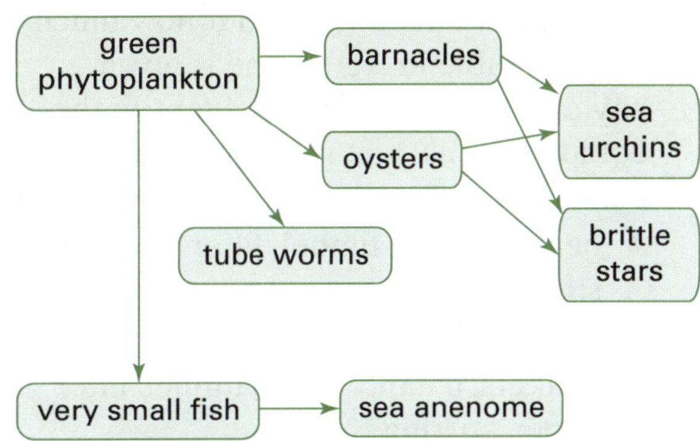

Figure T1.4 Food web

   Barnacles and small fish can be classified as

   A producers.
   B first-order consumers.
   C second-order consumers.
   D decomposers.

8. Rabbits are responsible for serious soil erosion in Australia. This occurs because

   A rabbits are not native animals.
   B rabbits eat the native plants whose roots bind the soil.
   C the faeces of the rabbits kill the plants.
   D of competition between rabbits and bilbies.

9. When filtering large volumes of suspensions, why do you think it is important that the mixture be allowed to settle and then most of the clear liquid

decanted through the filter paper before the remaining suspension is poured into the filter?

A Most of the liquid will pass quickly through the filter.

B No residue is lost during the decanting process.

C The residue will flow over the top of the filter paper.

D Filter papers are too small.

10. Which of the listed procedures could be used to separate cream from milk?

A filtration

B chromatography

C froth flotation

D centrifugation

11. Gabriella wants to work out how much soil is soluble in water. She is given any equipment she needs. Identify the equipment that she will need to use to perform the experiment.

A beakers, filter funnel, filter paper, stirring rod

B beakers, balance, filter paper, stirring rod

C beakers, balance, filter funnel, filter paper, stirring rod

D measuring cylinder, beakers, balance, filter paper, stirring rod

12. Which one of the following sets contains only household appliances which make use of the principle of filtration?

A water purifiers, washing machines, television

B food mixers, vacuum cleaners, swimming pool chlorinators

C vacuum cleaners, clothes dryers, water purifiers

D vacuum cleaners; microwave ovens; electric toothbrush

13. Around 21 December in southern Australia, the Sun

A is low in the northern sky at midday.

B moves along an arc which is low in the western sky.

C has reached its most southerly point in its annual motion.

D rises in the north-western sky and sets in the north-eastern sky.

14. The Earth's seasons are caused by the

A Earth's axis being tilted at an angle of about 23.5° from the orbital plane.

B position of the Moon in its orbit around the Earth.

C position of the Earth in its orbit around the Sun.

D differences in gravitation at different locations in the orbit of the Earth.

15. At which location would a 5-kg mass have the greatest weight?

A in orbit around the Earth

B on the Moon's surface

C at a distance of 3 km underground in a mine

D on the Earth's surface

16. Omar Khayyám was a Persian astronomer who

A discovered the moons of Jupiter.

B measured the Earth's year as 365.24 days.

C observed the rings of Saturn through a telescope.

D proposed the Earth-centred model of the universe.

17. Figure T1.5 shows three views of Venus (from the Earth) throughout the year. These demonstrate that Venus has

A seasons.

B gravity.

C phases.

D variable cloud cover.

Figure T1.5 Venus

18. Arrange the following steps of water purification in the correct order.

A filtration; fluoridation; settling; flocculation

B sedimentation; filtration; chlorination; acidity adjustment

C chlorination; acidity adjustment; flocculation; sedimentation
D flocculation; filtration; sedimentation; aeration

19. Figure T1.6 is a particle diagram showing a liquid being gently heated below its boiling point. Identify the process (X).

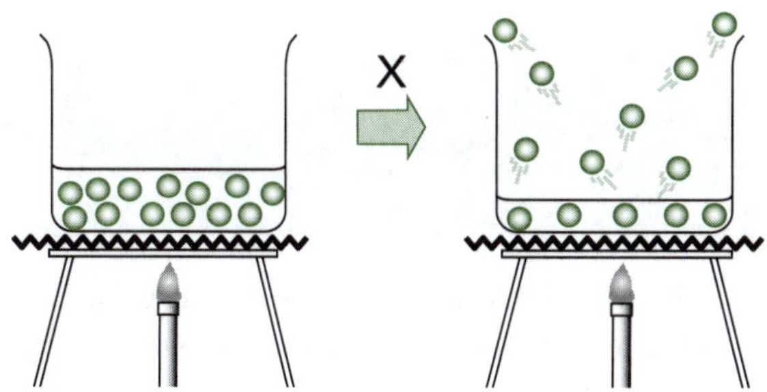

Figure T1.6 Process X

A condensation
B distillation
C fusion
D evaporation

20. Which of the following is a renewable resource?
A fossil fuels
B aluminium
C wind energy
D uranium

21. Identify a resource that was not available to humans 250 years ago.
A hydroelectric power
B coal
C copper
D biomass

22. Identify an advantage of geothermal power.
A It does not contribute to global warming.
B All rocks are suitable.
C Hot rocks are very close to the surface and are easy to access.
D Hot rocks are widespread across Australia.

23. Which of the following promotes cloud formation and rain?
A increases in solar radiation
B mountain ranges
C deserts
D cold, dry air rising

24. Figure T1.7 shows three paper clips hanging from the end of a steel screwdriver. Identify the force that is opposing gravity in this example.

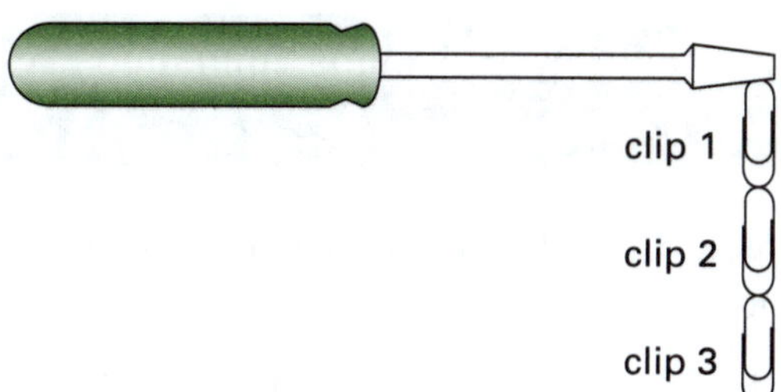

Figure T1.7 Opposing force

A electrostatic force
B contact force
C magnetic force
D frictional force

25. The mass and weight of a steel ball was measured on Earth and on the Moon. The results are shown in Figure T1.8.

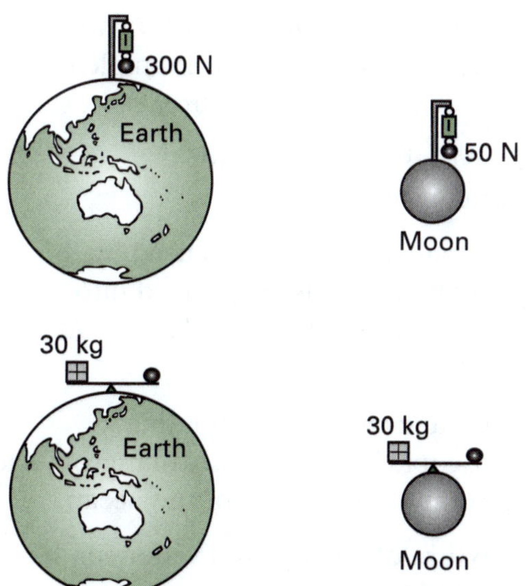

Figure T1.8 Mass and weight of a steel ball

What conclusion can be made from these experiments?
A The steel ball has a greater mass on Earth due to its stronger magnetic field.
B The weight of a body depends on the strength of the gravitational field whereas the mass of a body does not.

C The mass experiment on the Moon must be wrong as the mass did not change.

D The strength of the gravitational field on the Moon is the same as that of the Earth as the body had a weight of 30 kg in each place.

Part B: Restricted-response questions

(15 marks—1 mark for each question)

Complete the following restricted-response questions using the appropriate word.

26. When a body moves at constant speed the forces are ....

27. The tension in a rope supporting a heavy weight is an example of a .... force.

28. A .... charged glass rod will attract an ebonite rod rubbed with flannel.

29. The echidna is a member of the mammalian class known as ....

30. The name *Thylacinus cynocephalus* (or Tasmanian tiger) is an example of .... nomenclature.

31. Herbivores are also called ....-order consumers.

32. Myxomatosis (a disease caused by the myxoma virus) was introduced into Australia in 1950 to control .... populations.

33. Phytoplankton are free-floating microscopic .... that grow and live in the upper layers of the ocean.

34. A mixture of alcohol and water can be separated by ....

35. A mixture of fine sand and 1-cm diameter polystyrene balls can be separated by ....

36. Liquid honey can be separated from the honeycomb using ....

37. A .... is the partial shadow that surrounds the umbra.

38. In Australia, the phase of the Moon that immediately follows a full Moon is called the waning .... Moon.

39. Neptune is about 30 .... units from the Sun.

40. Copernicus and Galileo both supported the ....-centred model of the universe.

Part C: Knowledge, skill and processing data questions

(60 marks)

41. The strength of various glues (A, B, C, D and E) used to join two plastic blocks together was compared using the apparatus shown in Figure T1.9. The results are shown in Table T1.2.

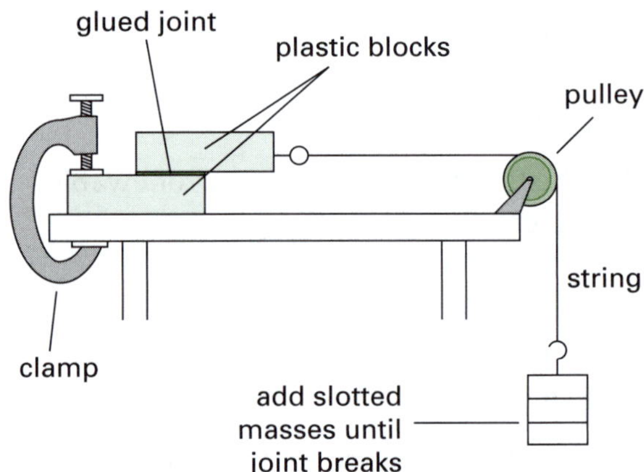

Figure T1.9 Glues

Table T1.2 Glue and joints

Glue brand	Number of weights required to break the joint
A	3
B	5
C	7
D	5
E	4

a) Which glue provides the greatest binding strength? *(1 mark)*

b) When testing glue B, the student found that the force of four weights was not sufficient to break the glued joint. At this point in the experiment, what can be said about the following?

i) the tension (strain force) in the string and the force of the four weights *(1 mark)*

ii) the tension in the string and the opposing force of the glued joint *(1 mark)*

42. Figure T1.10 shows how temperature affects the solubility of a slightly soluble solid, S, in water. Ten grams of the solid was added to 100 mL of water at a particular temperature and the mixture stirred. The weight of solid that had dissolved was determined every 2 minutes. The experiment was repeated at three different temperatures.

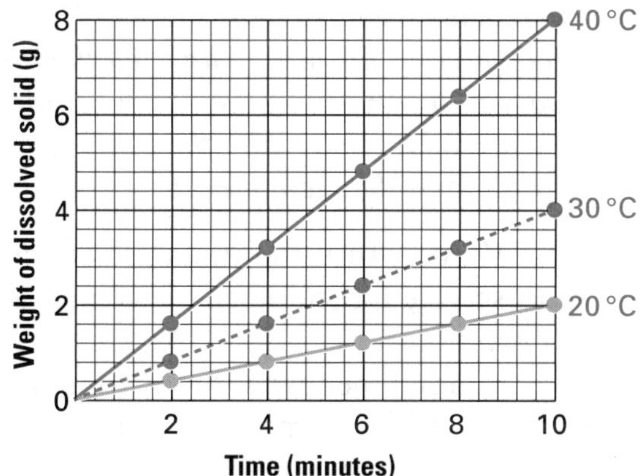

Figure T1.10 Weight of dissolved solid

a) What does this experiment tell you about the effect of temperature on the solubility of solid S? *(1 mark)*

b) How much of solid S has remained undissolved after 4 minutes in the experiment conducted at 40 °C? *(2 marks)*

c) Suggest a practical method for determining the weight of dissolved solid at a particular time (say 6 minutes) in the experiment at 30 °C. *(3 marks)*

43. Jonathon had just finished varnishing his new desk and painting his bedroom. He chose a blue matt plastic paint for the walls. He wet a cloth under the tap and used it to clean a few splashes of paint from the window pane. He washed the paint brush in a different solvent than that which he used to clean the varnish from his other brush. Jonathon chose the blue plastic paint because of its fast-drying properties and the ease with which the brushes can be cleaned.

a) What solvent is likely to be suitable for this paint? *(1 mark)*

b) The blue paint is a very fine suspension of tiny particles. What liquid is used to make this suspension? *(1 mark)*

c) Jonathon had to stir the paint before use. Why? *(1 mark)*

d) Suggest a suitable solvent for Jonathon's varnish brush. *(1 mark)*

44. Elise and Roberto decided to measure the salt content (salinity) of their local coastal river as part of a science project. Elise and Roberto live at Calinda, 50 km from the mouth of the river at Caledonia Bay. They took water samples at the first site and then each 10 km along the river from their home until they reached the bay. The collected samples were then taken back to the school laboratory for analysis.

a) How would you advise Elise and Roberto to collect the water samples? Give them some advice on control of variables. *(5 marks)*

b) A litre of water was collected at each site. Suggest a practical method that the students could use to find the total salt content in each of their samples. *(3 marks)*

c) Their results were tabulated as shown in Table T1.3.

Table T1.3 Salt content of river

Site	A	B	C	D	E
Salinity (mg/L)	8	300	600	3000	16000

i) Which site probably corresponded to Caledonia Bay? *(1 mark)*

ii) Which site probably corresponded to Calinda? *(1 mark)*

iii) Draw a bar graph of this data. *(4 marks)*

iv) Why is the salinity so high at sites D and E? *(1 mark)*

d) Farmers living near site B would like to be able to use the river water on their crops as well as for drinking, but found that it was too muddy. Can you suggest a practical, large-scale method that the farmers could use to help clear the water of the muddy particles? *(1 mark)*

45. A glass marble is allowed to fall vertically through a column of oil (see Figure T1.11).

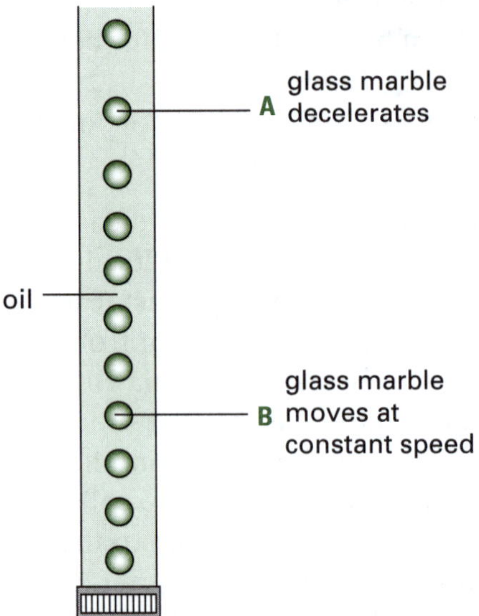

Figure T1.11 Glass marble experiment

a) What opposing forces act on the marble at point A? Copy the diagram and draw different arrows to show the relative sizes of the opposing forces. *(2 marks)*

b) At point B, the marble is falling at constant speed. How do the opposing forces compare? Copy the diagram and draw different arrows to show the relative sizes of these opposing forces. *(2 marks)*

46. Changes of state can be classified using an appropriate word from the list supplied. Select a word from the list to classify each change: condensation; evaporation; fusion; solidification.

a) Melissa hangs up her wet washing and allows it to dry in the sun. *(1 mark)*

b) Marcus places some solid beeswax in a test tube and allows it to melt into an oil as the tube is heated gently. *(1 mark)*

c) Marcia notices that steam from hot toast forms droplets of water on her cold plate. *(1 mark)*

d) Mervyn was annoyed when the water in his car's radiator turned to ice on a very cold winter night. *(1 mark)*

47. Words that are opposites of the correct word are hidden in the passage below. Identify these incorrect words and replace them with the correct antonym. *(7 marks)*

Carlo could feel the sweat dripping down his body as he played football for his school team. At half-time he warmed up with a cold drink and placed a damp cloth over his face so that condensation would supply some of the heat from his aching body. He was sure his feet must have contracted from the heat as they felt very loose in his boots. He put an ice block in his mouth and felt the hot sensation of solidifying ice down his throat. He was ready now to give the other team a real hiding in the second half.

48. In the 17th century, the Italian scientist Francesco Redi conducted experiments on the decomposition of food. At the time it was commonly believed that rotting meat spontaneously generated new life, often in the form of maggots. The hypothesis he wished to test was the following:

Maggots won't grow on meat if flies have no access.

Figure T1.12 shows the experiments he conducted with different types of meat, some of which were covered and some of which were not.

a) What was Redi's control? *(1 mark)*

b) Why did Redi choose to use a number of different types of meat? *(2 marks)*

c) Did the results confirm Redi's hypothesis? Explain. *(2 marks)*

d) Although no maggots were present, the meat still began to go rotten. What could have caused this? *(1 mark)*

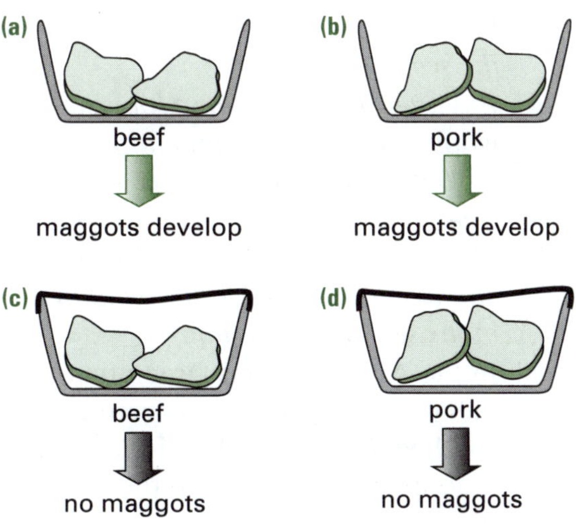

Figure T1.12 Redi's experiment

49. Use the key for sharks and rays in Figure T1.13 to identify each shark or ray described below.

a) This fish has two dorsal fins but the first one is shorter. A pelvic fin is also present. Its body is not kite shaped and its head is narrower than its body. *(1 mark)*

b) This fish has a kite-shaped body and no horned-shaped structures on its head. It has no dorsal fin. *(1 mark)*

50. A student set up a miniature ecosystem in a large fish tank. She added sand, various water plants, water snails, a few fish, brine shrimp and water fleas to it. Then she sealed the tank and left it out of direct

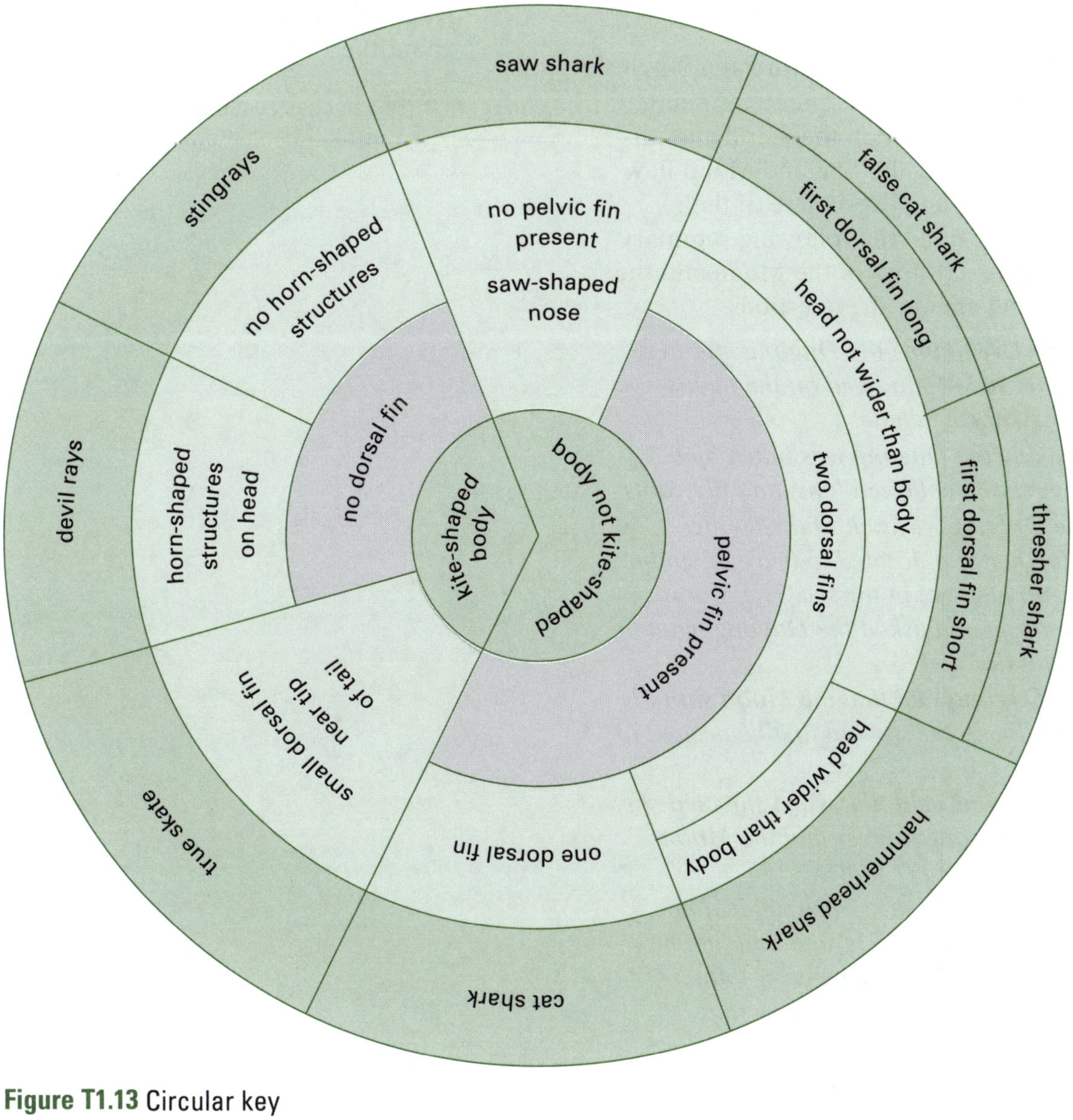

Figure T1.13 Circular key

sunlight. Several months later she found that her fish and plants looked very healthy.

a) Though the tank was sealed, the fish still needed oxygen to survive. Where did this oxygen come from? *(1 mark)*

b) What would the fish eat? *(1 mark)*

c) What happened to the wastes and dead plants at the bottom? *(1 mark)*

d) Where did the plants get the nutrients they need? *(1 mark)*

e) Where did the energy come from to keep this ecosystem going? *(1 mark)*

f) Her sister thought she could increase the number of fish in the tank without putting in any more plants or other organisms. Would this be wise? Explain. *(1 mark)*

51. In Australian Aboriginal culture, the Sun is seen as a woman and the Moon as a man. In the Dreamtime when the world and all on it were created, the Sun and Moon flew up into the sky and lived there in their present form. Read the following summary of an Aboriginal story of the Moon and the Dugong and answer the questions.

- *In the Dreamtime the Dugong and her brother the Moon lived on the plains near Arnhem Land.*
- *One day the Dugong was bitten by a leech as she collected lotus and lily roots.*
- *The Dugong was sick of the leeches biting her and decided to leave the plain and go and live in the sea. The Moon was upset and asked the Dugong what he should do.*
- *The Dugong said that the Moon should go to live in the sky but first he needed to die.*
- *The Dugong said that when she died she would be gone forever and the Moon could pick up her bones.*
- *The Moon said that when he died he would come back. He said that first he would get sick and grow very thin.*
- *He then said he would follow the Dugong down to the sea and then he would be so thin that he would only be bones and he would throw himself into the sea and die.*
- *But after three days the Moon said he would come alive again and gradually get better as he ate the lotus and lily roots.*
- *The Dugong agreed that this was a good plan and said that her brother should go and live in the sky but he should visit her in the sea once a month.*

a) In the Arnhem Land Dreamtime story, the Moon is described as getting sick and becoming very thin. Relate this description to a phase of the Moon. *(1 mark)*

b) Each month the Moon would return to the sea to visit his sister the Dugong. Relate this to the motion of the Moon about the Earth. *(1 mark)*

Go to pp. 214–216 to check your answers.

Course test 2

Go to p. v for *Tips for tests and examinations*

This test covers material from all the chapters in this book. It consists of the following.

- Section A: 25 multiple-choice questions *(1 mark for each)*
- Section B: 15 restricted-response questions *(1 mark for each)*
- Section C: 15 knowledge, skill and processing data questions *(total 60 marks)*

Total marks for this test: 100

Time: 2 hours

Part A: Multiple-choice questions

(25 marks—1 mark for each question)

1. In biological classification, the missing ranks in Figure T2.1 from top to bottom are:

kingdom

species

Figure T2.1 Classification

A phylum, class, order, family, genus.
B class, phylum, family, order, genus.
C genus, family, order, class, phylum.
D family, genus, phylum, class, order.

Use the food web in Figure T2.2 to answer Questions 2 and 3.

2. If the corn and oats were completely removed from this food web, which organism would be most affected?
A grasshoppers and crickets
B hawks and owls
C mice and rats
D raccoons and snakes

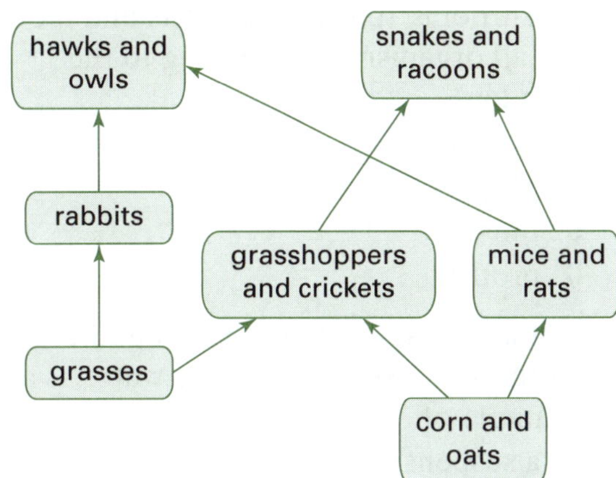

Figure T2.2 Food web

3. In Figure T2.2, there are a number of different producers as well as first- and second-order consumers. An example of a producer and a second-order consumer is
A grasses and crickets.
B corn and grasshoppers.
C oats and snakes.
D grasses and rabbits.

4. A container is filled with 100 mL of water and placed in a freezer. There it freezes at 0 °C. A second container contains only 90 mL of water and is placed in a second freezer. The temperature at which the water in the second container freezes is
A –10 °C.
B –1 °C.
C 0 °C.
D 10 °C.

5. On cold days your breath is often visible when you exhale. The process occurring to make this visible is
A evaporation.
B melting.
C boiling.
D condensation.

6. In which of the following situations is a chemical reaction occurring?
   A salt dissolving in water
   B a nail rusting
   C a bottle breaking
   D ice melting

7. Mushrooms, toadstools, yeasts and moulds are all organisms that belong to the kingdom
   A monera.
   B plants.
   C fungi.
   D protists.

8. Salt and water, when mixed together so all the salt dissolves into the water, is an example of
   A a suspension.
   B a solution.
   C an emulsion.
   D a colloid.

9. Which of these relationships among organisms is considered parasitic?
   A bees carrying pollen from the flower
   B ticks feeding on a dog
   C birds eating ticks from the back of a hippopotamus
   D a lion eating a deer it has just killed

10. Winnowing is an agricultural method developed by ancient cultures for separating grain from chaff. Threshed wheat was tossed in the air and winds would carry the chaff away. The wheat grain would fall to the ground (see Figure T2.3).

Figure T2.3 Winnowing

Separating wheat from the chaff relies on their difference in
   A weight.
   B colour.
   C size.
   D shape.

11. Which of the following is not a mixture?
   A hot chocolate made with milk
   B coffee made with hot water
   C a boiled sweet
   D a pressurised bottle of helium.

12. The Indian myna bird (see Figure T2.4) is an introduced pest. Its population has increased noticeably in the last few decades. Which of the following is a reasonable hypothesis for this population growth?

Figure T2.4 Myna bird

   A An organism that relies on similar food sources has migrated into the area.
   B Drought conditions have decreased water levels in local watering places.
   C The myna's main competitor has been eliminated, as the myna is a more aggressive bird.
   D Competition for food has increased among mynas.

13. Which kingdom contains organisms that are generally multicellular, have no chlorophyll and absorb nutrients from decaying tissue?
   A fungi
   B plants
   C protists
   D animals

14. One way to separate mixtures is by allowing a liquid to soak and move through another material containing the mixture. How the components of the mixture behave allows them to be separated. The technique described is

A filtration.
B distillation.
C chromatography.
D decanting.

15. Figure T2.5 shows a method of separating which substance?

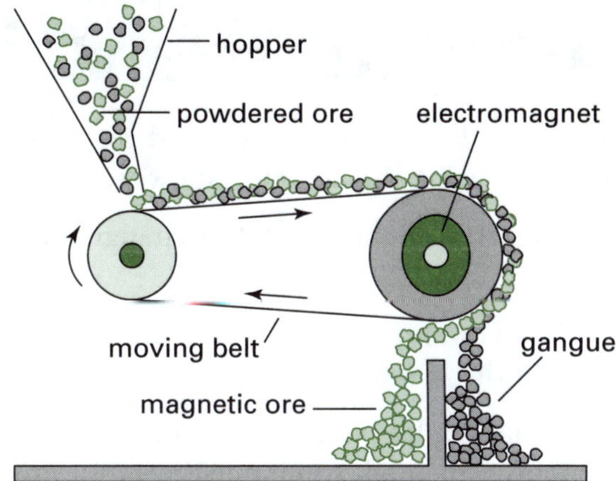

Figure T2.5 Separation

A iron
B copper
C aluminium
D zinc

16. Pouring water off after boiling potatoes is an example of

A decanting.
B filtering.
C sedimentation.
D evaporation.

17. Many introduced plants and animals often increase their numbers quite rapidly. Which of the following reasons could account for this?

i) The introduced species is resistant to pesticides.
ii) There is a large under-utilised food source in the new environment.
iii) The introduced species has few natural predators in its new environment.

A i) only
B ii) only
C ii) and iii) only
D i) and iii) only

18. Sea lions are known to eat salmon. The fishing industry wanted to determine what impact sea lions were having on the salmon populations. Which of the following studies would you deem the *least* urgent?

A study the sea lion population size
B study the sea lion stomach size
C study the percentage of salmon predators in the sea lion diet
D study the sea lion's daily energy requirements

19. Figure T2.6 shows the classification of matter. Four words were accidentally erased. These have been marked (i), (ii), (iii) and (iv). In order, they are

A heterogeneous, elements, impure substances, liquids.
B liquids, impure substances, elements, heterogeneous.
C liquids, elements, impure substances, heterogeneous.
D elements, liquids, heterogeneous, impure substances.

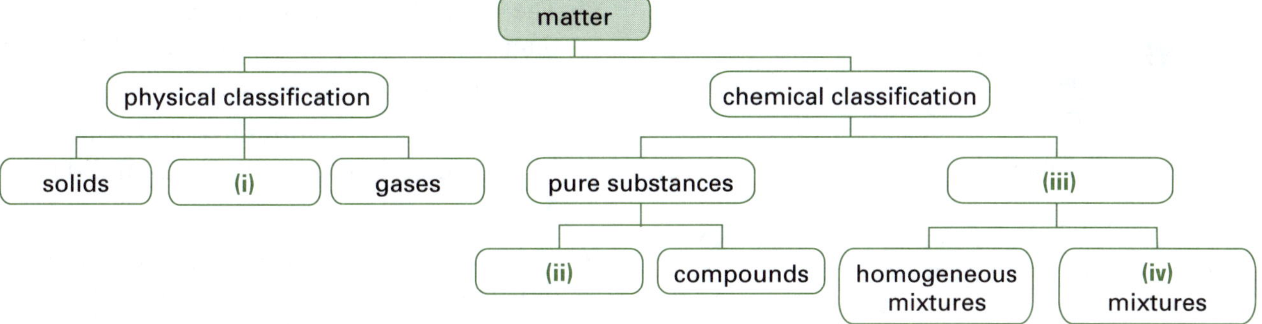

Figure T2.6 Matter classification

20. Which of these forces can be described as a contact force?
    - **A** gravitational force
    - **B** magnetic force
    - **C** electrostatic force
    - **D** frictional force

21. Which energy source is not renewable?
    - **A** oil
    - **B** solar
    - **C** wind
    - **D** tidal

22. As Earth's temperatures rise, due to the greenhouse effect
    - **A** the amount of water in the atmosphere will decrease.
    - **B** more islands will appear around the world.
    - **C** there will be adverse health effects on people.
    - **D** farmers will find that growing crops becomes easier.

23. Raindrops fall at a constant speed of 11 to 29 km/h in still air. Which of the following determines the speed of the raindrop?
    - **A** the height from which it falls
    - **B** the type of cloud that produced it
    - **C** the size of the raindrop
    - **D** the amount of rain that falls

24. In which position of Figure T2.7 is a full moon seen?

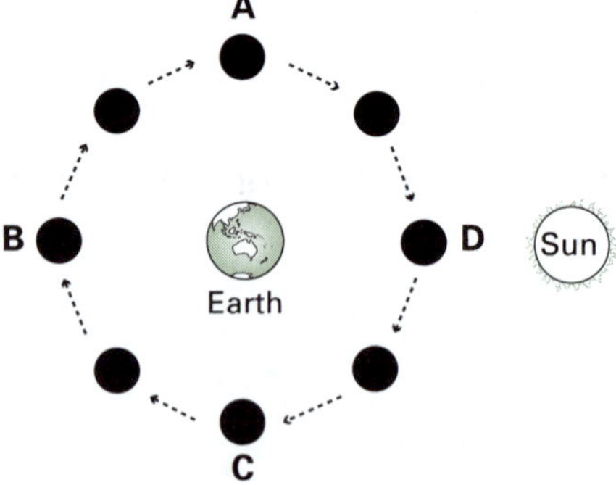

Figure T2.7 Full moon

25. Two forces, labelled F_1 and F_2, act on a box. Figure T2.8 shows a plan view (i.e. looking down from above the box). In which direction will the box move?

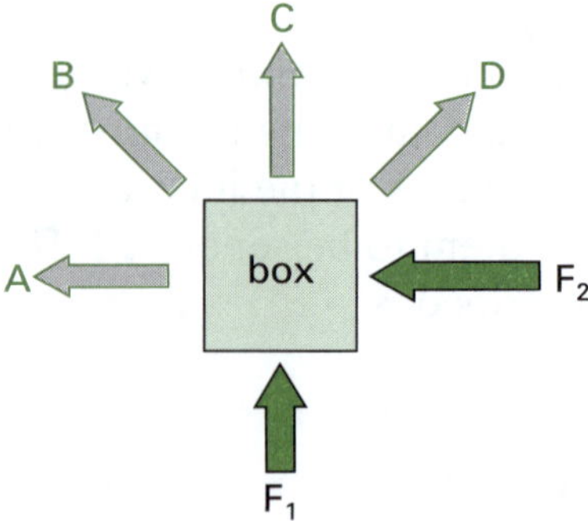

Figure T2.8 Where will the box move?

Part B: Restricted-response questions

(15 marks—1 mark for each question)

Complete the following restricted-response questions using the appropriate word.

26. Evaporation means matter changing from a liquid to a ....

27. Dew is water .... out of the air onto the leaves of plants.

28. In making lip balm, the cooked mixture is poured through some muslin (fabric) to .... it.

29. .... mixtures consist of two or more parts (phases), which have different compositions.

30. The constituents of any .... can be separated on the basis of their differences in their physical and chemical properties such as particle size, solubility or the effect of heat.

31. A .... organism is one that has escaped from domestication and returned, partly or wholly, to a wild state.

32. The Earth has a greater gravitational field than the Moon because the Earth has a greater ....

33. A .... eclipse happens when the Moon passes between the Earth and the Sun.

34. The .... occur because of the movement of the Earth around the Sun and the tilt of the Earth's axis.

35. Figure T2.9 shows a method of separation called ....

Figure T2.9 Separation

36. .... is the scientific study of the relation of living organisms with each other and their surroundings.

37. The calicivirus was introduced into Australia to help control .... plagues.

38. In the food chain shown in Figure T2.10, if all the snakes were killed this would cause an increase in the numbers of ....

Figure T2.10 Food chain

39. Figure T2.11 shows how water particles leave the surface of water during ....

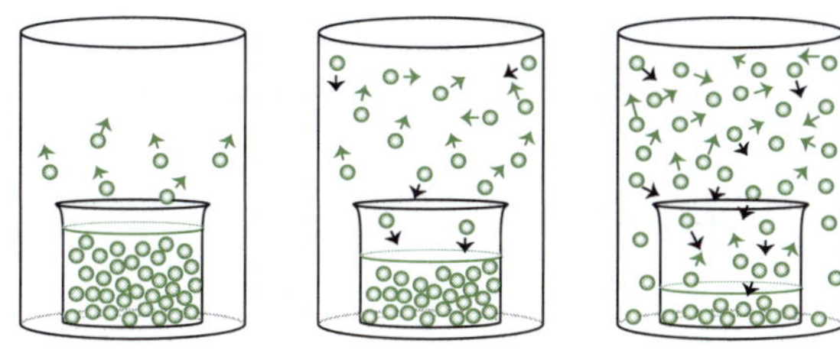

Figure T2.11 Water

40. A .... is an instrument, sometimes made by arranging lenses, mirrors or both, that is used to detect and observe distant objects.

Part C: Knowledge, skill and processing data questions

(60 marks)

41. Figure T2.12 shows the age of Australian mothers giving birth in a particular year.

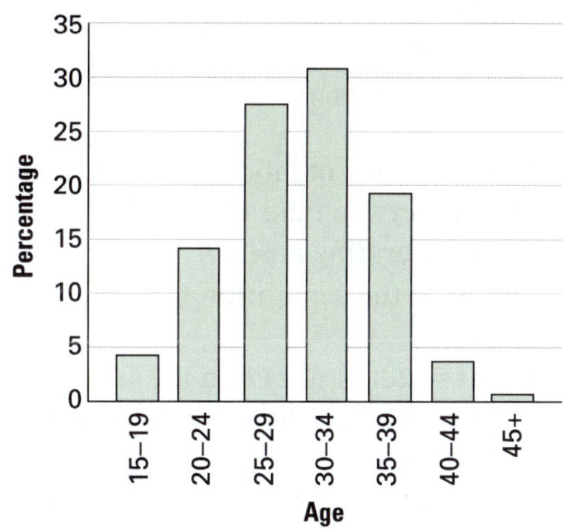

Figure T2.12 Age of Australian mothers giving birth

a) What percentage of babies was born to mothers aged 35–39? *(1 mark)*

b) Which two age groups accounts for around 60% of babies born? *(1 mark)*

c) The 25–29 year age group gave birth to 80 863 babies. Use this information to calculate the total number of babies born during the year. *(2 marks)*

d) Each group covers an age of 5 years. Explain why the last group is given as 45+. *(1 mark)*

42. a) What is sedimentation? Why is it useful in chemistry? *(2 marks)*

b) Adding a small quantity of alum can sometimes speed up the sedimentation process. What do you think the purpose of the alum is? *(1 mark)*

The following questions relate to Figure T2.13.

Figure T2.13 Evaporation

43. a) Name a piece of laboratory apparatus placed between the Bunsen burner and the evaporating basin. *(1 mark)*

b) Describe the separation method shown. *(1 mark)*

c) If salt water is placed in the basin, what will be left if all the water is boiled away? *(1 mark)*

d) If the water was to be recovered from the salt solution, what method of separation should be used? *(1 mark)*

44. a) Classification is important to scientists. Give a reason for this. *(1 mark)*

b) What is the meaning of the term *species*? *(1 mark)*

c) A donkey and a horse can mate to produce a mule, which is often sterile. Does this mean a donkey and a horse belong to the same species? Explain. *(1 mark)*

45. Figure T2.14 shows some buttons that need to be classified. One way to classify them is seen in Figure T2.15. This lists the five characteristics of the buttons. An example is shown for button F.

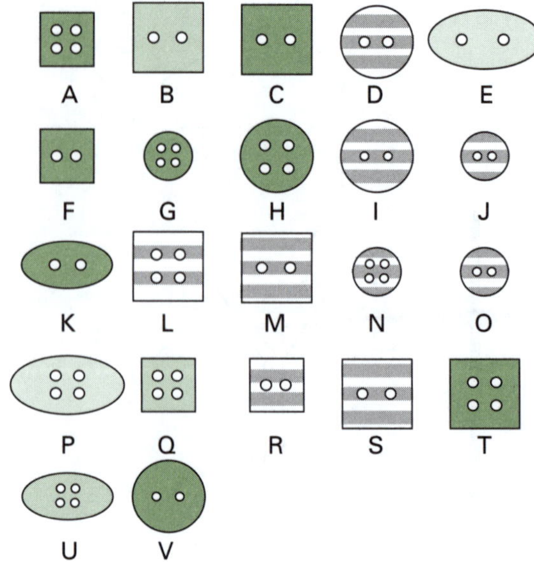

Figure T2.14 Buttons

a) Which button is described by the dashed line? *(1 mark)*

b) Copy the key in Figure T2.15 without the joining lines. Now draw lines on it (using different colours) to describe the following buttons.

i) E *(1 mark)*

ii) R *(1 mark)*

c) List two ways you can distinguish between buttons C and L? *(1 mark)*

d) You are told a button is small and square with four holes.

i) Which buttons fit this description? *(1 mark)*

ii) What other information do you need to distinguish between these buttons? *(1 mark)*

e) Which two pairs of buttons have all the same five characteristics? *(1 mark)*

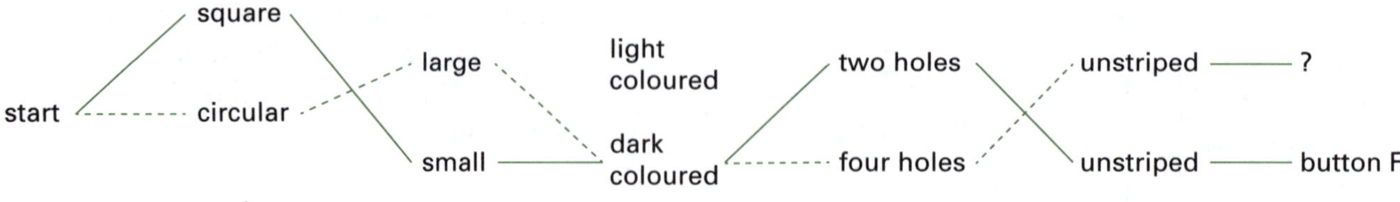

Figure T2.15 Button F

46. Figure T2.16 shows a man trying to move a rock using a long pole.

Figure T2.16 Lever

Identify the class of lever and explain why it works. *(2 marks)*

47. A crate has to be lifted so a system of pulleys is used as shown in Figure T2.17.

Figure T2.17 Pulley system

a) Does this system have a mechanical advantage? Explain. *(2 marks)*

b) Through how many metres does this man need to pull the rope to raise the crate 1 m off the ground? Explain. *(2 marks)*

48. A house had solar panels installed and kept a record of the percentage of electricity they could generate for their use. The results are shown in Figure T2.18. For how many months of the year could the panels provide for over 75% of the household's electrical needs? *(1 mark)*

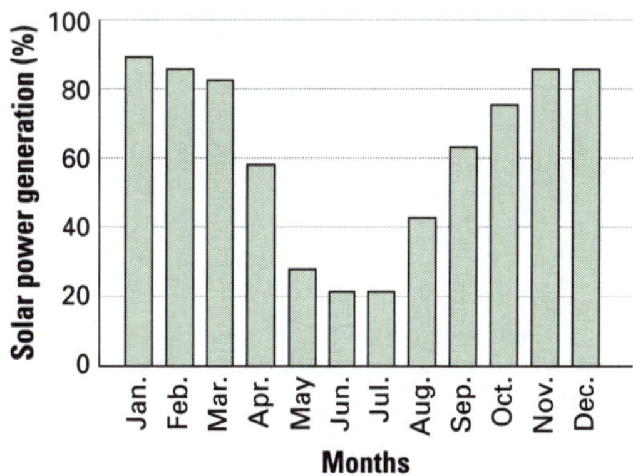

Figure T2.18 Percentage of energy generated by solar panels

49. The sector graph in Figure T2.19 shows the average water use in Australian households.

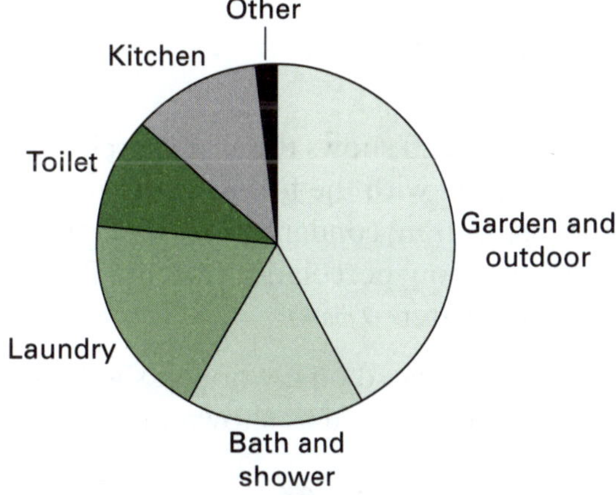

Figure T2.19 Average Australian water use

a) Which two areas use approximately the same amount of water? *(1 mark)*

b) Approximately what percentage of the household water use is for garden and outdoor? *(1 mark)*

c) The average household uses 300 kilolitres (thousand litres) of water each year. Approximately how many litres are used for the bath and shower? *(1 mark)*

d) Suggest one way the average household can conserve water. *(1 mark)*

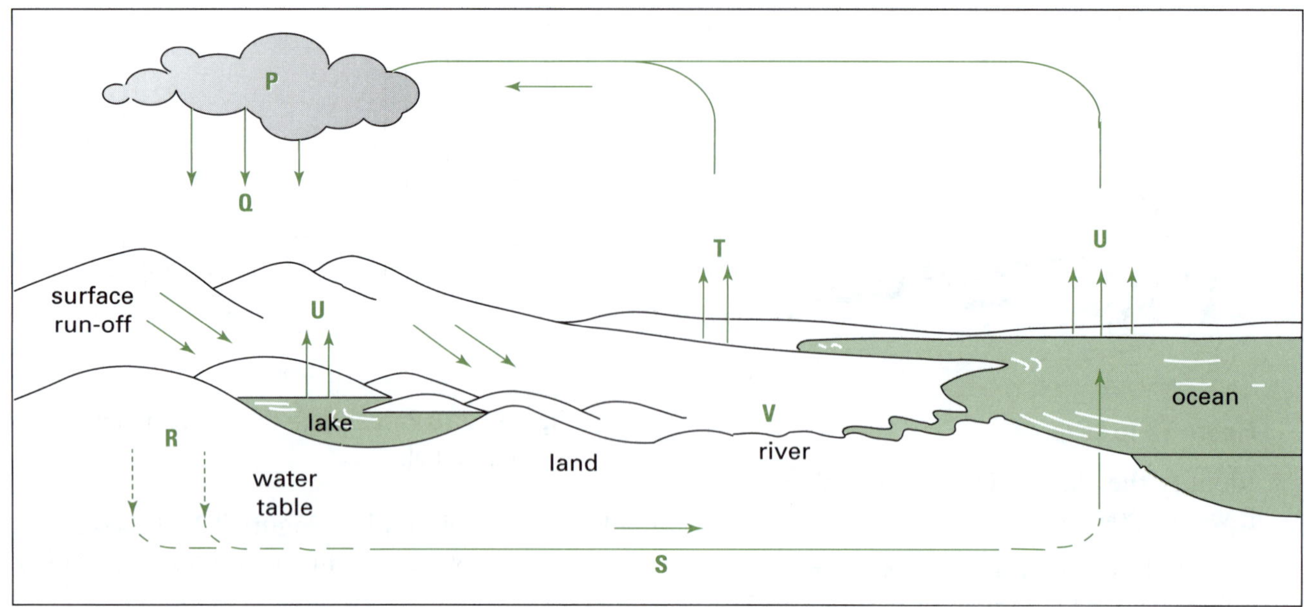

Figure T2.20 Water cycle

50. Figure T2.20 shows the water cycle. Replace the letters with the following words: transpiration; condensation; precipitation; evaporation; percolation; stream flow; and groundwater. *(7 marks)*

51. A box is placed on a wooden shelf. The shelf is then tilted as shown in Figure T2.21.

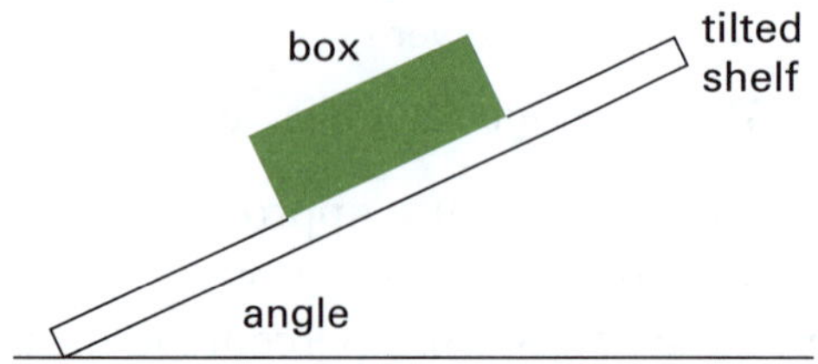

Figure T2.21 Tilted shelf

When the angle reaches 30°, the box begins to slide down the slope.

a) What was preventing the box from sliding before the angle reached 30°? *(1 mark)*

b) When the shelf is polished, the box begins to slide before the angle reaches 30°. Why is this? *(1 mark)*

c) Some people find that new shoes are slippery on polished floors so they rough up the soles. How does this help them walk better? *(1 mark)*

52. The Moon has no atmosphere. If a meteorite (Figure T2.22) crashed on the Moon nearby an astronaut would see it, but wouldn't hear it.

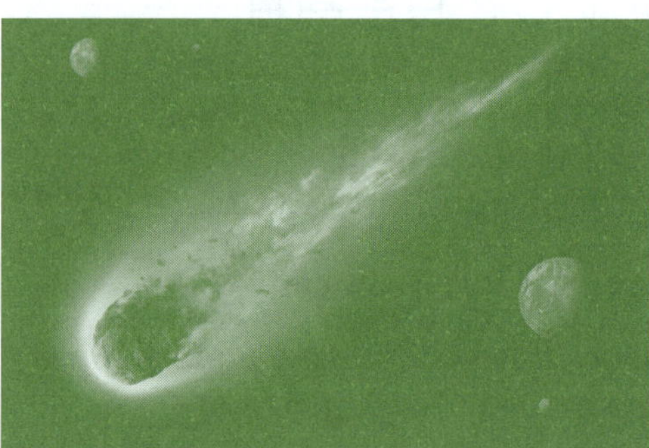

Figure T2.22 Meteorite

a) Explain why the astronaut sees the event but does not hear it. *(2 marks)*

b) Is there any other way the astronaut could detect the crashing meteorite? *(1 mark)*

53. A simple formula links mass and weight together:

weight = mass × acceleration due to gravity

Weight is a force and is measured in newtons. Mass is measured in kilograms. The acceleration due to gravity is 9.8 m/s^2.

a) A man has a mass of 72 kg. What is his weight on Earth? *(1 mark)*

b) On the Moon the acceleration due to gravity is only 1/6 as much as that on Earth. What would be the weight of the man on the Moon? *(2 marks)*

c) A 5-kg rock was brought back to Earth from the Moon. Which would change: its mass or its weight? *(1 mark)*

54. Figure T2.23 shows the location of an oil well. The three layers contain water, gas and oil.

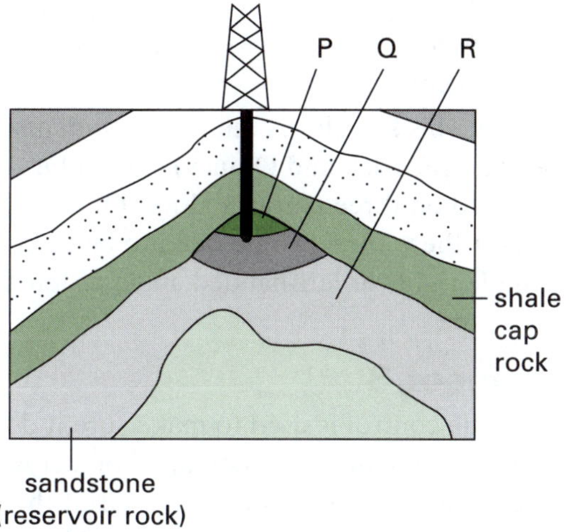

Figure T2.23 Oil well

Identify the material that is present in the three layers labelled P, Q and R. *(3 marks)*

55. Laboratory safety signs are important to remember. Figure T2.24 contains a jumbled set of haz-chem symbols. Match up each sign to its correct description and meaning. *(7 marks)*

Go to pp. 217–219 to check your answers.

Symbol	Description	Meaning
A 	**H** toxic	**O** can give off harmful radiation
B 	**I** oxidising agent	**P** can easily burn
C 	**J** harmful	**Q** poisonous
D 	**K** flammable	**R** can help fires get worse or spread
E 	**L** radioactive	**S** may explode
F 	**M** corrosive	**T** can affect health if breathed in or absorbed through the skin
G 	**N** explosive	**U** can burn skin and clothes

Figure T2.24 Jumbled haz-chem signs

Chapter 1 answers

Experiment 1

Analysis:

1. See Table A.1.

   Table A.1 Groupings based on hair colour

Black hair	Brown hair	Blonde hair
Mary, Raymond, Romeo	Catherine, Frederick, Penny, Quincy	Amy, Juliet, William

2. See Table A.2.

   Table A.2 Groupings based on ear lobes and eye colour

Free lobes/ blue eyes	Free lobes/ brown eyes	Fixed lobes/blue eyes	Fixed lobes/ brown eyes
Amy, Romeo	Catherine, Frederick, Quincy, William	Juliet, Raymond	Mary, Penny

Conclusion: The 10 students were able to be sorted into groups based on three characteristics. Additional characteristics would be required to sort all 10 students so that each could be individually identified.

Experiment 2

Analysis: There may be more than one answer. See Figure A.1 below.

Conclusion: This key shows that three students (Catherine, Frederick and Quincy) cannot be distinguished on three features alone. Other features for these three would need to be recorded; for example right- or left-handed, male or female, and eye shape.

Experiment 3

Analysis: The control is used to make sure it does not grow mould without coming into contact with the original mouldy fruit. The piece of fruit that was touched with the mouldy fruit should become infected and mould colonies should start to grow. The fungal and bacterial decomposers should cause decomposition of the fruit. The control will eventually decompose as microscopic decomposers are normally present in the air.

Conclusion: Mould spores can be transferred from an infected fruit to an uninfected fruit via contact.

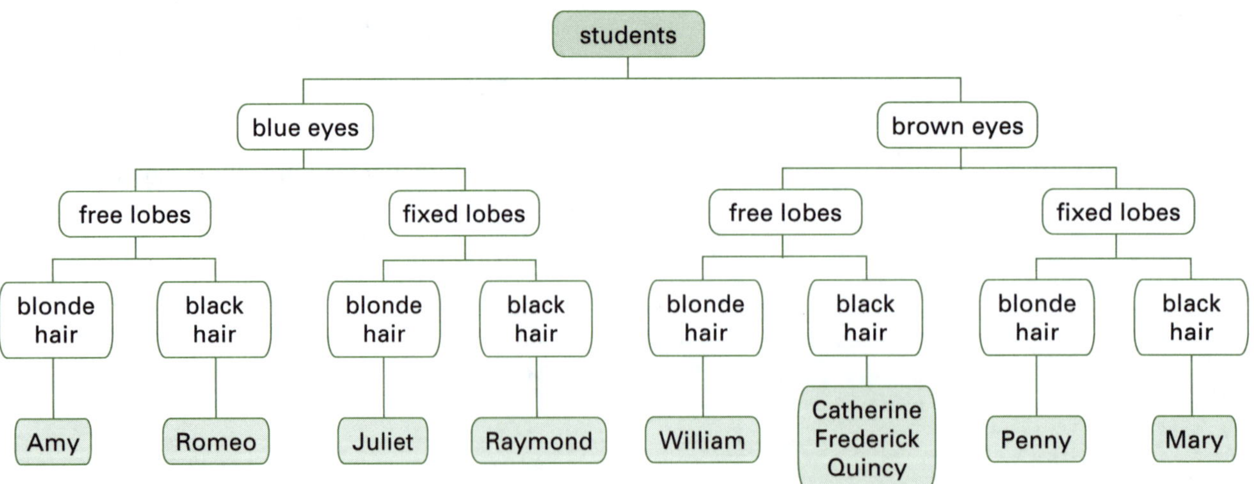

Figure A.1 Student flow chart

Test yourself 1

Part A: Knowledge

1. **B** ✓ Fungi are multicellular and have no chlorophyll. A is wrong as plants have chlorophyll; C is wrong as protists are unicellular; D is wrong as monera are unicellular.
2. **C** ✓ They are all types of fungus. A is wrong as monerans are bacteria; B is wrong as none of the named organisms are protists such as euglena or amoebas (this group contains protists); D is wrong as the named organisms do not have chlorophyll.
3. **A** ✓ They both have fur and mammary glands; B is wrong as birds have feathers and not fur; C is wrong as amphibians have slimy skin; D is wrong as reptiles have scaly skin
4. **C** ✓ They have dry scaly skin and lay eggs with soft shells; A is wrong as amphibians have slimy skin; B is wrong as fish have gills and goannas have lungs; D is wrong as dinosaurs are extinct.
5. **B** ✓ A wombat has a pouch; A is wrong as placentals suckle their young at the teat; C is wrong as a wombat does not lay eggs; D is wrong as vertebrates are not a mammalian group.
6. **a)** flow ✓; **b)** dichotomous ✓; **c)** gills ✓; **d)** brain ✓; **e)** soft ✓
7. A/I ✓; B/J ✓; C/F ✓; D/G ✓; E/H ✓

Part B: Skills

8. A = Camposaurus ✓; B = Iguanodon ✓; C = Triceratops ✓; D = Stegosaurus ✓; E = Nodosaurus ✓
9. **a)** A and C are more closely related as they both belong to the class Insecta; B is a mammal. ✓
   b) A ✓ and C ✓ are insects; B is a mammal ✓. (While you probably can't figure this out from the information alone, A is the Monarch butterfly, B is the domestic cat and C is the Colorado potato beetle.)
10. K = fly ✓; L = grasshopper ✓; M = ladybeetle ✓; N= dragonfly ✓
11. **a) i)** Any two of: has 6 legs; two antennae; three body parts; two pairs of wings; hatches from egg; compound eyes. ✓✓
    ii) Any two of: thin, hairless body; antennae have swellings on end; wings upright when at rest; wings usually colourful; most active during day. ✓✓
    iii) Any two of: wide furry body; antennae thick and feathery; wings horizontal when at rest; wings usually dull; most active at night. ✓✓
    b) Yes, because the description says wings are usually colourful. ✓ This does not mean that they always must be. ✓
    c) butterfly wings upright when at rest ✓
    d) clubbed antennae ✓
12. **a)** C and D ✓; **b)** D, E and F ✓; **c)** B ✓; **d)** A ✓; **e)** F ✓
13. bird ✓
14. Figure A.2 is one answer. Other keys are possible.

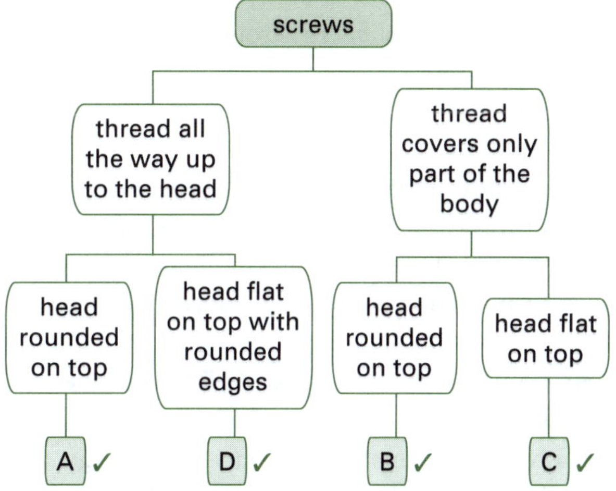

Figure A.2 Screw flow chart

Test yourself 2

Part A: Knowledge

1. **B** ✓ A herbivore is a plant-eater. A is wrong as producers are plants; C and D are wrong as second- and third-order consumers are carnivores.
2. **B** ✓ Whether shown or not, a decomposer is at the end of the food chain to ultimately return all nutrients to the environment. A is wrong as producers start a food chain; C and D are wrong as they are found after the producer.
3. **A** ✓ The main source of energy entering a food chain is from the sun. B is wrong as plants absorb energy from the sun; C is wrong as mineral salts are needed for life functions but are not a source of energy; D is wrong as

fats and carbohydrates store energy but are secondary sources of energy.

4. **A** ✓ The other animals were all introduced into Australia. B, C and D are not native animals.
5. **D** ✓ It doesn't matter who lights the fire, the effect is still the same. A is wrong as plant adaptations have a large effect on the outcome of a fire; B is wrong because the final outcome will depend on how much vegetation is burnt; C is wrong as large fires will kill everything whereas small fires will not.
6. **a)** ecology ✓; **b)** population ✓; **c)** community **d)** ecotourism ✓ (short for ecological tourism); **e)** cane toad ✓
7. A/I ✓; B/J ✓; C/H ✓; D/F ✓; E/G ✓

Part B: Skills

8. e ✓; c ✓; f ✓; a ✓; d ✓; b ✓
9. **a)** the transfer of energy from one organism to another ✓
   b) The plant (and its seeds) obtains energy from the sun. ✓
   c) the flow of energy ✓ (another way to think of it is to say the arrow means 'is eaten by')
   d) A food chain is one line of events. A food web is made up of many interconnected food chains. ✓
10. **a)** **i)** bulrush ✓; **ii)** muskrat ✓; **iii)** muskrat ✓; **iv)** owl ✓
    b) No, only a small part of the sun's energy is ever used by plants in the production of food. ✓ Of the energy-rich material stored in the plant, the plant itself uses some for its life processes. ✓ Not all of the material eaten by the herbivore is used either. ✓ For example, undigested food material passes out ✓ of the animal's body as waste and is utilised by decomposers.
    c) It is assumed that the sun is part of every food chain. ✓ Also, a food chain shows the energy links between living organisms. ✓
11. Sunlight is reflected off the silvery leaves keeping the leaves cool. ✓ Hanging vertically, the leaves are not in the path of much of the sun's rays ✓ and this also helps in preventing moisture loss. ✓
12. **a)** grasshopper ✓; **b)** crab ✓
13. **a)** A desert environment. ✓ Soils are not very fertile and rainfall is rare.
    b) Grassy plains. ✓ The red kangaroo needs grasses and similar plants to feed off and open areas for it to move through quickly. These plains can become quite hot during the day, so feeding at night is preferred.
    c) Dry environments. ✓ The bilby is suited to dry environments such as arid, semi-arid and relatively fertile areas. It makes its home in a burrow that spirals down, making it hard for its predators to get in.
    d) Dry areas with plenty of sunshine. ✓ Spinifex survives on the poorest, most arid soils in Australia. It is spinifex that has prevented our deserts from becoming a world of bare, shifting sand.
14. **a)** The food pyramid shows the loss of energy along a food chain. ✓
    b) Each feeding level is smaller than the preceding level. ✓ There must be a greater mass ✓ (not necessarily greater in numbers) of plants (producers) than plant eaters (herbivores), and more plant eaters than meat eaters (carnivores). The amount of energy available to the next higher level is only a fraction ✓ of the energy of the previous level.

Chapter test

Part A: Multiple-choice questions

1. **A** ✓ All these animals are herbivores and require large molar teeth to chew the grass; B is wrong as none of these animals are marsupials; C is wrong as all these animals are warm blooded; D is wrong as all the animals are herbivores.
2. **B** ✓ Stella, because they have four features in common. Zoe (A) has only two features in common; Linh (C) has only one feature in common; Lauren (D) has three features in common with Julia.
3. **B** ✓ Eye colour, because group 1 do not have brown eyes. A is wrong as Julia and Stella have different hair colour; C is wrong as Linh and Emily have different skin colours; D is wrong as Zoe has lobed ears but Linh does not.
4. **D** ✓ Lobed ears, because you cannot change this without plastic surgery. A, B and C list characteristics that are easy to change.
5. **C** ✓ Goannas are lizards that have four legs and dry, scaly skin. A is wrong as frogs have wet,

slimy skins; B is wrong as snakes have no legs; D is wrong as sharks have no legs.

6. **D** ✓ The snail moves on a large muscular foot. The octopus has many smaller muscular feet that are formed like suction caps. A is wrong as the octopus has no shell; B is wrong as these animals are invertebrates; C is wrong as these animals are herbivores.

7. **A** ✓ Plants are producers as they absorb sunlight to make food; the remaining animals are all consumers.

8. **B** ✓ All these animals eat grass. A is wrong as sparrows and frogs are not first-order; C is wrong as frogs and snakes are not first-order; D is wrong as frogs are second-order consumers.

9. **C** ✓ Mice eat grass and seeds. A, B and D are wrong as mice eat plants and so are herbivores.

10. **A** ✓ The people in these countries are poor and need income from planting palms to extract the oil. B is wrong as no large cities are being built; C is wrong as mosquitoes are controlled by chemicals; D is wrong as oil and gas wells take up very little land.

Part B: Short-answer questions

11. **a)** mammals ✓ **b)** order ✓ **c)** kingdom ✓ **d)** phylum ✓ **e)** plant ✓

12. A/H ✓; B/F ✓; C/I ✓; D/J ✓; E/G ✓

13. flat-head wood screw (choices are; 1B; 7A; 8A) ✓

14. set screw (choices are: 1A; 2B; 4B) ✓

15. A = air ✓; B = land ✓; C = water ✓

16.
- They live in places that are hard to get to. ✓
- They are found only in a small area. ✓
- There are not many of these trees growing in that area. ✓
- They look like a species that has already been discovered. ✓

17. See Figure A.3. *(1 mark for each choice level correctly drawn)*

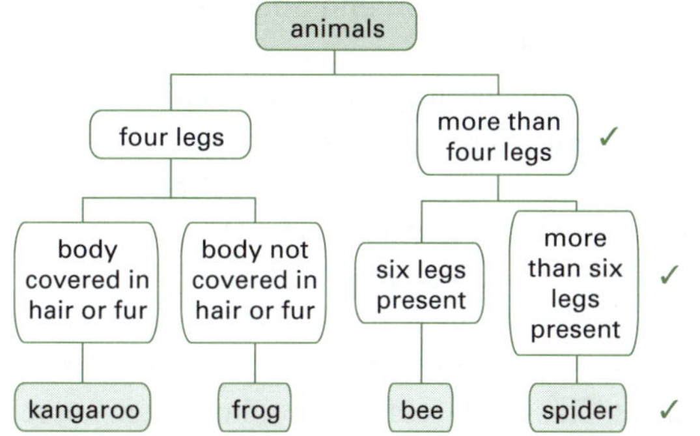

Figure A.3 Animal flow chart

18. See Table A.3. *(1 mark for each correct column)*

Table A.3 Similarities and differences between the platypus and other placental mammals

Differences ✓	Similarities ✓
A platypus lays eggs.	The body of a platypus is like an otter's.
Mammary glands have no teat.	The tail of a platypus is like a beaver's.
A platypus has a beak like a duck.	The body of a platypus is covered with fur.
A platypus has no teeth.	Mammary glands are present.
Milk trickles onto a skin patch.	

19. **a)** Katya has brown or blonde hair, ears with lobes and is right-handed. ✓
    b) Lily has black hair, blue eyes and glasses. ✓
    c) Ivy has black hair whereas Lauren's is blonde or brown. ✓
    d) Wearing glasses, and hair curly or straight—these characteristics are readily changed. ✓

20. **a)** \$4.40/pair (choices are: 1B; 3B; 7A; 10A) ✓
    b) The most expensive eyes are the black, leather, shoe-button eyes in the size range 7 mm–12 mm (choice 11b). ✓

21. The word backbone suggests one long bone is present. In fact the back is made of many smaller vertebra linked together to form a vertebral column. ✓

22. Plants have chlorophyll in their leaves ✓; plants have cellulose cell walls ✓.

23. Class = Mammal ✓ Order = Primate ✓
    Any two of: chimpanzee; gorilla; orangutan; gibbon; monkey; lemur; macaque. ✓✓

24. **a)** *Eucalyptus* = genus ✓; *citriodora* = species ✓
    b) *Eucalyptus alba* ✓ because the same genus name makes them more closely related ✓

25. humidity D ✓, Q ✓; wind B ✓, S ✓; rainfall A✓, R ✓; temperature C ✓, P ✓

26. All scientists around the world speak in different languages but the scientific name never changes. ✓

27. Statement N is correct. ✓ Many of the nutrients they need are in the upper layers of the soil. They can also access a greater area to soak up rainwater. In dry forests little rain falls and often doesn't soak deeply into the soil.

By having shallow roots covering a large area, these trees can take advantage to collect as much moisture as possible. ✓

28. Statement E is correct. ✓ Without deep roots there is not much anchorage and trees can fall easily. This is especially true for large trees and with strong winds. ✓ (Note: spinifex grasses also have roots spread widely, but they are usually less than a metre tall. Spinifex roots go down a long way, approximately 3 m. They need this in the desert sands to anchor themselves against shifting sands.)

29. Statement G is correct. ✓ To protect the seeds from predators while it has a long wait before the rains come. A tough protective case makes it difficult, though not impossible, for animals to eat the enclosed seeds. It also protects the seed from wind abrasion and from drying out. This gives the seed a greater chance of surviving until the rains come. ✓

30. Statement M is correct. ✓ Without fire, many of the hard seed casings do not break to allow the seeds to germinate. Consequently the number of trees relying on fire to begin their life cycle would dwindle. The woody fruit of some plants protects the seed from desiccation and from fires. Some of these plants need fire to open the fruiting bodies and release the seed into nutrient-rich ash. ✓

31. **a)** grass → mouse → snake → hawk ✓

b) algae → tadpole → long-necked tortoise → tiger snake ✓

c) cabbage leaf → snail → blue-tongue lizard → cat ✓

32. **a)** seal or killer whale✓; **b)** small fish ✓; **c)** small fish ✓; **d)** phytoplankton ✓

33. In a food chain, energy passes from one link to another. When a herbivore eats plants, only a fraction of the energy becomes new body mass. The rest of the energy is lost as waste or used up by the herbivore to carry out its life processes (such as movement, digestion and reproduction). ✓ So when a carnivore eats a herbivore, only a small amount of total original energy passes to the carnivore. Of the energy transferred from the herbivore to the carnivore, some energy will be lost or 'used up' by the carnivore. Therefore, the carnivore then has to eat many herbivores to get enough energy to grow. ✓

34. Decomposers break down dead plants and animals. When plants and animals die, they become food for decomposers like bacteria, fungi and earthworms. ✓

Decomposers:

- recycle dead plants and animals into chemical nutrients such as carbon and nitrogen that are released back into the soil, air and water
- break down the waste (faeces) of other organisms
- are very important for any ecosystem because if they weren't there, the plants would not get essential nutrients, and dead matter and waste would build up. ✓

35. **a)** The number and variety of organisms (plants and animals) found within a specified geographic region, such as a forest; the variety of life forms within a given ecosystem ('bio' = life, 'diversity' = difference). ✓

b) There are many different kinds of animals and plants living side by side, or on top of each other, in a forest. ✓ In a rainforest, many of these plants and animals make their homes in or near the treetops and rarely come to the ground. While a field of wheat or corn contains many plants, they are all of the same kind. There is no variety. (Biologists call this a monoculture.) ✓

36. Putting back vegetation that has been removed in past decades:

- helps to restore the environment to the condition it was before ✓
- will take some time as trees and shrubs grow back slowly and animals adopt these as their new homes. ✓

In order to determine how effective this re-vegetation is, detailed data is needed to show:

- where the actions are taking place
- which vegetation and animals are being targeted
- how the actions affect the relationships of organisms in habitats
- how populations are responding
- how changed water flow may be restored in some locations. ✓

The best strategy is to encourage farmers and others to preserve remaining forests on their lands. This costs less than replanting in some landscapes. ✓

37. Forest fragmentation occurs when large, continuous forests are divided into smaller blocks by roads, clearing for agriculture, urbanisation or other human development. Fragmentation can:
   - harm many forest animals by making them easy prey for predators ✓
   - allow more predators to gain access to the woodland ✓
   - produce smaller or more isolated forest patches that will lose species faster than those that are larger or less isolated ✓
   - lead to habitat destruction and the inability of individual forest fragments to support viable populations, especially of large vertebrates. For example, migration paths of organisms may be disrupted causing them to move over open areas where they may more easily be taken by predators, or having to cross roads, thereby contributing to road kill. This is true of koalas which need to move from one stand of eucalypts to another in search of food or a mate. ✓

38. The Murray-Darling Basin river system forms the largest and most developed river system in Australia. The changes faced by people living in the region as they adapt to the physical limits of water availability are challenging. This is continually being researched by scientists to find better ways of managing the finite volumes of water in this fragile ecosystem. Issues of concern include:
   - securing and maintaining water entitlements and allocations for irrigation of agriculture and other activities ✓
   - the health of important wetlands, floodplains and the estuary ✓
   - the quality of water and keeping salt levels low ✓
   - the jobs and livelihoods of local people. ✓

39. a) An animal that was once domesticated but has reverted to the wild state. There are many feral animals in Australia: pigs, camels, cats, horses, goats and foxes. ✓

   b) plenty of food, able to compete successfully against native animals, good climate, suitable terrain, few or no predators ✓

   c) the hot top end of Australia ✓

   d) Conventional control measures require a lot of people to organise on a continuous basis. With hundreds of millions of rabbits, this was an impossible task. ✓

   e) Species specific means that it will affect only the organism it is aimed at. ✓ If myxomatosis affected other animals besides rabbits this could cause more problems than it solves. ✓

40. a) No. It was thought it would control beetles affecting sugar cane (but it didn't). Not much thought was given at the time that it could affect other species. ✓

   b) Cane toads eat almost anything they can swallow, including pet food, carrion and household scraps, but most of their food is living insects. ✓ They can poison larger animals that threaten them or try to eat them. ✓

   c) Wear gloves and goggles to protect you from the poison. Preferably scoop them up or use tongs to place them in large containers, such as plastic garbage bins, or use toad traps. ✓ Physical removal can never stop the cane toad invasion. This is because you have to take them out faster than they can replace themselves. That isn't too hard if you are dealing with an animal that reproduces slowly, but unfortunately cane toads reproduce at an amazing rate. They can reach maturity in just a few months. And, of course, toads can be found in hard-to-reach places or where there are no people to catch them. ✓

   d) It is easy to confuse cane toads with harmless native frog species, so it is best to have someone who knows what to look for examine them first. ✓

   e) Biological control is a method of controlling pests that relies on predators, parasites, diseases or other natural mechanisms. It can be an important component of integrated pest management programs. However, selecting the appropriate biological control agent is a difficult task as it must be specific to the organism you are trying to control and not spread to other organisms. ✓

Chapter 2 answers

Experiment 1

Analysis:

Part A: The mixture becomes more well-mixed with grinding but it still remains heterogeneous under the high power of the microscope.

Part B: This rock is heterogeneous even when viewed with the naked eye.

Part C: These results confirm that a true solution is homogeneous.

Conclusion: In this experiment we found that a solution is homogenous and that mixtures of solids or crystals are heterogeneous.

Experiment 2

Conclusion: Oily substances tend to dissolve in kerosene whereas salt and baking soda do not. They dissolve in water. Oily substances do not dissolve in water.

Experiment 3

Analysis: The chalk is insoluble in water whereas the copper sulfate is soluble. The chalk forms the residue in the filter paper. The copper sulfate solution forms the filtrate. This filtrate can be evaporated in an evaporating basin to obtain the copper sulfate crystals.

Conclusion: A solid mixture of chalk and copper sulfate were separated by filtration as the chalk was insoluble in water and the copper sulfate was soluble.

Test yourself 1

Part A: Knowledge

1. **D** ✓ Methylated spirits is a solution of methanol in ethanol. A is wrong as methylated spirits is a mixture; B is wrong as it is a homogeneous mixture; C is wrong as the methanol is the solute.
2. **A** ✓ Carbon dioxide contains two oxygen atoms and one carbon atom; it has a fixed composition. B and C are wrong as carbon dioxide is not a mixture; D is wrong as the gas is not dissolved in any solvent.
3. **B** ✓ Dilute solutions contain a large amount of solvent and not very much solute. A is wrong as this describes a concentrated solution; C is wrong as the solvent is always in the greater amount; D is wrong as a solvent is part of a solution.
4. **C** ✓ Milk is not clear and so is not a solution (D) or a homogeneous mixture (B). A is wrong as milk is a mixture and not a pure substance.
5. **A** ✓ Various separation methods can be used to separate mixtures. B is wrong as mixtures can be separated physically; C is wrong as mixtures may only contain two components; D is wrong as solid–solid solutions are not clear.
6. **a)** constant (or fixed) ✓; **b)** same ✓; **c)** variable ✓; **d)** pure ✓; **e)** magnet ✓
7. A/I ✓; B/F ✓; C/J ✓; D/H ✓; E/G ✓

Part B: Skills

8. Yes, tap water is a mixture. ✓ Tap water is produced from water that has collected in dams. This water is treated in a variety of ways to ensure it is fit to drink. Apart from water, tap water contains some dissolved air as well as small amounts of chlorine (used to kill dangerous microbes) and small amounts of fluoride for our teeth. ✓
9. Substances A, B, D and E are all considered pure substances because there is only one type of particle present. ✓ Substances C and F are mixtures as there is more than one type of particle in each. ✓
10. The answer is C. ✓ Each particle in diagram C is two different types of atoms and they are in the correct ratio of 1:2. ✓
11. All substances ✓ are pure as they have a constant composition. ✓
12. Convert the information for solution Y to the same volume of water. Thus, 2 g of salt dissolved in 20 mL of water is the same as $2 \times 5 = 10$ g of salt dissolved in $20 \times 5 = 100$ mL of water. ✓ Therefore, solution Y is more concentrated as 100 mL of that solution contains 10 g of salt whereas solution X contains only 5 g of salt. ✓
13. **a)** liquid ✓; **b)** solid ✓; **c)** liquid–solid ✓
14. **a)** oil ✓; **b)** unleaded petrol ✓; **c)** mower fuel ✓

Test yourself 2

Part A: Knowledge

1. **B** ✓ After sedimentation the supernatant liquid above is poured off to reduce the volume of liquid to be filtered. A is wrong as a sieve is

used to collect solids; C is wrong as the solids must be allowed to settle; D is wrong as this process happens before the mixture is decanted.

2. A ✓ The solid salt remains as it has a high boiling point. B is wrong as salt is soluble; C is wrong as density is irrelevant; D is wrong as salt is a pure substance and not relevant.
3. D ✓ The liquid that condenses after distillation is called the distillate. A is wrong as the solute is the sugar component of the original solution; B is wrong as a filtrate is the liquid that passes through a filter paper; C is strong as the solution was the sugar solution.
4. D ✓ Red cells are much heavier. A is wrong as this is the liquid part of the blood; B is wrong as these cells are not as numerous as red cells and not as dense; C is wrong as these are fewer in number and lighter.
5. A ✓ The oil combines with the steam and is removed from the leaves. B is wrong as this method is used to separate minerals from rocks; C is wrong as this is used to collect crystals after evaporation of a solvent; D is wrong as this method is used to separate a mixture of soluble substances.
6. a) non-renewable ✓; b) smoke ✓; c) filtered ✓; d) centrifugation ✓; e) residue ✓
7. A/G ✓; B/H ✓; C/I ✓; D /J ✓; E/F ✓

Part B: Skills

8. The liquid is poured down the glass rod which acts as a guide and prevents splashing. ✓
9. One use is the production of table salt from sea water in large flat pans (e.g. in South Australia). ✓
10. X will form the distillate ✓ as it has the lower boiling point and vapourises more readily. ✓
11. a) Yellow dye ✓ travels down the column very slowly and so it is the least soluble dye.
    b) Blue dye ✓ travels down the column the fastest and so it is the most soluble dye.
12. a) The literal meaning is 'flee the centre'. ✓
    b) Yes. In the centrifuge, the suspended material is pushed to the end of the tube, which is the point furthest away from the centre. ✓
13. The crushing of the large lumps releases the metallic minerals from the rocky minerals that surround them. Once freed, the metallic minerals can adhere to the air bubbles in the froth flotation tank. ✓
14. volume of pure alcohol $= 12 \div 100 \times 750$
    $= 90$ mL ✓

Chapter test

Part A: Multiple-choice questions

1. B ✓ Solutions cannot be filtered or centrifuged. A is wrong as dissolved substances pass through a filter; C is wrong as gases cannot be separated in this way; D is wrong as solutions cannot be centrifuged.
2. C ✓ Milk is a mixture of water, fats, proteins and other nutrients. A and B are wrong as they are pure elements; D is wrong as this is a pure compound.
3. A ✓ Fruit pulp is suspended in the water. B is wrong as it is a mixture; C is wrong as it is a suspension; D is wrong as only the pulp sediments.
4. D ✓ Solutes dissolve in solvents to produce solutions. A is wrong as the liquid is the solvent; B is wrong as there is no sediment; C is wrong as the ink is the solution.
5. C ✓ This is the correct reading from the graph. A, B and D are all incorrect readings.
6. B ✓ Alcohol has a lower boiling point than water. It is more volatile and will escape. A is wrong as the wine would just change composition; C is wrong as there is no such thing as grape crystals; D is wrong as removing the alcohol would make it more dilute.
7. B ✓ There are numerous additives in tap water (e.g. chlorine). A, C and D are all pure substances.
8. B ✓ The zinc is present in the smaller amount. A is wrong as the brass that forms is the solution; C is wrong as the copper is the solvent; D is wrong as there are no suspended particles.
9. C ✓ The mineral sticks to the bubbles which rise to form froth. A is wrong as lead is not magnetic; B is wrong as two solids cannot be separated this way; D is wrong as the boiling points are far too high.
10. D ✓ The salt is left behind and fresh water is the distillate. A is wrong as the materials are not magnetic; B is wrong as a solution will not separate on centrifugation; C is wrong as this method is used to obtain minerals from rock.

Part B: Short-answer questions

11. **a)** suspension ✓; **b)** solute ✓; **c)** centrifugation **d)** distillation ✓; **e)** transparent ✓
12. A/I ✓; B/F ✓; C/H ✓; D/J ✓; E/G ✓
13. **a)** sediment ✓; **b)** solution ✓
14. Showing a sediment ✓; showing particles correctly located. ✓ See Figure A.4.

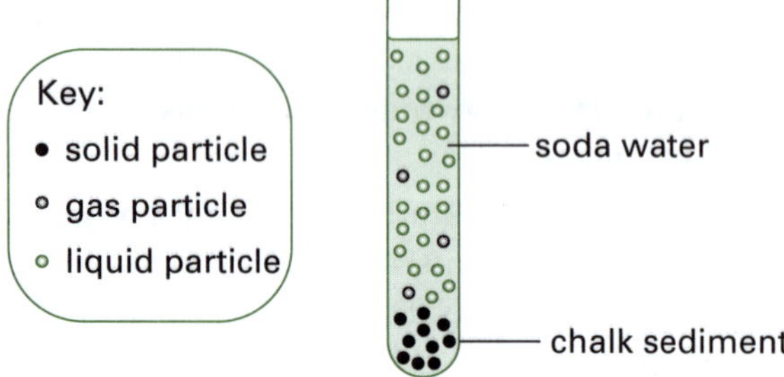

Figure A.4 Sediment and particles

15. The force acting on the wet clothes during the spinning causes the water to be flung outwards to the sides of the tub where it can be removed. ✓
16. **a)** Designs will vary: showing the positions of the bore water bowls ✓; showing the position of the bowl that collects the distilled water ✓; showing how the perspex sheets are positioned and clamped ✓. See Figure A.5 in the next column.

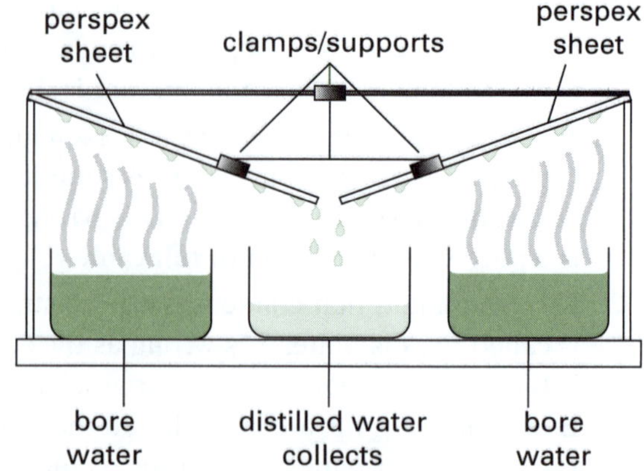

Figure A.5 Solar distillation plant

b) The solar rays pass though the perspex and warm the water which evaporates. ✓ The water vapour condenses on the underside of the perspex. The water droplets roll down the sloping sheet into the collection bowl. ✓

17. **a)** method in correct order: 4, 6, 7, 5, 2, 1, 8, 3 ✓
    b) See Figure A.6. Flask correctly clamped ✓; air condenser correctly placed ✓; ice bath ✓.
    c) The smell will be much stronger than in the flowers as the chemicals have been concentrated. ✓
    d) The ice bath helps to condense the vapours as the air condenser is not as efficient as the water condenser. ✓
18. **a)** The breezes in the air blow away the chaff. ✓
    b) grain separated from chaff ✓

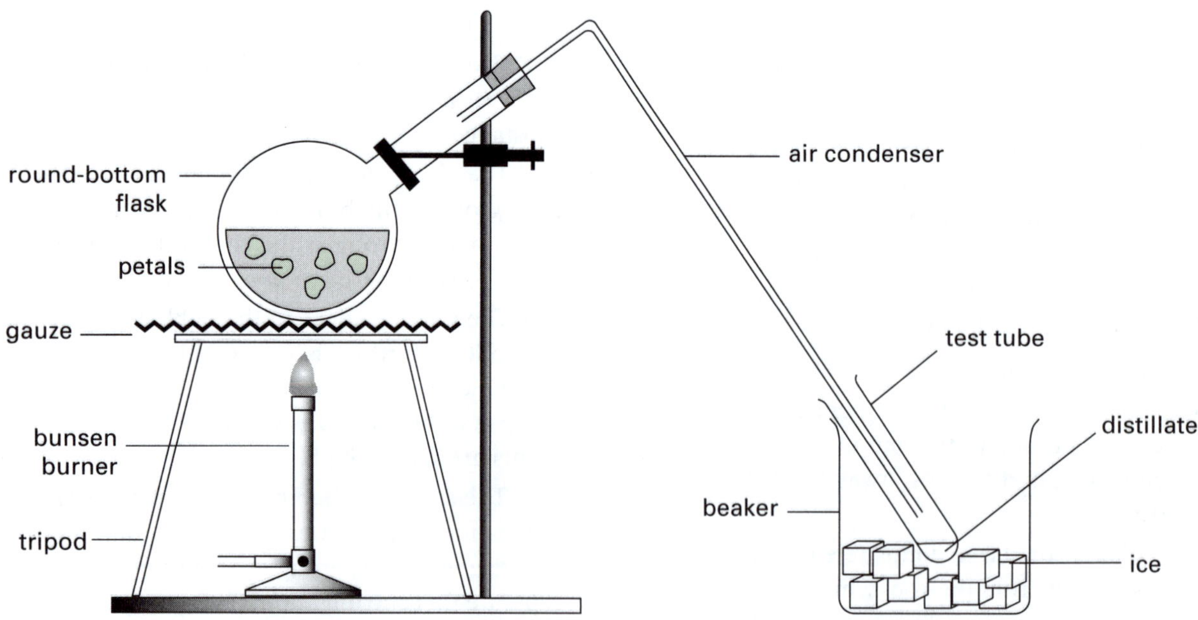

Figure A.6 Equipment used to extract perfume

19. a) i) Tube 1: The X component of the mixture dissolves in the petrol and the Y part sinks to the bottom. ✓
Tube 2: The X component sinks to the bottom and the Y component dissolves in glycerine. ✓
ii) filtration ✓
iii) All X would dissolve but only part of Y would dissolve, leaving the rest of Y to sediment. ✓
b) Yes. ✓ X is insoluble, Y is soluble in water. Separate by filtering and then evaporate to dryness. ✓
20. a) filtration ✓; magnetic separation ✓
b) Wrap Strapmag around a beaker containing oil/metal particles. Observe whether the metal pieces move to the sides near the Strapmag. ✓
21. a) Olive oil forms the upper layer as it is less dense than water. ✓
b) Pour the oil and water mixture into the separating funnel with the tap initially closed. Allow the two layers to form. ✓ Open the tap and run the lower water layer into a clean beaker and close the tap before any olive oil escapes. Remove the beaker of water and replace it with a clean beaker. Open the tap and run the olive oil into the new beaker. ✓
22. Hold a magnet over the bowl. The iron nails will be attracted to the magnet whereas the brass nails will not as brass is not magnetic. ✓
23. D, E, C, A, B ✓
24. Centrifuging separates the blood components. Some patients only need plasma and others the blood cells. ✓
25. a) The water or air on the outside of the condenser reduces the temperature so that the distillate condenses back. ✓
b) The top of the condenser always has cold water leaving and has the most heat to be removed. ✓
26. solute = iodine ✓; solvent = alcohol ✓; solution = tincture of iodine ✓
27. a) Eucalyptus oil acts as a solvent in removing grease stains on clothes. ✓
b) A hot solvent generally dissolves a greater quantity of solute than a cold solvent. ✓
28. a) F ✓; b) T ✓; c) T ✓; d) T ✓; e) T ✓; f) F ✓; g) T ✓
29. You need to know the physical properties of the substances. ✓
30. a) Some of the particles are small enough to pass through the pores in the filter paper. ✓
b) The second paper was finer. It filtered out the smaller particles. ✓
31. Solutions have particles so small that they cannot be seen. They pass through the pores of a filter paper. ✓ It does not separate in a centrifuge because the solute is dissolved. ✓
32. See Figure A.7. The main idea of chromatography is that different solutes have different degrees of solubility in any solvent. During chromatography the different solutes move at different rates through the column. ✓

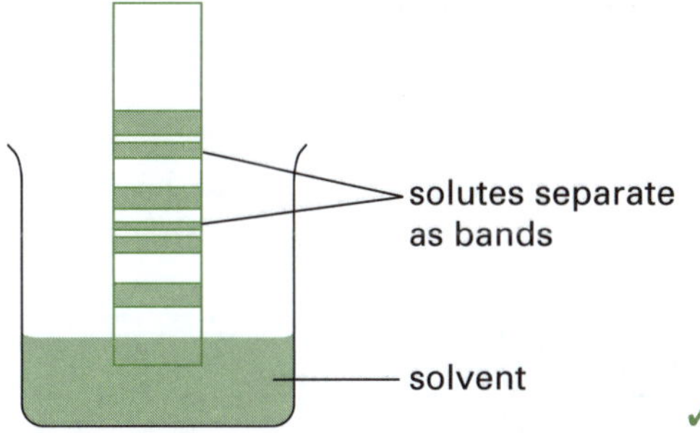

Figure A.7 Chromatographic separation

33. a) The literal meaning is colour writing. ✓
b) Chromatography is the separation of mixtures by colour using their different solubility in the solvent. ✓
34. (Answers will vary.) In a *centrifuge*, the *spinning* separates the *lighter* particles from the *heavier* ones, which *settle* out. ✓
35. • Containment and skimming. Long floating booms are used to surround the oil slick and then the contained oil is skimmed from the surface. Skimming can also involve the use of polyethylene mop-like pads that absorb surface oil. ✓
• Dispersants. Detergent-based chemicals mix with the oil and convert the slick into tiny droplets which can then mix with the water and be absorbed into the aquatic environment. While this method works to some extent in the open ocean, studies have shown that this method is more devastating to corals than the untreated oil slick. ✓
• In-situ burning. Oil is flammable and on some occasions the oil has been contained

with fire-resistant booms and then set alight. The sea must be calm for this to work. ✓

36. Steps:

1. Rubbish is placed on a conveyor belt and plastic bags and non-recyclables are removed. ✓
2. A rotating machine sorts waste by weight. ✓
3. Paper and plastic are light and are optically or manually sorted. ✓
4. Glass and metal are heavy. The iron/steel is removed magnetically, other metals are removed electrically and glass is hand-sorted. ✓

37. See Figure A.8. The clay particles are extremely fine and remain suspended. The sand is heavy and sinks to the bottom. Mud particles are larger than silt particles.

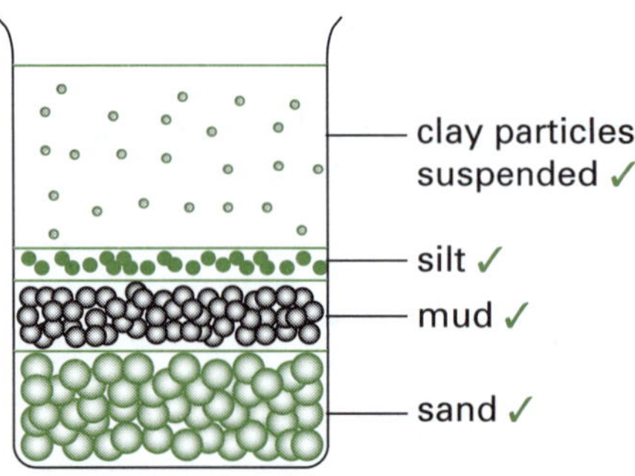

Figure A.8 Layers of sediment and suspended material

38. **a)** chromatography ✓; **b)** distillation ✓; **c)** evaporation ✓; **d)** filtration ✓

39. The iron filings will stick to the magnet ✓ and he will have a lot of trouble removing them. Using the paper as a barrier, ✓ he can easily remove the magnet and the iron filings will immediately fall into a container. ✓

40. Crush the igneous rock into fine particles ✓ then use a strong magnet to separate the magnetic ore from the rock fragments. ✓

Chapter 3 answers

Experiment 1

Part A

Analysis: The eight observations are related to the following sequence of Moon phases: new moon; waxing crescent; first quarter; waxing gibbous; full moon; waning gibbous; third quarter; and waning crescent.

Conclusion: This experiment demonstrates the phases of the Moon in the correct order as seen from the southern hemisphere.

Part B

Analysis: A = summer; B = autumn; C = winter; D = spring.

Conclusion: The tilt of the Earth's axis relative to the plane of the ecliptic allows us to explain the four seasons of the year in the correct order.

Experiment 2

Analysis: The Sun's energy has evaporated the water to form water vapour which has condensed on the underside of the plastic. The water droplets have then dripped into the small beaker. This model does not show all aspects of the water cycle.

Conclusion: The experiment demonstrates the evaporation of salty water from oceans and the condensation of pure water that falls to the Earth and forms our freshwater.

Experiment 3

Analysis:

1. See Figure A.9.

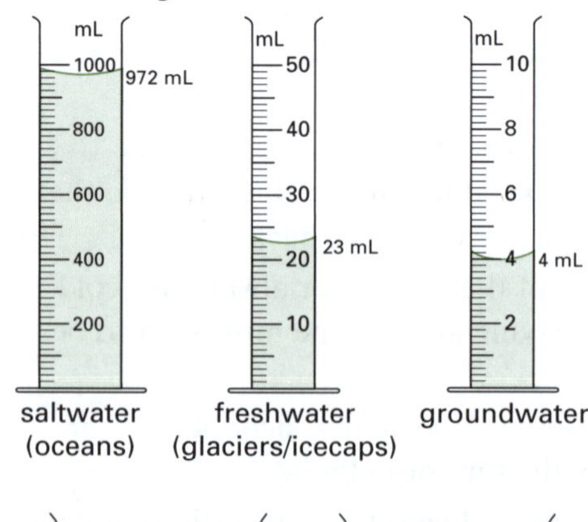

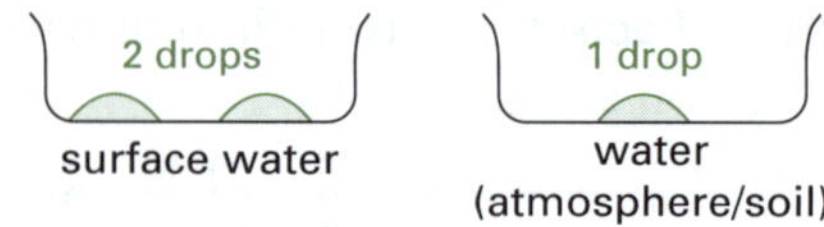

Figure A.9 Water distribution experiment

2. See Figure A.10.

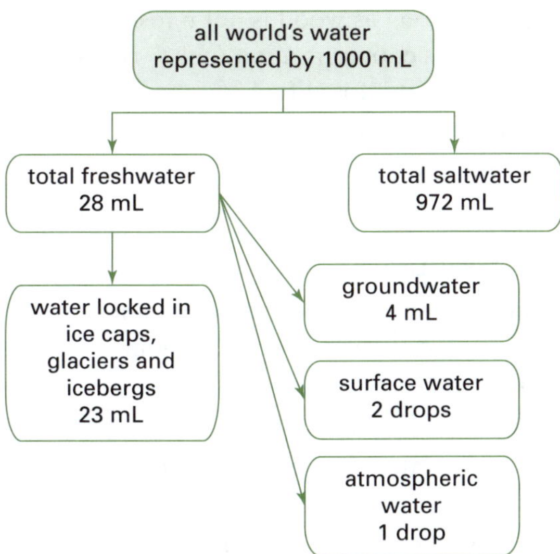

Figure A.10 Flow chart showing how water is distributed on Earth

Conclusion: Most of the Earth's water is found in the oceans, polar caps, glaciers and groundwater.

Test yourself 1

Part A: Knowledge

1. A ✓ All three objects need to be lined up for an eclipse to occur. B is wrong as no shadow is cast on the Earth or Moon; C will not occur; D produces a solar eclipse but does not account for the lunar eclipse.
2. A ✓ For only about ¼ of the moon to be lit up, as viewed from Earth, the Earth, Moon and Sun need to be in the positions shown. At B the bright crescent is at the left; at C there is a gibbous moon; at D the Moon is full.
3. C ✓ The solstices occur in the summer (longest daylight) and winter (shortest daylight). A is wrong as it only refers to a season; B is wrong as it refers to spring; D is wrong as it refers to equal length days and nights.
4. B ✓ Each season must be 3 months long to get four seasons in the year. Thus A, C and D are wrong.
5. C ✓ If the Earth didn't have a tilt on its path around the Sun, there would not be any seasons. A and B are wrong as they have nothing to do with the seasons; D is wrong as it is the tilt of the Earth relative to the orbital plane that causes the seasons.
6. **a)** solar ✓; **b)** day ✓; **c)** gravity ✓; **d)** Galileo ✓; **e)** Moon ✓; **f)** mass ✓
7. A/I ✓; B/H ✓; C/F ✓; D/G ✓; E/J ✓

Part B: Skills

8. **a)** R ✓ Draw a line from Earth through the third image from the bottom of the planet, and it passes through the position marked R.
   b) Between successive snapshots, there is a greater distance ✓ between the positions of Earth than for the planet. The Earth is moving faster. ✓
   c) Retrograde motion is the apparent backwards movement of a planet across the background of stars, as seen from Earth. ✓ Of course, the planet doesn't move backwards. Retrograde motion arises from the combined motions of the planets and the Earth. The diagram shows that as the faster-moving Earth overtakes the planet, the planet appears to move backwards. ✓
9. **a)** No. ✓ Plants can sense the Earth's gravitational field. ✓ Roots bend and grow downwards towards gravity, while shoots bend away from the direction of gravity and grow upwards. ✓
   b) Jim can grow seedlings entirely in darkness, such as in a dark cupboard. ✓ No matter how the seeds are planted, the root will grow downwards and the shoot upwards. ✓ As the plant is in complete darkness, the explanation for this cannot be light. (Of course, once the shoot has broken the soil it will need sunlight to provide further nourishment, having exhausted the supply of food provided within the seed.)
10. **a)** Drop two similarly sized, but different weight, objects from a balcony. ✓ (Be careful not to hit anyone below.) Have someone on the ground film the result ✓, then play it back slowly to show that the objects do indeed land at the same time ✓.
   b) No. ✓ Besides gravity pulling objects down, there is air resistance on Earth that can slow the downward path of objects ✓. The two pieces of paper are each being pulled downwards due to the force of gravity. When initially dropped, both pieces of paper begin to accelerate (gain speed). As they gain speed, they both encounter the

upward force of air resistance. The amount of air resistance depends on the speed of the falling object and the surface area of the falling object. Based on surface area alone, it is safe to assume that (for the same speed) the flat sheet would encounter more air resistance than the scrunched paper. Thus the flat sheet does not fall as fast as the scrunched sheet.

11. The telescope made it possible for Galileo to:
    - discover the largest moons of Jupiter ✓
    - see the mountains and valleys on the Moon ✓
    - observe the phases of Venus ✓
    - observe the rings of Saturn ✓.

12. **a)** winter ✓; **b)** spring ✓; **c)** autumn ✓; **d)** summer ✓

13. **a)** It is summer in the southern hemisphere. ✓ The sun here is high overhead during the day and the rays of light fall at a steep angle ✓, concentrating them on small areas of land ✓.

    b) No ✓, in the northern hemisphere it is winter ✓ at this time of year. The sun's rays strike at a shallow angle ✓, and the same quantity of sunlight needs to warm a greater area of land ✓.

14. A = summer night ✓; B = winter night ✓; C = spring day ✓

Test yourself 2

Part A: Knowledge

1. **D** ✓ Many synthetic resources are derived from non-renewable resources, such as plastics. A, B and C are wrong because they are reasons why synthetics are good alternatives.

2. **B** ✓ The question describes a non-renewable resource. A is wrong as renewables can be replaced quickly; C is wrong as natural resources such as timber are quick to replace; D is wrong as the chemical industry can rapidly replace various resources.

3. **A** ✓ You can always breed more silkworms and obtain more silk. B and D are wrong as silk is renewable; C is wrong as silk is natural.

4. **A** ✓ Clay has been used for thousands of years to make bricks, tiles and china. B and C are wrong as clay is not used as a fertiliser; D is wrong as clay cannot be turned into copper.

5. **D** ✓ Coal, crude oil and natural gas are all fossil fuels. A is wrong as wood is not a fossil fuel; B is wrong as copper is not a fuel; C is wrong as hydrogen is not a fossil fuel.

6. **a)** nickel ✓; **b)** bauxite ✓; **c)** steel ✓; **d)** resources ✓; **e)** fuel ✓

7. A/H ✓; B/J ✓; C/F ✓ ; D/I ✓; E/G ✓

Part B: Skills

8. **a)** The filament becomes white hot and tungsten still remains solid without melting. ✓

   b) X = 100 – 61 – 27 – 7 = 5% ✓

9. **a)** air ✓; **b)** water ✓; **c)** earth ✓; **d)** air ✓; **e)** earth ✓

10. **a)** C ✓ and D ✓; **b)** A ✓ and B ✓; **c)** A ✓ and C ✓

11. **a)** Jane is correct. ✓

    b) Fish numbers are finite. ✓

    c) true ✓

    d) Fishing needs to be controlled so that no more are taken out than replacement can sustain. ✓

12. **a)** A ✓ and B ✓

    b) to protect itself from the climate in high, dry and windy plateaus of their homeland ✓

    c) to make clothes ✓

    d) A cashmere jumper would be expensive, ✓ considering how much wool goes into producing it and how long it takes one goat to produce ✓ enough for this purpose.

13. **a)** A ✓ and D ✓

    b) Hypothesis: These minerals act as fertiliser so red algae can grow in great quantities. ✓

    c) **i)** less ✓; **ii)** no ✓; **iii)** If the underwater plants can't synthesise then no oxygen is produced and so fish will die. ✓

    d) Soil is not bound by root systems when trees are cut down. The topsoil can blow away, making the ground barren. In monsoon areas where plenty of rain falls, dirt is turned into mud that can flow down hillsides smothering everything in its path. ✓

14. **a)** A, B, E and F ✓

    b) No. ✓ If farmers do not look after their land, they will not have a sustained future in the industry. ✓

Test yourself 3

Part A: Knowledge

1. **C** ✓ Water vapour turns to water liquid when heat is removed. A is wrong as liquid water particles are not tightly packed, they are free to move past each other; B is wrong as heat would need to be removed; C is wrong as it would expand.
2. **B** ✓ The leaf pores are called stomata and the evaporation of water from these pores provides a force that drags water up from the roots. A, C and D are incorrect according to the above definition.
3. **A** ✓ When the humid air rises, it cools and water vapour condenses to droplets of water. B is wrong as it is a warm not a cold air mass that moves from the ocean; C and D are wrong as they also refer to cold air masses that contain very little water vapour.
4. **C** ✓ Soil absorbs water whereas concrete does not. A is wrong as paved surfaces are not porous; B is wrong as they increase the amount of storm water reaching the sea; D is wrong as global warming has nothing to do with paved surfaces.
5. **B** ✓ Wetlands are a complex system where contaminants can be removed from the water without using chemicals. A is wrong as fish do not remove wastes; C is wrong as wetlands are not used for that purpose; D is wrong as phosphates are not selectively absorbed from blackwater in such locations. Greywater is treated.
6. **a)** suspended ✓; **b)** fertilisers ✓; **c)** sea ✓; **d)** percolates ✓; **e)** transpiration ✓
7. A/F ✓; B/J ✓; C/G ✓; D/I ✓; E/H ✓

Part B: Skills

8. (i)/(b) ✓; (ii)/(a) ✓; (iii)/(c)✓
9. **a)** The higher the temperature, the greater the moisture-holding capacity of the air. ✓
   b) 23 g/m^3 ✓
   c) 11 °C ✓
   d) At 15 °C the moisture-holding capacity is 13 g/m^3. Doubling this is 26 g/m^3. Therefore, the temperature is 27 to 28 °C . ✓
10. **a)** cools; **b)** condenses; **c)** moisture; **d)** rain; **e)** permeable; **f)** surface; **g)** groundwater; **h)** sun; **i)** evaporation; **j)** fresh *(½ mark for each answer)* ✓
11. F, A, C, B, D, E ✓
12. W = condensation; X = freezing; Y = evaporation; Z = melting *(½ mark for each answer)* ✓
13. **a)** water vapour ✓
    b) liquid water ✓
    c) Water vapour cools and condenses to liquid water on the cold surface of the clock glass. ✓
    d) The ice in the clock glass melts or undergoes fusion. ✓
14. See Figure A.11.

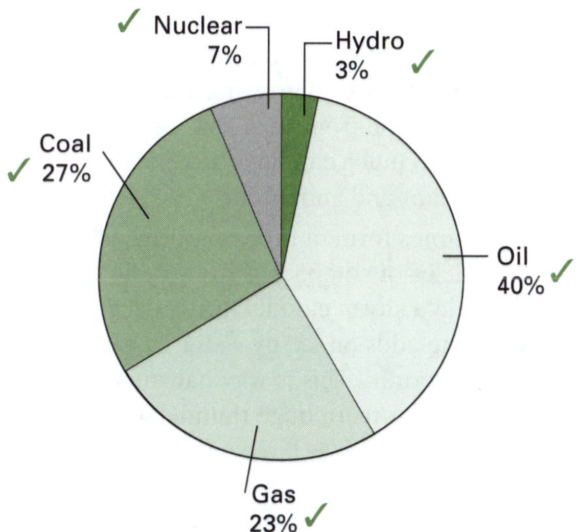

Figure A.11 World energy consumption of various energy sources

Chapter test

Part A: Multiple choice

1. **B** ✓ An equinox occurs twice a year, when the tilt of the Earth's axis is inclined neither away from nor towards the Sun. The centre of the Sun is in the same plane as the Earth's equator. This occurs around 20/21 March and 22/23 September each year. Answers A, C and D refer to the solstices.
2. **B** ✓ 1.023 × 60 × 60 = 3682.8 km/h. Multiply by the number of seconds in an hour.
3. **D** ✓ A is summer; B is spring; C is winter. In Australia, the seasons begin at the beginning of the month.
4. **D** ✓ Answers A, B and C are incorrect because they are the remains of organisms that were once living.

5. **D** ✓ There is only some much uranium, or other fissionable material, in the ground to produce sustained nuclear fission to generate heat and do useful work. Sources A, B and C are all renewable.
6. **A** ✓ Answers B, C and D are incorrect because the amount of power that can be derived comes from the difference between high and low tides. There are always tides every day, and power can be transferred great distances to where it is needed.
7. **A** ✓ B, C and D are incorrect because water from any source can leach or remove minerals from the soil. Irrigation is the artificial application of water to the land or soil and is used mainly to assist in growing agricultural crops. It can come from various sources, not just bores, and too much can have a damaging effect on native plant and animal life.
8. **C** ✓ Hailstones form in thunderstorms when small particles of compacted snow cycle up and down in a storm cloud. During each cycle the hailstone adds on ice by water freezing during the updraft. This is why hailstones have a layered structure. In huge thunderstorms hailstones may grow as large as 8 to 10 cm in diameter before finally falling out of the cloud. Therefore answers A, B and D are incorrect.
9. **B** ✓ A is solid to liquid; C is gas to liquid; D is liquid to gas.
10. **A** ✓ When a parcel of air containing water vapour gets to a mountain it is forced upwards, since the mountain is blocking its way. This air expands and cools. As it cools it becomes saturated because the amount of water it can hold decreases. So the water vapour begins to condense into clouds and finally falls as rain. Therefore answers B, C and D are incorrect.

Part B: Short-answer questions

11. A lunar eclipse (an eclipse of the Moon) is perfectly safe to watch with the naked eye; you're only looking at the Moon, at night, which is quite safe. ✓ A solar eclipse is potentially dangerous, however, because this involves looking at the Sun, which can damage your eyesight. Sunglasses do not provide adequate protection, as they do not block out the wavelengths of light that are likely to damage your eyes, nor sufficiently reduce the intensity of the visible light. ✓ Even when almost the entire surface of the Sun is obscured during the partial phases of a total eclipse, the remaining crescent is intensely bright and can cause eye damage.
12. **B** ✓ He would see the rim of the Sun surrounding the Moon.
13. a) A full Moon ✓ is the only time when a lunar eclipse is possible. That is when the Moon may move through the shadow cast by the Earth.

    b) Because of the tilt of the Moon's orbit around the Earth relative to the Earth's orbit around the Sun, the Moon may pass above or below the shadow. ✓ Hence a lunar eclipse does not occur at every full Moon. ✓
14. February, with only 28 days, can lack a full moon as one can occur just before and just after February. ✓ The cycle of the phases from full moon to full moon takes 29.5 days. (Some years in which February lacked, or will lack, a full moon are 1809, 1847, 1866, 1885, 1915, 1934, 1961, 1999, 2018, 2037, 2067 and 2094.)
15. Looking down at the South Pole, the Earth's rotation appears clockwise. ✓
16. speed = $\frac{40\,000}{24}$ = 1670 km/h
    (correct to the nearest 10 km/h) ✓
    Even at this huge speed it takes one full day to spin once on its axis.
17. a) C: 20 days ✓; G: 7 days ✓; E: 5 days ✓; I: 3 days ✓

    b) Moons are not visible when behind the planet. ✓

    c) I, E, G, C ✓
18. full; B; D; E; C; ✓ then G; F; A ✓
19. a) Uranus was too far away for it to be observed by the naked eye or simple telescopes. ✓

    b) A = Venus ✓, B = Mars ✓
20. The telescope has allowed scientists to explore the heavens far more thoroughly. It has revealed unsuspected phenomena and allowed them to examine and change theories about the origin of the universe and the objects within it. Only so much can be seen with the unaided eye. Using telescopes has allowed scientists to see further and examine more obscure objects in the universe. ✓

21. Other answers are possible.
    - Electricity ✓ flowing freely through power lines as we know it today was unknown, although the beginnings of using a simple battery to produce current were established. Most appliances in the home today use electricity, and it is a major energy resource we can't do without.
    - Plastics ✓ were unknown, although today they are found everywhere with a large number of varieties and uses available.
    - Rubber ✓ remained a curiosity in the Western world. But Central American tribes had extracted latex from a type of rubber tree in the area as early as 1600 BC. This was mixed with the juice of a local vine, creating an ancient processed rubber. It wasn't until the 19th century that rubber could be turned into something more useful through a process called vulcanisation. A vast array of products is now made with vulcanised rubber including tyres, shoe soles, hoses and balls.
    - Although petroleum ✓ oozing out of the ground was in use for thousands of years, it wasn't until the 19th century that a method was found to distil the components from it. Today oil meets about 90% of vehicle fuel needs. Petroleum also provides around half of the total energy consumed in Australia.
    - Many metals ✓ we take for granted today were not purified until the 19th century. These include aluminium, chromium, sodium and uranium.

22. Renewable resources are ones that can be replaced or reproduced easily. ✓ For instance air, wind and sunlight are continuously available and their quantities are not affected by human consumption. While many renewable resources can be depleted by human use, they may also be replenished, thus maintaining a flow. Some, like agricultural crops, take a short time for renewal. Others, like water, may take a longer time. And there are others, like forests, which take even longer. This is where human management is needed to ensure our sources of renewable resources are maintained. ✓
    Non-renewable resources are formed over very long geological periods. ✓ This includes minerals and fossil fuels. Since they are formed extremely slowly, they cannot be replenished once they are depleted. While metallic minerals can be recycled and re-used, and so extend the availability of these minerals, others like coal and petroleum cannot be recycled. ✓

23. There is only a finite quantity of non-renewable resources and we must learn to use them wisely. Recycling is an important part of this. ✓ Most councils now provide recycling bins for paper, metal, glass and other materials that can be recycled cheaply. For example, the process in recycling aluminium involves simply re-melting the metal. This is far less expensive and less energy intensive than creating new aluminium. Recycling aluminium uses about 5% of the energy required to produce aluminium from its ore. This preserves our natural resources. We can also reduce the amount we consume. ✓ Asking 'do I really need that?' before buying something only to soon discard it. We can also save energy (which is often made from non-renewable sources) by turning off appliances when not in use. ✓

24. a) Most cans were used in 2005 ✓, and this was about 6.8 billion. ✓
    b) In 2001 about 0.5 billion cans were recycled out of 4 billion used. This is around
    $\frac{0.5}{4.0} \times 100 = 12.5\%$. ✓
    c) In 2010 about 2.0 billion cans were recycled out of 4.7 billion used. This is around
    $\frac{2.0}{4.7} \times 100 = 43\%$ (rounded). ✓
    d) On these values alone, recycling rates should continue to gradually increase. ✓

25. a) Aluminium is one-third the weight of steel, ✓ resists corrosion ✓ and is easy to recycle, ✓ and there is an enormous range of applications.
    b) Demand is growing by
    $\frac{3}{100} \times 24 = 0.72$ million tonnes each year. ✓
    (Of course, this number will continue to grow.)

26. a) See Table A.4.

Table A.4 Composition and mass of Australian coins

Coin	Metal composition	Mass (g)	Mass of copper present (g)
5 ¢	75% Cu, 25% Ni	2.83	2.1225 ✓
10 ¢	75% Cu, 25% Ni	5.65	4.2375 ✓
20 ¢	75% Cu, 25% Ni	11.30 ✓	8.4750
50 ¢	75% Cu, 25% Ni	15.55 ✓	11.6625
$1	92% Cu, 6% Al, 2% Ni	9.00	8.2800 ✓
$2	92% Cu, 6% Al, 2% Ni	6.60 ✓	6.0720

b) i) $\frac{20}{100} \times 13.28 = 2.656$ g ✓

ii) There would be over 10 g of silver in the coin and this is more than the coin is worth. ✓

27. a) See Figure A.12.

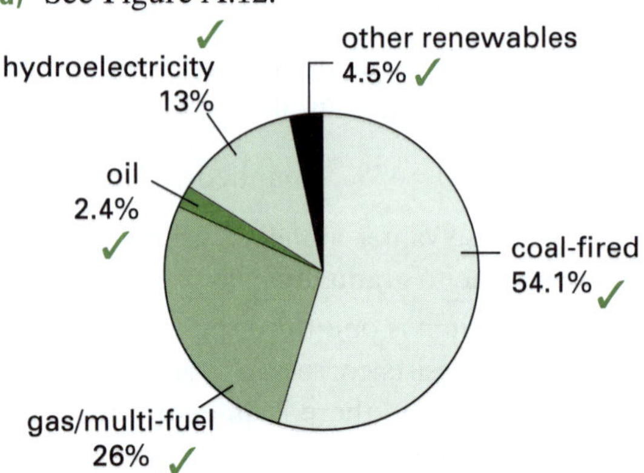

Figure A.12 Sources of electrical energy in Australia, 2008–09

b) The largest sector is the coal sector. ✓

28. a) In the 45 years between 1965 and 2010, coal consumption grew from about 65 million GJ (gigajoules) to around 220 million GJ. ✓ So each year the growth was about $\frac{220-65}{45} = 3.44$ million GJ/year. ✓ (This value is approximate. Your value could be different, though close.)

b) From 2010 to 2030 is 20 years. So coal consumption would increase by $20 \times 3.44 = 68.8$ or by about 69 million GJ. Adding this on to the 2010 value, $220 + 69 = 289$ million GJ of coal energy should be used by 2030. ✓

c) This value is based on trends between 1965 and 2010, and is an estimate. ✓ While extending the graph (extrapolating) for a few years is pretty safe, who knows what might occur between now and 2030? Maybe coal might be seen as an undesirable energy source and alternative energies developed. Australia might invest in nuclear power, with the coal influence being lessened. On the other hand, the demand for energy might increase faster than the graph is currently showing, in which case more coal will be consumed.

29. a) hydro-electric ✓; **b)** nuclear energy ✓;
c) hydro-electric ✓; **d)** solar hot water ✓;
e) wind power ✓; **f)** solar electricity ✓;
g) wind power ✓; **h)** geothermal power ✓;
i) wind power ✓; **j)** tidal power ✓;
k) geothermal power ✓

30. a) Suitable areas for Australian wind power resources lie mainly on a narrow strip of sites around the southern coastline of the mainland, ✓ and half of Tasmania. ✓

b) These are areas of high wind for much of the time, so locating turbines here would be of benefit in harnessing this energy. ✓ Wind resource drops sharply with increasing distance from the coast and inland Australia is, generally, unsuitable for wind power. ✓

c) These areas overlap with some of the most scenic and populated areas of Australia. ✓ It is therefore natural for people in these regions to be against siting wind farms virtually in their back yard. They would be concerned by the visual and noise impact on the natural environment. ✓

31. No. ✓ Water will evaporate at any temperature but the rate at which this happens increases as the temperature increases. As liquid water is heated, the particles near the liquid surface can escape into the air. The more energy they have the easier it is to escape from the surface. ✓ This is why a bucket of water left outside for several days slowly evaporates.

32. Plant roots have fine hairs that absorb groundwater. ✓ This water then moves into the stem of the plant where it is then drawn upwards in a continuous column by the forces caused by transpiration from the leaves. ✓

33. No, the steam is not water vapour. ✓ The clear region near the spout of the kettle is where hot water vapour is found. Water vapour cannot be seen. As the hot vapour moves further away, it starts to cool down and some water vapour condenses into fine droplets of water. This is the whitish region we call steam. Thus steam is a mixture of liquid water and water vapour. ✓

34. a) Average percentage contribution of cloud water droplets = 100 – (51 + 17 + 6 + 5)
= 21% ✓

b) See Figure A.13.

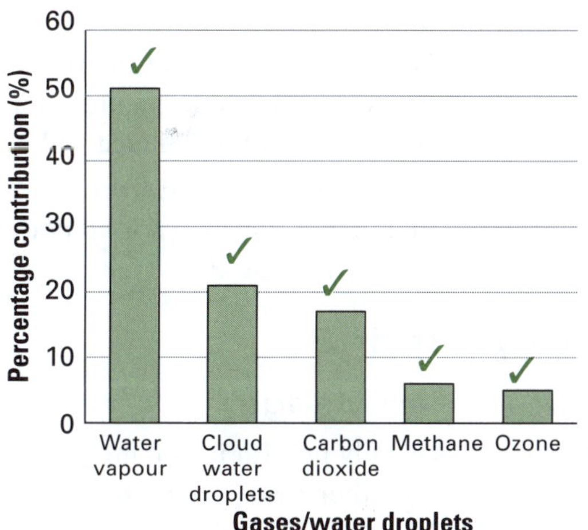

Figure A.13 Average contribution of gases/water droplets to the greenhouse effect

35. The correct order is: C ✓; A ✓; D ✓; B ✓. The colloidal particles are suspended in the water. The alum is added and starts to clump the particles together. As the clumps stick together, they get heavier and fall under gravity to the bottom of the tank.

36. a) washbasin (45.5 mg/L) ✓

b) laundry (1.5 + 10.5 = 12.0 mg/L) ✓

c) mass of phosphorus = 1.5 mg/L × 300 L
= 450 mg ✓

37. The Australian Aborigines cared for waterways sustainably. They used the resources of the waterways to the minimum extent that allowed them to survive as they were hunter/gatherers. They did not degrade the waterways or overexploit them. This was the tried and tested way of doing things; they were not early conscientious environmentalists in the modern-day sense. ✓

The farming practices of early European settlers often led to waterway degradation as they did not have a culture of sustainability. This practice led to pollution of the waterways and the collapse of river systems by removing too much water for irrigation. ✓

38. are ✓; by ✓; for ✓; be ✓; that ✓

39. A/G ✓; B/F ✓; C/E ✓; D/H ✓

40. After the water passes through the sand/gravel filter ✓ and enters the tank to be chlorinated, it should be clear and colourless as the solid particles should all be removed by the filter.

Chapter 4 answers

Experiment 1

Analysis: The neutral pieces of paper were attracted to the positively charged rod but on touching them they gained some of the positive charges. Both the rod and paper were now positively charged so they repelled each other and the papers jumped off the rod.

Conclusion: Neutral objects can be attracted to charged objects as long as they are very light. On contact with the charged object, the neutral object becomes charged with the same charge as the original charged object.

Experiment 2

Analysis: As the effort distance increases, the effort (E) required to balance the lever decreases. As the effort is getting smaller, the last experiment produces a lever with the greatest mechanical advantage as very little effort is required to raise the load.

Conclusion: In order to achieve the greatest mechanical advantage in a first-class lever, the load must be close to the fulcrum and the effort as far away from the fulcrum as practical.

Experiment 3

Analysis: It takes much less effort to raise the load using a 2 × 2 pulley system than a 1 × 1 pulley system. Both systems have a mechanical advantage as the effort is less than the load but the 2 × 2 system has the greater mechanical advantage.

Conclusion: The greater the number of pulley wheels in a pulley system, the greater is the mechanical advantage.

Test yourself 1

Part A: Knowledge

1. A ✓ A mass experiences a force in a gravitational field and this is called the weight force. B is wrong as no contact is made on a surface; C is wrong as no frictional forces are involved because there is no relative movement; D is wrong as the Earth's magnetic field does not affect the weight of a body unless the body is magnetic.
2. D ✓ When a net force operates, the body accelerates. A and B are wrong as the direction depends on the direction of the net force; C is wrong as the forces need to balance to move with constant speed.
3. A ✓ The floating boat is not moving up or down and so the two forces balance. B is wrong as an unbalanced force would lead to motion; C is wrong as the Earth's magnetism is irrelevant; D is wrong because a greater buoyancy force would lead the boat to continue rising out of the water.
4. C ✓ We cannot see the magnetic field, only feel its effect on magnetic bodies. A is wrong as like poles repel; B is wrong as unlike poles attract; D is wrong as the poles are of equal strength.
5. A ✓ The silk rubs off some of the surface electrons, leaving the glass rod positive. B is wrong as the rod will be negative; C is wrong as repulsion will occur; D is wrong as attraction will occur.
6. **a)** motion ✓; **b)** field ✓; **c)** gravity ✓; **d)** mass ✓; **e)** accelerate ✓
7. A/G ✓; B/J ✓; C/I ✓; D/H ✓; E/F ✓

Part B: Skills

8. The weight will decrease ✓ because of the buoyancy force which acts upwards. ✓
9. Spray water on the plastic sheet. ✓ The water will reduce friction between the children and the plastic as they slide. ✓
10. Neutral objects are attracted to charged bodies. ✓ In the case of a positive body the electric field around it attracts the negative charge in the cardboard as close as possible to the charged object. ✓
11. Graph (b) is correct. ✓ This graph shows no change in speed over time. Graph (a) is wrong as the speed is decreasing; graph (c) is wrong as the speed rises and falls.
12. **a)** Decelerate means to slow down or decrease in speed. ✓
    b) Graph (a) is correct. ✓ The speed decreases rapidly with time and then the speed decreases less rapidly as it gets closer to landing. Graph (b) is wrong as the speed increases; graph (c) is wrong as the speed decreases at a constant rate.
13. **a)** The weight will increase with the magnets present as the steel is magnetic and is attracted to the magnet. ✓ Thus the total force increases. ✓
    b) The distance (D) will affect the reading on the balance. If D is made smaller then the force registered on the balance will rise. ✓
14. The acceleration due to the Earth's gravitational field is constant for all bodies regardless of their mass. ✓

Test yourself 2

Part A: Knowledge

1. D ✓ This is only helpful if a change of speed is required. In other cases a mechanical advantage is needed when the load is very heavy. A, B and C are wrong because these are the three purposes of machines.
2. D ✓ The load is the nut, the fulcrum is at one end and the effort is applied to the handles. A is wrong as this is first class; B and D are wrong as this is third class.
3. B ✓ Pliers are levers. A, C and D do not have the structure of a lever.
4. D ✓ Less damage is done to the head and brain if the impact force is reduced by spreading the energy of the impact over a longer time. A is wrong as this is the function of the soft inner lining; B and C are wrong as this is the function of the rigid outer shell.
5. C ✓ The bicep contracts when the weight is raised. A and D are wrong as these are leg muscles; B is wrong as this muscle relaxes during this exercise.

6. a) hinge ✓; b) tricep ✓; c) firmly ✓; d) longer ✓; e) first ✓
7. A/J ✓; B/I ✓; C/F ✓; D/G ✓; E/H ✓

Part B: Skills

8. second-class lever ✓; A = effort ✓; B = load ✓; C = fulcrum ✓
9. a) 60 × 160 = L × 40 ✓
   L = 60 × 160/40 = 240 kg ✓
   b) first-class lever ✓
10. See Figure A.14.

a)

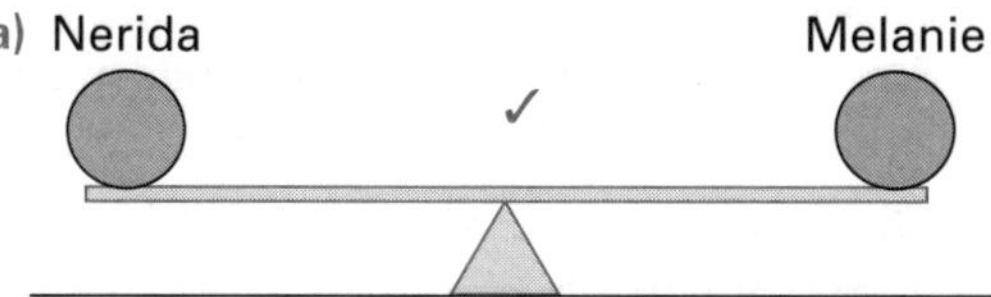

They sit the same distance on each side of the fulcrum

b)

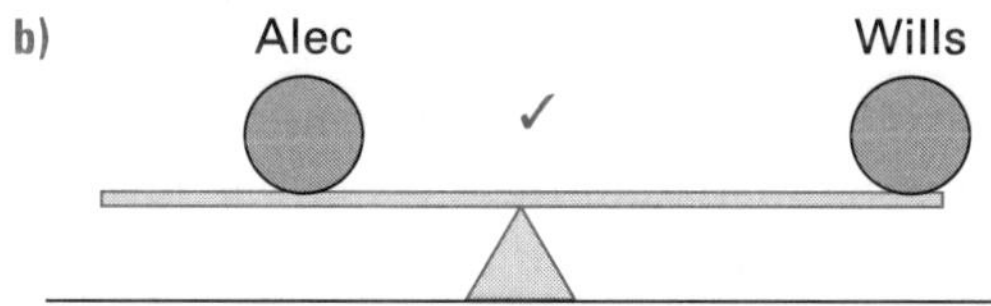

Alec has to sit 1.5 m from the fulcrum when Wills sits 3 m from the fulcrum

Figure A.14 Where each child should sit on the see-saw

11. The door is a lever pivoting about its hinge. The further from the pivot point, the more efficiently we can use our energy to move the door open and shut. ✓ If the handle is too close to the hinge (pivot), the effort required to open the door will be too great. ✓
12. a) third class ✓
    b) No. ✓ The advantage of this lever is that the load of soil can be moved rapidly from one location to another, but the effort required is greater than the weight of the soil. ✓
13. Yes, it has a mechanical advantage ✓ as there are two ropes supporting the load. The supporting bar assists in raising the load. The effort is about half the weight of the load. ✓
14. The cheetah has a greater sprinting speed as the ratio of the length of its front foot and toes to the length of its heel is greater than that for the lion. ✓ The loss of mechanical advantage is compensated for by an increase in speed. ✓

Chapter test

Part A: Multiple-choice questions

1. **B** ✓ Weight is the force that pulls you towards the centre of the Earth in the gravitational field. A is wrong as no surfaces make contact; C is wrong as mass is not a force; D is wrong as this is the unit of force.
2. **D** ✓ Two surfaces make contact and if they are rough then friction results. A, B and C are examples of field forces.
3. **D** ✓ Uncharged objects have an even distribution of positive and negative charges. They do not attract other uncharged objects. A, B and C are correct statements about electrostatic forces.
4. **C** ✓ The swim bladder can inflate or deflate to change the buoyancy. A is wrong as gravity will pull the fish down; B is wrong as there is no magnetic force involved; D is wrong as friction will act in all directions.
5. **D** ✓ All objects are not moving and so there must be balanced forces for all four. While A, B and C are true, only D is completely correct.
6. **B** ✓ A is wrong, as a fishing rod is a third-class lever; C is wrong, as scissors are first-class levers; D is wrong as your effort is applied upwards.
7. **B** ✓ There is very little friction between the ice blade and the ice. In A, C and D, friction is quite large. You cannot walk or write without some friction. D is an example where friction is high.
8. **C** ✓ Considerable rocket thrust is required to escape Earth's gravity and the astronauts experience very high forces. The other choices are incorrect since in A the gravity on the Moon is much less than on Earth; in B the force here is their normal weight; in D the thrust force acting on a lander is much less than the rocket taking off from Earth.
9. **A** ✓ Retrorockets apply a reverse thrust that opposes gravity and the craft's speed decreases so it lands safely. B is wrong because there is no air on the Moon to allow a parachute to be used; C is wrong because gliding is not possible without an atmosphere; D is wrong because they do not store gases on board.
10. **B** ✓ Stainless steel is magnetic. A, C and D are wrong as they are not made of iron or steel.

Part B: Short-answer questions

11. a) balanced ✓

b) See Figure A.15.

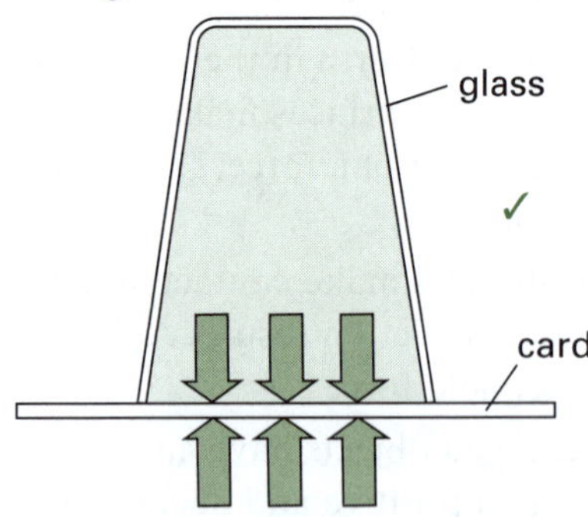

Figure A.15 Arrows show the gravitational force acting on the mass of water

c) air pressure ✓ (The gravitational force acting on the mass of water is actually balanced by the upward force due to air pressure acting on the card and water. Therefore, there is no motion as forces are balanced.)

12. See Figure A.16. *(6 marks, 1 for each force arrow)*

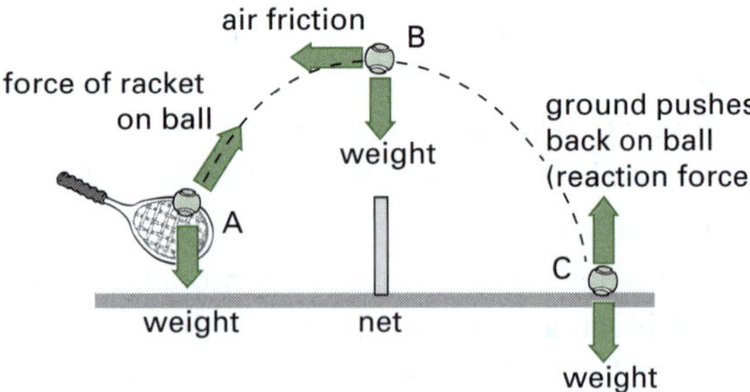

Figure A.16 Arrows show the forces acting on the ball

13. The ball bearings are highly smooth (low friction) and only touch a very small area of the wheel (also low friction). ✓

14. a) See Figure A.17.

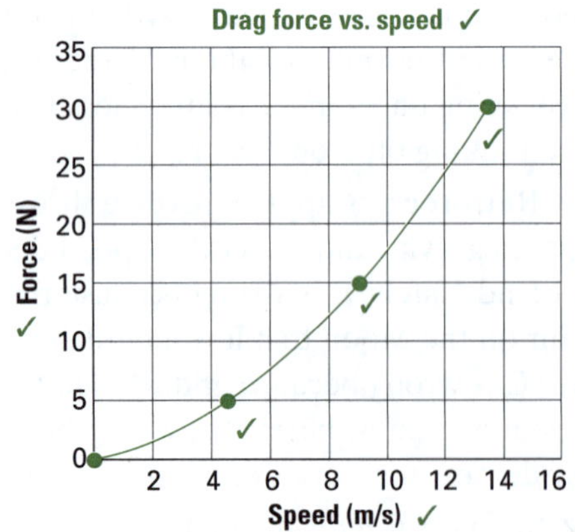

Figure A.17 Graph showing the interaction between force and speed

b) As speed increases, drag increases. ✓

c) Make the bike rider as smooth as possible to reduce friction with the atmosphere (drag). ✓ Examples include: shiny, smooth racing suites; helmets designed to cut through the air; aerodynamic bikes. ✓

15. a) a magnet ✓; **b)** diagram A ✓

16.
- Apply a pull on the force meter until the magnet is just removed. ✓
- Record the value. ✓
- Repeat five times and record the values. ✓
- Average the values. ✓
- Since there are two poles of equal size and equal force, divide this value by 2. ✓

17. a) The reaction force while running is reduced by 500 N with shoes on. ✓

b) The reaction force is greater than the action force when the athlete is running. When he stood still they were balanced. This could cause problems to the joints. ✓

18. A = third-class lever. The jaw muscle is placed between the load of the jaw and the pivot that is where the lower and upper jaw meet. ✓
B = second-class lever. The nut is the load that is between the fulcrum and the effort. ✓
C = second-class lever. The fulcrum in this case is the oar tip in the water. The load is applied at the rowlock which is attached to the boat. Thus the load is between the effort and the fulcrum in the water. ✓

19. a) 300 g ✓

b) 150 g ✓

c) Pulley system (2) has the mechanical advantage as the effort is less than the load. ✓

20. a) using a force meter ✓

b) The 20-kg weight ✓ has the greatest weight because it is attracted to the Earth by the greatest gravity force. ✓

21. The body falls towards the Earth with increasing speed. ✓ Eventually it will reach a terminal speed and then fall to the surface at that constant speed. ✓

22. a) Similarities:
- There is an area of influence (field) around the magnet and the Earth. ✓
- They extend into space in three dimensions. ✓
- They are invisible. ✓

b) Differences:

- Magnetic fields only attract iron/iron alloys. Gravity fields attract all matter. ✓
- Magnetic fields can attract as well as repel. ✓

23. a) *massa*: composed of lumps—composed of small particles such as atoms or molecules ✓

b) massive: lots of lumps—very big, contains a lot of matter ✓

c) Gravity is the force which pulls matter towards the Earth and gives matter its weight. ✓

24. a) gravitational ✓; b) extension ✓

25. a) The body will move (accelerate) to the right as F_1 is greater than F_2. ✓

b) The body will accelerate to the left as F_1 is less than F_2. ✓

c) The body will remain at rest or in constant motion, as there is no net force. ✓

26. As the body is not moving, the tensional force equals the weight force in size but in opposite directions. ✓

27. The coefficient of friction is very small between waxed skis on snow. Therefore, there is much less friction and the skier will travel further and faster. ✓

28. a) The pole of magnet B ✓ must also be north as it repels the north pole of magnet A. But the pole ✓ of magnet C is south as it is attracted to the pole of B.

b) In each case the north or south pole induces an opposite pole in the iron bar and then attraction occurs. ✓

c) Copper is not affected by a magnetic field. It is not magnetic. ✓

29. We know force by its effects on objects (push, pull, twist, deformation and so on). ✓

30. By lubricating the chain, he has reduced the friction between the metal parts. ✓

31. a) Each 100 g attached to the spring results in a 3-cm stretch in the spring. ✓

b) 24 cm ✓

c) 1200 g ✓

d) No, either the coils could be stretched so far that damage is produced ✓ or the spring uncoils totally so it no longer acts like a spring ✓.

32. a) positive ✓

b) The rod on the clock glass will be repelled and start to move away. ✓ This happens because like charges repel. ✓

33. There may be other suitable terms that are not covered here. Any one of the suggested answers will score a mark.

a) speeding up; going faster; increasing speed ✓

b) initially; at the beginning; at the start ✓

c) constant speed (velocity); covering equal distances in equal time; no acceleration; keep the needle on the specified speed (Note that acceleration = 0 does not necessarily mean stationary.) ✓

d) stopped; stationary; not moving; at rest ✓

34. e) In any road crash there are really two collisions. ✓

h) The first is the car's collision where the car, hitting something, buckles and bends and then comes to a stop. ✓

f) The part of the vehicle that receives the first impact of the collision stops abruptly. ✓

a) The second, and more important, collision is the human collision, where an occupant hits some part of the car. ✓

g) In a crash, unrestrained occupants keep moving inside the passenger compartment during the time it takes the car to stop. ✓

c) They are still moving forwards at their original speed when they slam into the steering wheel, windscreen or some other part of the car. ✓

b) But the passenger compartment comes to a more gradual stop as some of the impact is absorbed in the crushing of, say, the engine bay or boot. ✓

d) The net result is that the passenger compartment often remains relatively undamaged. ✓

35. a) easy ✓; small ✓; would ✓ ; big ✓; harder ✓; did not ✓

b) As the mass of the trolley and its contents increases, the force required to accelerate it increases. ✓

36. Concrete blocks are rigid and can cause serious trauma to the occupants of any car that hits them at speed. ✓
The Brifen wire-rope safety fence not only prevents head-on collisions, but also absorbs the impact of vehicle accidents and minimises

injury to people. The time of the collision is increased and the energy and impact forces decreased. ✓

37. Without such a support, the whiplash during a crash can cause extensive damage to the neck area. ✓ The supports are not there to rest your head, and possibly go to sleep, while driving.
38. diagram (b) ✓ because it has a greater mechanical advantage
39. When the electromagnet is switched on and attracts the iron bar, the pointer on the scale moves to the left. ✓ The lever is then balanced back to horizontal with small weights added to the pan. ✓ The number of weights added is equal to the force needed to balance the electromagnet at that setting of the voltage. The balancing weight force equals the strength of the electromagnet. ✓
40. the Moon ✓

Chapter 5 answers

Experiment 1

Analysis: If a liquid, such as water, is heated continuously, vaporisation takes place in all parts of the liquid. This liquid undergoes a complete change of state into gaseous state, a process called boiling.

There are many observations you should be able to make. Dozens have been recorded for such experiments. You will see that bubbles start forming before the water boils. These bubbles are mostly of water vapour trying to escape. When the bubbles are large enough by combining with other small bubbles (coalesce), they come on the surface and escape as water vapour. The hot water vapour cools to form steam.

Here are some questions you should be able to answer.

- When and where do bubbles begin to form?
- Are there more bubbles on the bottom of the beaker than on the sides? Are bubbles distributed evenly around the glass surface?
- What is the shape and size of the bubbles while they are still attached to the glass, and when they begin to rise to the surface?
- Do bubbles grow as they rise to the surface?
- How long do they take to get to the surface? Is this the same time for all bubbles?
- Do all bubbles travel at the same rate? Is the speed constant, or do they accelerate to the top?
- Do bubbles travel in straight lines, or zigzag as they rise?
- What is the shape and size of the bubbles when they burst at the surface?
- Is the surface of the water flat before, during and after boiling?
- Once the Bunsen has been turned off, do any more bubbles form?
- Are any sounds heard? Are there any other movements?

Conclusion: In boiling a beaker of water there are many processes occurring within the water.

(This experiment only involved observation. Scientists often try to explain these observations.)

Experiment 2

Analysis:

1. a) There may be a difference in the rate of melting caused by the different volume/shape of the beaker. The key words here are 'may be'. So keeping them the same eliminates this possible difference.
   b) As the ice melts, water is cooled. This would slow the rate of melting of more ice. For a good comparison, both volumes should be kept the same.
   c) The initial temperature of the waters should be the same. The warmer the water, the faster will be the rate of the ice melting.
   d) More ice will take longer to melt. This, then, won't provide a good comparison between the two.
2. The quantity of salt in the water. The two beakers should be set up exactly the same except for the salt in one beaker. Then you know that any difference would be due to the salt.
3. The time it takes for the ice cube to melt.

Conclusion: The hypothesis is incorrect. The ice cube in fresh water melts faster than the ice cube in salty water.

(There are three possible outcomes: either the ice cube melts faster in fresh water, or slower in fresh water, or there is no difference. In this last possibility, the salt would make no significant difference to the ice melting.

While this is the end of this experiment, scientists often need to know why. As the ice cube melts in tap water, the cold water sinks to the bottom of the beaker, creating convection currents, with the warmer water from below moving up to take its place. This constant movement keeps the ice cube warmer than the cube in the salty water. Salt water is much denser than fresh water. So the water melting off the ice cube in a beaker of salt water doesn't sink at all. Without any convection currents to carry the cold water away from the ice cube, the ice cube melts much more slowly.)

Experiment 3

Analysis: The graphs you print out will depend on the motions you or the trolley make, and will be different from student to student.

Figure A.18 has sample graphs you may use to help determine what each of the curves you obtain show.

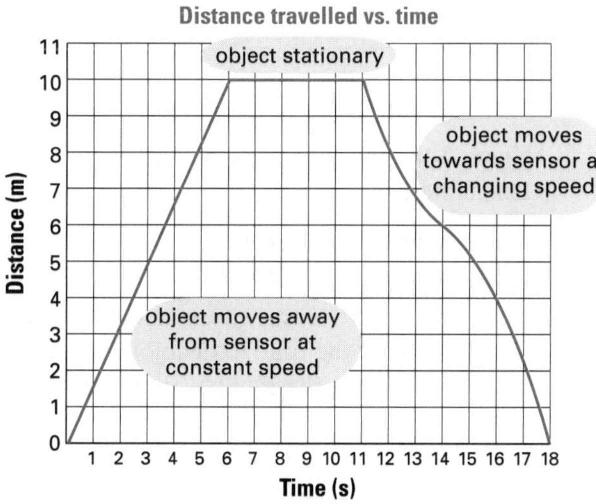

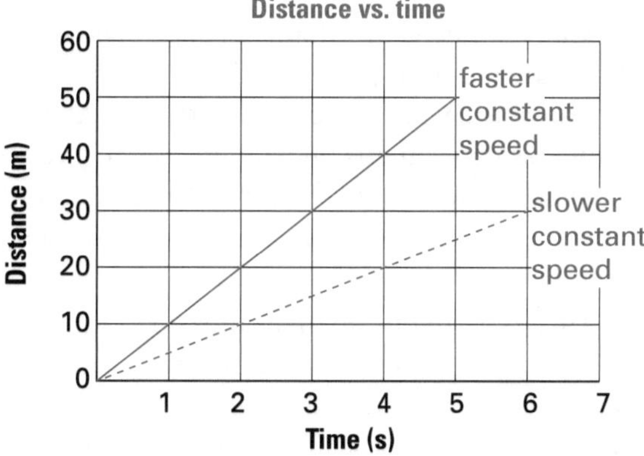

Figure A.18 Sample graphs

Conclusion: A data logger can accurately determine the motion of an object and represent it graphically.

Test yourself 1

Part A: Knowledge

1. D ✓ D is correct as an inference is one possible explanation of an observation. The inference must be reasonable and not frivolous. A is wrong as an inference may not be correct when further testing is done; B is incorrect as no prediction is involved; C is incorrect as theories are developed after much experimentation.
2. A ✓ The independent variable is also called the manipulated variable. B is wrong as this is the dependent variable; the control is the experiment in which no changes are made and so C is not correct; D is incorrect because the value is changed.
3. A ✓ This is an opinion and can vary from one person to another. B, C and D are all scientific hypotheses as they can be tested.
4. B ✓ Predictions suggest possible future experiments and their potential results. A is not correct as predictions are made after an observation is made; C is wrong as theories are developed after many different experiments are analysed; D is wrong as explanations are not involved in framing a prediction.
5. D ✓ It is poor science to discard the results that do not support the hypothesis. A, B and C are all factors that must be taken into account in designing experiments.
6. a) support ✓; b) true ✓; c) temporary ✓; d) variables ✓; e) independent ✓
7. A/G ✓; B/H ✓; C/I ✓; D/F ✓; E/J ✓

Part B: Skills

8. a) The southern side of the house is more moist and shady. ✓
   b) The boys all purchased and consumed contaminated food at the canteen during lunch. ✓
   c) Hydrogen peroxide breaks down (decomposes) as the temperature increases. ✓
   d) Kerosene is less dense than water. ✓
   e) The children's parents have pale skin and blue eyes. ✓
9. At 50 °C they will take 25 seconds to react together (time decreases by half for each 10-degree rise). ✓
10. a) the temperature of the acid ✓

b) the time for the magnesium to dissolve ✓

c) volume of acid; ✓ concentration of acid; ✓ rate of stirring; ✓ size of magnesium pieces; ✓ surface area of magnesium pieces; ✓ reaction vessel ✓

11. a) scales to weigh quantities of soil; ✓ measuring cylinder to measure volumes of water ✓ poured in and collected; pots with perforated bases for the soils to be placed into; ✓ container to collect water passing through soil; light gauze to be placed in the bottom of the pots ✓ to trap soil and only allow water to pass through

b) the quantity (mass) of soil in each case ✓; the amount of water poured through the soil samples ✓; the size of the pots ✓; temperature ✓

c) If the soil already contains some moisture, ✓ then it can hold less water and give a false impression that its water-holding capacity ✓ is less than it really is. Soils should be dried first in a drying oven. ✓

d) Weigh a given quantity of soil ✓ in each case and place it in a pot, ✓ pressing the soil down firmly. Pour in more water than the soil can hold, ✓ so the excess passes out through the bottom ✓ and is collected. Measure the volume of water that passes out and compare it ✓ to the quantity poured through. The difference is the amount of water retained ✓ by the soil.

12. Any two of: work faster; one person prompts the other to work to ability and on time; competitiveness; different skills/abilities brought to the team by each person. ✓✓

13. a) *(2-mark maximum: any two answers)* ✓✓
- temperature of air
- weather conditions (rain, wind, and so on)
- size and shape of balloon
- altitude of the weather station

b) *(2-mark maximum: any two answers)* ✓✓
- mass of the car
- condition of the road surface
- speed of the car (when the brakes are applied)

c) *(2-mark maximum: any two answers)* ✓✓
- microscopic plants, animals in water
- suspended materials such as colloids and other solids
- dissolved materials (i.e. water purity)
- smoothness of the water surface

14. a) *(3-mark maximum: any three answers)* ✓✓✓
- amount of blue solution
- amount of mild bleach
- size and shape of the container
- time to mix the solutions

b) See Figure A.19.

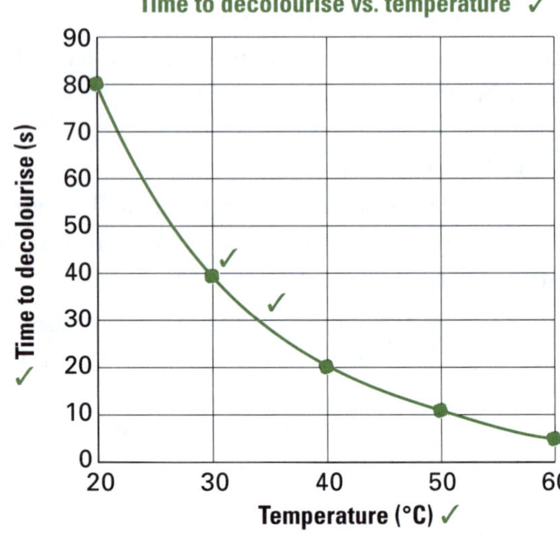

Figure A.19 Time to decolourise versus temperature

c) speed of reaction increased as the temperature increased ✓

d) Yes. ✓ When the temperature changes from 20 °C to 30 °C, the time to decolourise halves. Therefore the rate has doubled. ✓

Test yourself 2

Part A: Knowledge

1. **B** ✓ Many councils or interest groups produce books showing native plant and animal species in the local area. A is incorrect as there may not be any fossil specimens and, in any event, you are trying to identify a living tree. Both C and D are too general to provide the answer to this question.

2. **D** ✓ This experiment involves measuring height and will depend on the accuracy of the measuring instrument (ruler, tape measure) and how it is used. A, B and C are incorrect as they have nothing to do with the activity of measuring height. Besides, A and B are easy to determine as you only need to count the students and ask each person's age in full years.

3. **C** ✓ This is not easy to measure. There are scales that can determine the degree of happiness, but these are subjective. A, B and D can all be measured and so are incorrect.
4. **A** ✓ Growth depends on the soil type; the soil type does not depend on growth. B, C and D are all incorrect as these are kept constant (the same) during the experiment.
5. **D** ✓ Only this is an observation. While A and B are true, they are not observations she could make. As for C, who knows what butterflies enjoy or whether they can feel this emotion.
6. **a)** control ✓; **b)** constants ✓; **c)** fair ✓; **d)** datum ✓; **e)** investigation ✓
7. A/J ✓; B/I ✓; C/H ✓; D/F ✓; E/G ✓

Part B: Skills

8. See Table A.5.

Table A.5 Incorrect and correct techniques

Mistake	Correction
a) Uses her hand to transfer chemicals	Use a spatula ✓
b) Looks down into the test tube	Point the test tube away from all students ✓
c) Leaves the desk and talks to friends	Remain at desk while completing the experiment ✓
d) Places nose directly over the bottle to sniff	Use hand to waft air containing the smell of the chemical towards her nose ✓
e) Places glassware close to the edge of the table	Set up glassware away from the edge of the table ✓
f) Thumb over the top of the test tube and the contents shaken up and down	Gently tap the test tube on the hand to mix the contents ✓
g) After boiling, picks up the beaker with hand	Allow the beaker to cool before picking it up and pouring the contents down the sink ✓
h) Pours solid wastes into the sink	Pour water down the sink and put solid wastes in the bin provided by the teacher ✓

9. At 10.00 am the temperature is 19.4 °C. ✓ At 4.30 pm the temperature is 21.3 °C. ✓ The rise in temperature = 21.3 – 19.4 = 1.9 °C. ✓
10. **a)** 19.7 mL ✓
    b) parallax error ✓
    c) by keeping the eye level with the bottom of the meniscus ✓ (The meniscus is the U shape made by a liquid in a narrow tube.)
11. **a)** Collaborative research is any research project that is carried out by two or more people. ✓
    b) The workload can be shared when a project is large or involved ✓; researchers can pool their areas of expertise ✓.
    c) It can result in more reliable and powerful results. ✓ These results can be obtained faster than they would if the research were done independently. ✓ Researchers can pool their knowledge and analyse each other's work before completing the research. ✓
12. **a)** Reliability has to do with the quality of the measurement. It refers to the consistency or repeatability of these measurements. ✓ For example, in a test a measure is reliable if a person's score on that test given twice is similar.
    b) Validity is the strength of your conclusions, inferences or propositions. In short, were you right? It generally refers to the extent to which a concept, conclusion or measurement is well founded and corresponds accurately to the real world. ✓
    c) Yes, it refers to the strength or degree to which the tool you used measures what it claims to measure. ✓
    d) **i)** not valid, not reliable ✓
    ii) valid and reliable ✓
    iii) not valid, but reliable ✓
13. You infer by drawing a conclusion that combines information you already know with new information. It can also refer to the process of arriving at some conclusion that could be true given the current information. ✓ For example, when you see that the sky appears with dark clouds you can infer that it is likely to rain. ✓ A prediction is the act of reasoning about the future based on past experience. ✓ For example, you can confidently predict that day will follow night. ✓ In science a prediction

is a rigorous statement forecasting what will happen under specific conditions.

14. a) Accuracy refers to how close a measurement of a quantity is to its actual (true) value. ✓ In other words, the extent to which a given measurement agrees with the standard value for that measurement.
    b) the measuring cylinder (iii) ✓
    c) At the narrow end the liquid level rises higher compared to the wide end when each millilitre is added. ✓ So if you need to take, say, 10 mL of medicine you can more accurately measure this out. ✓ A millilitre or two out for this volume can make a big difference. ✓ On the other hand, if you need to take, say, 30 mL of medicine then being out by a millilitre or two is not that critical. ✓

Chapter test

Part A: Multiple-choice questions

1. **A** ✓ This is the correct sequence, or words to this effect. B, C and D are all wrong as they are written out of sequence.
2. **C** ✓ The flowers on each plant depend on the humidity in the container. B is the independent variable, and should be different in each experimental plant set up. A and D should be the same for all plants.
3. **B** ✓ A survey among a variety of people, young and old, male and female, city and country, would be the best way to gauge attitudes. A and C are wrong as they may be biased in one particular direction and not giving a balanced view, although results of surveys are often reported in newspapers. D gives opinions from scientists but this may not be representative of the population.
4. **D** ✓ Many research questions are complex and require several scientists to help solve them. Some questions can remain unsolved for some time until devices are invented to allow the research to continue, or for scientists to think of other ways of attacking the problem. A is incorrect as earlier scientists were not any less capable of carrying out quality research. B is wrong as the information provided shows many researchers needed to pool their results. Of course experiments should be repeated, but simply repeating the same experiment should lead to the same results, so C is wrong.
5. **D** ✓ His measurements should be both accurate and precise, so C is wrong. If he is doing this correctly there should be no difference between readings, so A is incorrect. If there is any problem with the weighing scales they should not be used, so B is wrong.
6. **A** ✓ Newspapers present general information, sometimes biased, to average readers. They are not meant to be scholarly articles. B, C and D, if used properly, can be informative and accurate, so are not correct answers to this question.
7. **A** ✓ Like any muscles, the more you use them the stronger, and larger, they become. A weightlifter, for example, was not born with large muscles but developed them through exercise. So B is incorrect, since flying around strengthened the wing muscles of birds. C is wrong as caged birds originated from birds that flew freely and probably had strong muscles. D is also incorrect as caged birds are bred for their colourfulness, not their wing muscles.
8. **C** ✓ The control group is almost identical to the experimental group, with the experimental group changed by one key variable of interest. The control group remains constant during the experiment. A and B are wrong as they have no effect on calculations or repetitions; D is wrong, as tradition is not a reason for using control groups.
9. **A** ✓ She needs to compare germination of seeds that came through a fire and those that didn't. B, C and D are wrong, as they don't allow for this control group to be set up.
10. **B** ✓ Only this hypothesis links in with this information about the rate of growth. A, C and D might be true, but are not supported by the information concerning growth. There is no evidence to suggest more watering results in a better survival rate. In fact, many desert plants can be killed by too much water.

Part B: Short-answer questions

11. a) Celsius ✓
    b) Peter, 23.4 °C ✓; Paul, 22.5 °C ✓; Mary, 22.0 °C ✓
    c) Paul ✓
    d) Your sight should be in line with the meniscus of the liquid in the thermometer ✓

to get the correct reading, otherwise you could be reading under or over the true value ✓.

12. a) Inference. ✓ The observed crack could have been caused by many things (e.g. a stone, a ball). A change in temperature is just one possibility.
    b) Generalisation. ✓ Examination of a wide variety of copper compounds shows that most of them are blue and so this is a useful generalisation.
    c) Observation. ✓ The colour of the different feathers can be observed.
    d) Prediction. ✓ The statement suggests that the student has already tried the dilute acid and is now predicting that a more concentrated solution will react even faster.
13. a) This is a hypothesis. ✓ It is suggesting an answer to a problem and it can be experimentally tested. It predicts the outcome of the research. ✓
    b) This is not a hypothesis. ✓ It does suggest an answer to a problem. It cannot, however, be tested directly as there are no living Neanderthals or ancestors of modern humans. The finding of Neanderthal bones in caves does not provide sufficient evidence to prove the hypothesis. ✓
    c) This is not a hypothesis. ✓ It involves personal opinions and this will vary from one group to another. ✓
14. The independent variable is the time period. ✓ The dependent variable is the height the plant grew. ✓
    The height of the plant depends on the length of time it has been allowed to grow. Time does not depend on the plant's height. The student can make measurements over regular time intervals. For instance, he/she can measure the height of the plant every second day for a fortnight. Consider the absurd situation if time depended on the plant's height. When the plant reached its maximum height and stopped growing, does this mean that time now stands still?
15. Student B will obtain a more accurate answer as he/she is avoiding parallax error. ✓ Student A will obtain an answer that is too low due to parallax error.
16. Patrice made the best measurement. ✓ All measuring instruments are accurate within certain limits. There is a limit to the number of decimal places that you can read on a scale. The limit of reading is normally half a scale division. ✓ The scale is not that accurate for Nikolas to come up with the value he obtained. ✓ James could make a better estimate of the value. ✓
17. The measuring cylinder is more accurate than the beaker. ✓ The scale on the measuring cylinder is divided into 0.1-mL units whereas the beaker's scale is divided into 10-mL units. ✓
18. The volume of liquid is 96 mL. ✓ The volume is measured at the base of the curved meniscus.
19. See Figure A.20.
    a)

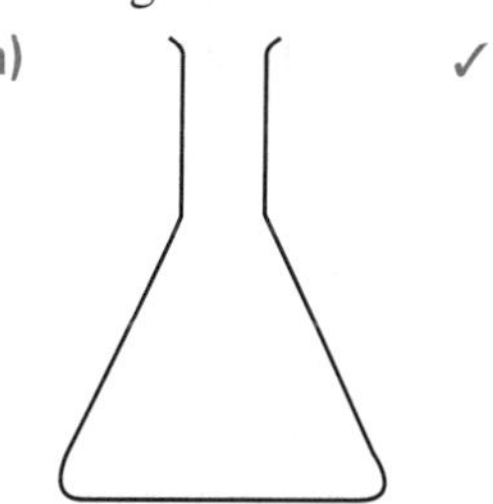

Figure A.20 a) Conical flask

b)

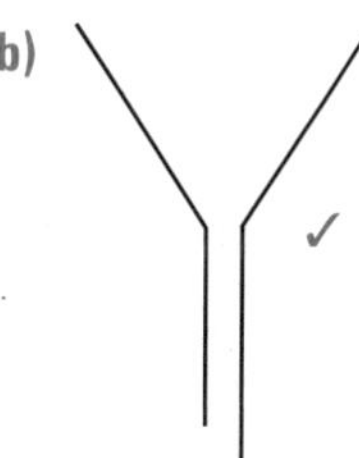

Figure A.20 b) Filter funnel

c)

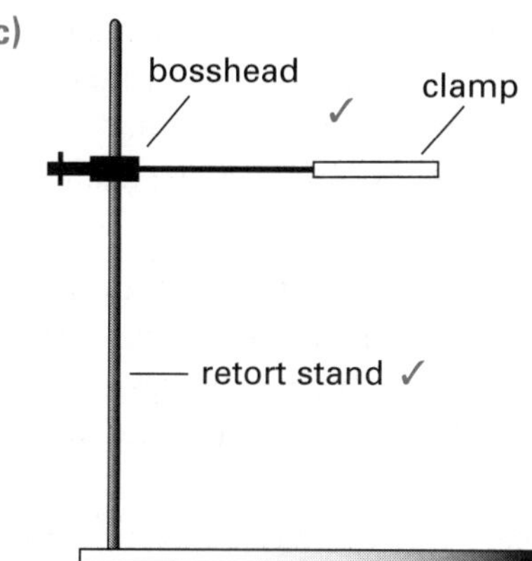

Figure A.20 c) Retort stand with attached bosshead and clamp

20. a) Grams. ✓ Given that there is a small box of throat lozenges in the pan, milligrams is too small a unit, while kilograms is too large a unit. ✓
b) 190 g ✓
c) 123 g ✓
d) To zero (or tare) the scales means to have the pointer show exactly 0 g ✓ when there is nothing in the balance pan. Sometimes, through use, the scale is not always zero when it is about to be used. This would lead to the reading you make to be either larger or smaller than the true value. ✓

21. a) test tube holder ✓; b) watch glass ✓;
c) test tube ✓; d) test tube rack ✓;
e) wash bottle ✓; f) conical flask ✓;
g) petri dish ✓; h) evaporating basin ✓;
i) funnel ✓; j) beaker ✓;
k) measuring cylinder ✓

22. a) A = pestle ✓; B = mortar (the bowl)✓
b) A mortar and pestle is a tool used to crush, grind and mix solid substances. ✓
c) Yes. The mortar is the vessel in which the pounding occurs ✓, while the pestle does the pounding. ✓
d) It must be hard enough to crush the substance and not be worn away by it. ✓ It cannot be too brittle or it will break during the pounding and grinding. ✓ The material should be cohesive so that small bits of the mortar or pestle do not get mixed in with the substances being crushed. ✓ Smooth and non-porous materials are chosen that will not absorb or trap the substances being ground. ✓ Material in the mortar and pestle must not react with substances placed in it.✓ It should be easy to clean without leaving stains. ✓

23. a) beaker ✓
b) These marks are not intended for obtaining a precise measurement of volume ✓ (there are more accurate measuring vessels such as graduated cylinders), but rather an estimate ✓ of the volume.
c) Borosilicate glasses do not expand very much when heated making them resistant to thermal shock, more so than any other common glass. Such glass is less subject to thermal stress and so will not crack when heated. ✓

24. a) While this is possible, some of the salt may spill outside of the test tube. ✓ The test tube is narrow and a heaped spatula will not neatly fit into the opening.
b) Some salts can stick to ordinary paper, especially if they are moist, and may not be completely transferred. ✓
c) The paper is opened at both ends when rolled into a cylinder. This way it won't tip out until it is required to. ✓

25. a) A burette is narrower and longer than a measuring cylinder. Therefore, a small volume added to it will show an appreciable change in height of the liquid. ✓ This means it is more sensitive in measuring changes in smaller volumes of liquids. Many burettes are graduated in tenths of millilitres, while measuring cylinders measure to the nearest millilitre.
b) This simple method prevents unnecessary spillage. ✓ Of course, you should use a clean beaker and never pour any unused liquid back into the reagent bottle.

26. Safety goggles: these help you avoid eye splashes, especially from chemicals that can do much damage to such delicate organs. ✓ As they say, 'foresight is better than no sight'.
Gloves: chemicals may be harmful, poisonous or corrosive. ✓ Some chemical reagents readily pass through the skin and into the bloodstream, which can cause serious health problems.
Apron or lab coat: some chemicals can eat through fabrics and damage clothing. An apron or lab coat can help protect your clothing. ✓
Tying hair: this is to prevent it from accidentally coming into contact with a flame or touching chemicals on the benchtop. ✓
Washing hands: some chemicals may inadvertently make contact with your hands. Often students go outside to eat and you could ingest these substances. ✓ So make it a habit to wash your hands after experiments.

27. a) A solution being filtered through filter paper. ✓
b) A = stirring rod ✓; B = filter funnel ✓;
C = filter paper ✓; D = retort stand ✓;
E = ring clamp ✓
c) Liquid runs down the side of the beaker and doesn't splash ✓, as would happen if the end

of the funnel ended in mid-air above the liquid level.

d) The stirring rod guides the liquid to be filtered into the filter paper. ✓ It is good practice to have the stirring rod just touch the side of the filter paper so no splashing occurs.

28. a) A liquid in an evaporating basin being heated to dryness, leaving the solids behind. ✓

b) The wire gauze could be set up on a tripod. ✓

c) Lower the heat and turn off the Bunsen burner when the liquid has almost evaporated. ✓ There should be enough heat in the evaporating basin to finish the process. ✓ If heating continues right to the end, there is a good chance of splattering and the basin cracking from the heat.

29. a) True. ✓ Should any accident occur, there would be no one around to help you if you get into trouble.

b) False. ✓ Once a chemical has been removed from the reagent bottle or jar, it could potentially be contaminated so dispose of any excess.

c) False. ✓ There should be a special receptacle for disposing of chemical wastes. Wastepaper baskets are for waste paper.

d) True. ✓ There is no problem with wearing contact lenses and goggles. Just don't assume contact lenses can substitute for goggles.

e) False. ✓ This will only localise heat and may cause liquid to spurt out of the test tube.

f) True. ✓ Liquid could run down the stopper and onto the bench.

g) True. ✓ An unexpected reaction might occur if two different substances are ground together.

30. The point is to measure the temperature of the water, so the bulb needs to be placed somewhere need the middle of this volume. ✓ If the bulb sat on the bottom it would be measuring the heat coming in through the glass, which is not the temperature of the water. ✓

31. a) A = rubber hose ✓; B = gas inlet ✓; C = air inlet ✓; D = barrel ✓

b) Q ✓

c) A safety flame is the highly visible, yellow sooty flame ✓ the Bunsen burner is turned to whenever it is not heating anything ✓. It is more highly visible than the almost non-luminous blue heating flame and is used for safety reasons.

32. As test tubes are often heated, you need something that will not conduct this heat to your hand. Wood is a poor heat conductor. ✓

33. a) As ships are loaded up, they sink further into the water. Some unscrupulous and greedy owners would overload their ships. The Plimsoll line prevented this from happening. ✓

b) i) salt water in winter ✓

ii) tropical salt water ✓

(The denser the water, the higher up the ship will be and so the Plimsoll line marker for that water will be lower.)

34. a) 60 km ✓

b) 5 hours and 12 hours after starting ✓

c) 60 – 20 = 40 km ✓

d) 20 and 25 hrs ✓

e) 60 + 40 + (210 – 20) = 290 km ✓✓
(Don't forget, the cyclist travelled 40 km in the homeward direction between $t = 12$ hours and $t = 20$ hours.)

35. a) number of fish hatched ✓

b) temperature ✓

c) 90 ✓

d) around 29 °C ✓

e) between 25 °C ✓ and 32 °C ✓

36. a) volume of blood lost each day

b) number of hookworms

c) in the left-hand column: 36 ✓, 56 ✓; in the right-hand column: 14 ✓, 20 ✓, 31.5 ✓

d) Each hookworm sucks around 0.5 mL of blood each day. ✓

37. a) 10 °C ✓; b) 95 °F ✓; c) 32 °F ✓; d) True ✓

38. a) 375 kg ✓

b) 12.0 hands ✓

c) 2.5 ✓

d) No. ✓ This chart was developed for horses only. A different chart would need to be developed for other farm animals. ✓

39. a) $\frac{78.1}{100} \times 360 = 281°$ ✓

b) $\frac{75.6}{360} \times 100 = 21.0\%$ ✓

c) $100 - (78.1 + 21.0) = 0.9\%$ ✓

40. (Note in Figure A.21 that the two scales do not need to start at 0.)

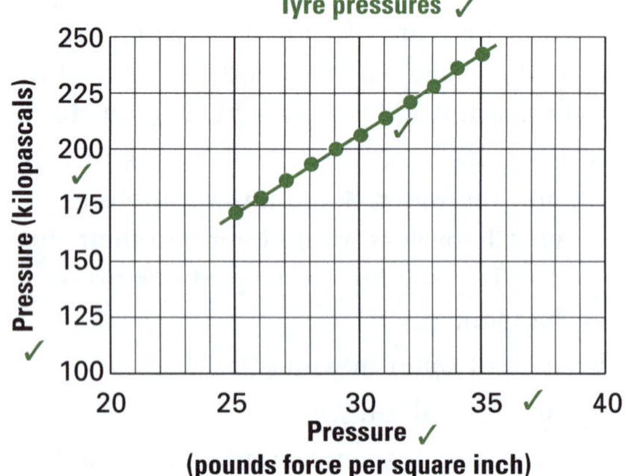

Figure A.21 Tyre pressures

Test 1 answers

Part A: Multiple-choice questions

1. **A** ✓ All birds have feathers but not all birds can fly. Penguin's wings have become flippers and they do not fly so B is wrong; Ostriches and emus have wings but are too heavy to fly so C is wrong; D is wrong as amphibians don't have teeth either.
2. **D** ✓ Shale is grey or black; sandstone has orange and white grains; granite is crystalline. A is incorrect as shale does not have orange and white grains; B is wrong as coal is not crystalline; C is wrong as marble does not have orange and white grains.
3. **C** ✓ A is wrong as it is a producer; B is wrong because sea snails are a first-order consumer; D is wrong as it is a third-order consumer.
4. **A** ✓ Decomposers break down dead organisms. B is wrong as carnivores eat herbivores; C is wrong as producers are plants; D is wrong as herbivores are plant eaters.
5. **D** ✓ The animal has hair and hooves and walks on four legs. A is incorrect as a pyebum has no hooves; B is wrong as an alflee walks on two feet; C is wrong as a slolum has hair.
6. **B** ✓ The animal walks on two legs and has many tail feathers. It has a long neck and long legs. A is wrong as a slolum has hair; C is wrong as a bintee has a short neck; D is wrong as an umee has hair.
7. **B** ✓ They are both first-order consumers as they eat producers. A is wrong as producers are plants; C is wrong as they are first-order consumers; D is wrong as they are not bacteria or fungi.
8. **B** ✓ Rabbits breed rapidly and eat native vegetation. The loss of plants and their roots leads to soil erosion. Rabbits are known to eat the plant, roots and all, so there is nothing left to grow back. A is an irrelevant answer as being non-native does not matter; C is wrong as faeces fertilise the soil; D is wrong as competition for food is irrelevant.
9. **A** ✓ The clear supernatant liquid will pass through the filter rapidly and this will only leave a small volume of suspension to filter. B is wrong as a little reside is inevitably lost if it is very fine; C is wrong as the residue remains at the bottom of the beaker; D is wrong as filter papers come in many different sizes.
10. **D** ✓ By spinning the cream and milk rapidly, the milk sinks to the bottom and the cream rises to the top. Cream is less dense than milk. A is wrong as it is too slow and incomplete; B is wrong as this separates soluble materials; C is wrong as this is used to separate minerals from rocks.
11. **C** ✓ She needs a balance to weigh the soil before and after filtration. A is wrong as there is no balance; B and D are wrong as she has no filter funnel.
12. **C** ✓ Vacuum cleaners need to filter the dust from the air entering the cleaner; a clothes dryer contains a filter to remove fluff; water purifiers filter impurities from water. A is wrong as televisions do not filter; B is wrong as chlorinators do not filter; D is wrong as microwave ovens and electric toothbrushes do not filter.
13. **C** ✓ In Australia 21 December represents the summer solstice. At this time the Sun has reached its most southerly point in the sky. A is wrong as the sun will be high not low; B is wrong as the arc is high in the sky from east to west; D is wrong as the sun rises in the east and sets in the west.
14. **A** ✓ Summer in the southern hemisphere occurs when this hemisphere is tilted towards

the sun and winter occurs when this hemisphere is tilted away from the sun. B is wrong as the Moon does not affect the seasons; C is wrong as the tilt is important, not the position; D is wrong as gravity hardly varies along the orbit.

15. **C** ✓ Weight increases as you get closer to the centre of the Earth. A is wrong as it would have no weight in orbit; B is wrong as gravity is less on the Moon; D is wrong as it is not as close to the Earth's centre as the underground mine.
16. **B** ✓ He built an observatory to measure the length of the solar year. A and C are wrong as Galileo did these; D is wrong as Copernicus proposed this model.
17. **C** ✓ Venus shows various phases like the Moon as the Sun's light shines on different parts of its surface due to the relative position of it and the Earth. A is wrong as this observations cannot detect seasons; B is wrong as gravity is constant; D is wrong as Venus is permanently covered in cloud.
18. **B** ✓ The flocs are allowed to sediment and then they are filtered. Chlorination occurs next to kill microbes and then the acidity is adjusted. A is wrong as fluoridation is the last step; C is wrong as chlorination is a much later step; D is wrong as aeration is an early event.
19. **D** ✓ Evaporation converts a liquid into a gas. A is wrong as condensation is the reverse process. B is wrong as no gas is being condensed. C is wrong as no melting is occurring.
20. **C** ✓ The absorption of solar energy by the atmosphere produces changes in air pressure and therefore wind. The other resources are fixed and may eventually run out as they are not renewable. Thus A, B and D are wrong.
21. **A** ✓ Electrical turbines were not invented until the 20th century. Answers B, C and D are wrong as all these were known for thousands of years.
22. **A** ✓ No combustion occurs and so no greenhouse gases are released. B is wrong as not all rocks are hot; C is wrong as hot rocks are not close to the surface; D is wrong as they are not widespread.
23. **B** ✓ Warm, moist air has to rise as it moves over a mountain range. As it rises, it cools and water droplets begin to form clouds. A is wrong as solar radiation is fairly constant; C is wrong as the air over deserts is not moist; D is wrong as the air is dry.
24. **C** ✓ Many screwdrivers are made from magnetised steel so that carpenters can hold steel screws on them while working. Some scissors, too, have magnetised tips for picking up pins dropped onto the floor. A is wrong as steel is not electrified; B is wrong as attraction does not result by contact forces; D is wrong as friction does not cause attraction.
25. **B** ✓ The weight of a body (object) depends on how strong gravity is. The Moon's gravity is much less than that of the Earth. The mass is a measure of the amount of matter in a body. This does not change with a change in location. A is wrong as mass is a property of the body, not its location; C is wrong as its weight changes on the moon and its mass remains the same as on Earth; D is wrong as Moon's gravity is weaker. Weight is measured in newtons not kilograms.

Part B: Restricted-response questions

26. balanced ✓
27. contact ✓
28. positively ✓
29. monotremes ✓
30. binomial ✓
31. first ✓
32. rabbit ✓
33. protists ✓
34. distillation ✓
35. sieving ✓
36. centrifugation ✓
37. penumbra ✓
38. gibbous ✓
39. astronomical ✓ One astronomical unit (1 AU) is the average Earth–Sun distance.
40. Sun ✓

Part C: Knowledge, skill and processing data questions

41. **a)** glue C ✓; **b) i)** same ✓; **ii)** same ✓
42. **a)** More dissolves at higher temperatures (solubility increases with temperature). ✓
    b) from graph, 7 g ✓; undissolved solid = 10 – 3 = 7 g ✓
    c) Remove a portion, say 100 mL, in a weighed beaker. ✓ Heat almost to dryness. Allow to cool and evaporate the last amount of

water. ✓ Reweigh the beaker. ✓ Calculate the mass of solute in grams.

43. **a)** water ✓

b) water ✓

c) The particles of solid would have sedimented and need to be remixed. ✓

d) methylated spirits, mineral turpentine or kerosene ✓

44. **a)** Collect 2-L samples from each position. ✓ Control variables: same amount ✓, same depth ✓, same time of day ✓ and same distance from shore. ✓

b) 1. Weigh the evaporating basin (X). ✓
2. Add 1 L of water.
3. Evaporate nearly to dryness using a drying oven or low-temperature electric hotplate. ✓
4. Allow the last water to evaporate naturally.
5. Reweigh basin (Y). ✓
6. Calculate the mass of salt (X – Y).

c) **i)** E ✓

ii) A ✓

iii) See Figure A.22.

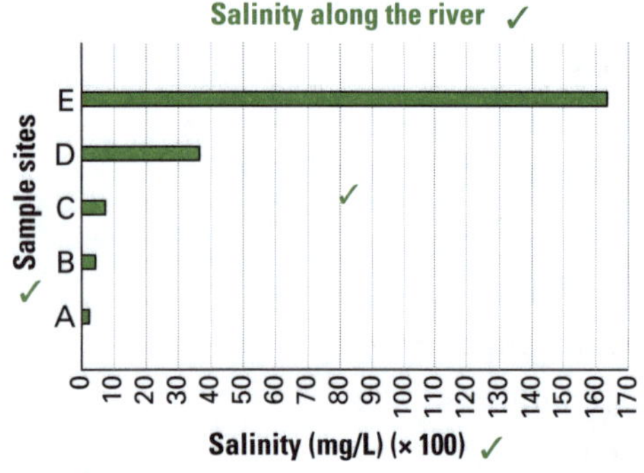

Figure A.22 Salinity along the river

iv) close to the ocean (river is tidal) ✓

d) Pass the water through a sand filter. ✓

45. **a)** Weight force acts downwards; buoyancy force acts upwards. ✓ See Figure A.23.

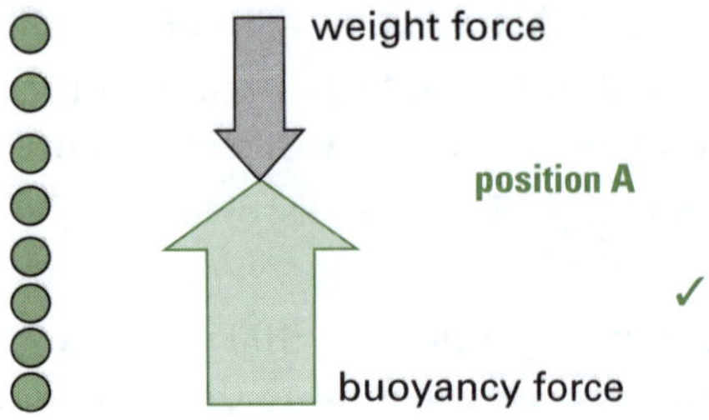

Figure A.23 Weight force and buoyancy force

b) Forces are equal. ✓ See Figure A.24.

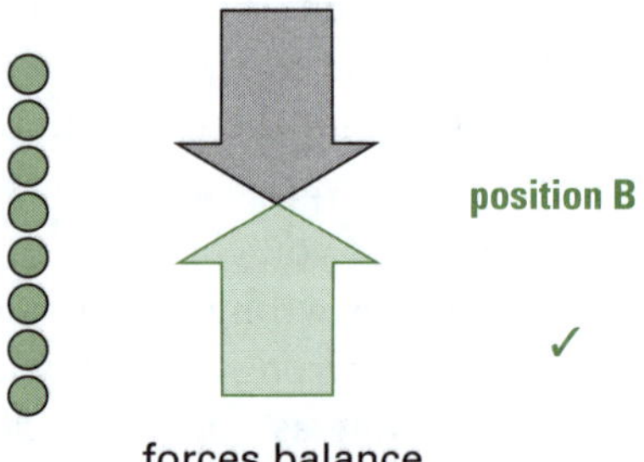

Figure A.24 Forces balance

46. **a)** evaporation ✓; **b)** fusion ✓;
c) condensation ✓; **d)** solidification ✓

47. warmed up = cooled down ✓; condensation = evaporation ✓; supply = remove ✓; contracted = expanded ✓; loose = tight ✓; hot = cold ✓; solidifying = melting ✓

48. **a)** diagrams C and D (meat covered) ✓

b) to show that maggots are not already on one type of meat ✓ and to show that his hypothesis is true in more than one case ✓

c) Yes. ✓ If flies had access to meat, then maggots did grow. ✓

d) Bacteria/microbes in air landing on the food or bacteria/microbes already in the meat. ✓

49. **a)** thresher shark ✓; **b)** stingray ✓

50. **a)** from the plants ✓

b) plants, brine shrimp, perhaps also some algal scum which may grow in the water ✓

c) They decomposed, releasing nutrients to cycle again starting through the plants ✓

d) from the decaying material (debris) in the tank, such as fish wastes and dead protists ✓

e) from the Sun ✓

f) Possibly not. At the moment the system is balanced. Putting in more fish may make greater demands on the plants than they can cope with, and the ecosystem could collapse. ✓

51. **a)** The Moon goes from small to a full moon and then to a thin crescent phase, then disappears (new moon). Life is similar: from child to adult to old age, then death. ✓

b) The Moon seems to come back each month. It disappears then reappears. On the coast the Moon rises out of the sea (east coast) or sets into the sea (west coast). ✓

Test 2 answers

Part A: Multiple-choice questions

1. **A** ✓ The correct order is kingdom, phylum, class, order, family, genus, species. B, C and D are wrong because they do not have phylum as the level under kingdom.
2. **C** ✓ According to this food web mice and rats feed exclusively on oats, whereas the other organisms can get their food from other sources. A is wrong as grasshoppers and crickets can eat grass; B is wrong as hawks and owls can eat rabbits that eat grass; D is wrong as racoons and snakes can eat grasshoppers and crickets that also can eat grass.
3. **C** ✓ Oats is a producer organism (green plant), while snakes are second-order consumers (they eat first-order consumers). A is wrong as crickets are first-order consumers; B is wrong as grasshoppers are first-order consumers; D is wrong as rabbits are first-order consumers.
4. **C** ✓ Regardless of the amount of water, pure water under these conditions always freezes at the same temperature. Therefore answers A, B and D are incorrect.
5. **D** ✓ Water as a vapour or gas condenses to fine droplets of liquid. A, B and C are wrong as no evaporation, boiling or melting occurs.
6. **B** ✓ A, C and D are physical changes.
7. **C** ✓ This is a large group of organisms that includes microorganisms such as penicillium and smut as well as the more familiar mushrooms and toadstools. A is wrong as these are bacteria; B is wrong as mushrooms are not green; D is wrong as mushrooms are multicellular.
8. **B** ✓ A solution is a homogeneous mixture of two or more substances. A is wrong as no solid particles are visible; C is wrong as a solution is transparent; D is wrong as light does not scatter when passing through salt water.
9. **B** ✓ In a parasitic relationship one organism benefits while the other is harmed in some way. A is wrong as both benefit; C is wrong as both benefit; D is wrong as this is a carnivore–prey interaction.
10. **A** ✓ It involves throwing the mixture into the air so that the wind blows away the lighter chaff, while the heavier wheat grains fall back down for recovery. B is wrong as the colour does not matter; C is wrong as some large objects can be lighter; D is wrong as the shape does not matter.
11. **D** ✓ Helium is an element and a pure substance. A, B and C are all examples of mixtures.
12. **C** ✓ All the other reasons (A, B and D) are reasons why the myna's population might decrease.
13. **A** ✓ This group of organisms includes mushrooms, yeast, and moulds. Fungi obtain energy by oozing digestive enzymes that decompose other biological tissues. B is wrong as it has chlorophyll; C is wrong as these are not multicellular; D is wrong as they do not absorb nutrients from decaying tissue.
14. **C** ✓ Chromatography is a process used for separating mixtures by virtue of differences in how well the components stick to the material they move through. A is wrong as filtration cannot separate two soluble materials; B is wrong as the mixture is not boiled; D is wrong as this method separates a sediment from a solution.
15. **A** ✓ Only iron is attracted to a magnet. In Figure T2.3, the electromagnets are automatically switched on near the top, attracting iron particles. Just past the bottom they are switched off, after gangue material has fallen off the conveyor belt. B, C and D are not magnetic.
16. **A** ✓ In chemistry, decanting is the process of carefully pouring a solution from a container in order to leave the precipitate (sediments) in the bottom of the original container. It is the mechanical transfer of the clear upper liquid without disturbing the settled solid particles. B is wrong as no sieve is used; C is wrong as the potatoes do not sink to the bottom; D is wrong as the water is not completely evaporated.
17. **C** ✓ Think of the dramatic increase in rabbit populations. A is wrong as pesticides are not used; B is wrong as the main reason is that there are none or few natural predators; D is wrong as it includes i).
18. **B** ✓ Regardless of the size of the sea lion's stomach, the question is to determine the effect sea lions have on salmon populations. This is why A, C and D are important to study.

19. **C** ✓ If you weren't sure, you could have worked it out by looking at the other boxes around them. A is wrong as (i) is not heterogeneous; B is wrong as (ii) is elements; D is wrong as (i) is not an element.

20. **D** ✓ Friction occurs when objects are in contact with each other. A, B and C are called field forces.

21. **A** ✓ Oil is a fossil fuel. B, C and D are all renewable.

22. **C** ✓ In order to make the others (A, B and D) true, take their opposites.

23. **C** ✓ The air buoys up the raindrop and slows its acceleration until it drops at a constant speed. The size of the drop will have an effect on this final speed. A is wrong as the same speed is reached from many different cloud heights. B is wrong as the type of cloud does not matter. The quantity of rain also does not matter, so D is wrong.

24. **B** ✓ D gives a new moon; A is at the first quarter; C is the third quarter.

25. **B** ✓ The two forces combine to move the box in the direction B because F_2 is greater than F_1. A is wrong as the force F_1 will move the box up; C is wrong as F_2 will force the box to the left; D is wrong as F_2 is larger than F_1.

Part B: Restricted-response questions

26. gas ✓
27. condensing ✓
28. filter ✓
29. heterogeneous ✓
30. mixture ✓
31. feral ✓
32. mass ✓
33. solar ✓
34. seasons ✓
35. filtration ✓
36. ecology ✓
37. rabbit ✓
38. frogs ✓
39. evaporation ✓
40. telescope *or* binoculars ✓

Part C: Knowledge, skill and processing data questions

41. (Numerical answers in this question depend on the accuracy of reading the graph.)
    a) between 18% and 19% ✓
    b) the 25–29 and 30–34 years age groups ✓
    c) The 25–29 year age group covers about 27% of mothers. ✓
    27% of mothers had 80 863 babies
    1% of mothers had $\frac{80\,863}{27}$ babies
    So 100% of mothers had $\frac{80\,863}{27} \times 100$ = 299 493 babies.
    Around 299 500 babies were born in that year. ✓
    d) There are so few babies born to mothers 45 years and over that they are placed in the same group for convenience. ✓

42. a) The coarse particles of the undissolved solid, being heavier than the liquid (usually water), settle down due to gravity. ✓ This is called sedimentation. This allows the clear upper layer of the liquid to then be gently poured out into another container, or though a filter paper without clogging the pores. ✓
    b) Alum causes the fine particles to clump together. Bigger particles settle down faster than the finer particles. ✓

43. a) any of tripod, wire gauze or pipeclay triangle ✓
    b) The solution contains a mixture of a liquid and a dissolved solid. This is heated gently in an evaporating basin. Gradually the solvent evaporates and the solution containing the dissolved solute becomes thicker. The semi-solid mass left on the evaporating basin is slowly heated to dryness leaving the solid behind. ✓
    c) salt ✓
    d) distillation ✓

44. a) Classification systems are an aid to memory. It is impossible to remember the characteristics of a large number of different things unless we can group them into categories, whose members share many characteristics. They greatly improve our powers of prediction. If, for example, we know that females of mammal species have

mammary glands to produce milk for their young, we can be quite certain that a newly discovered mammalian animal will also have functional mammary glands. They also improve our ability to explain relationships among things. ✓

b) A species is often defined as a group of organisms capable of interbreeding naturally in the wild and producing fertile offspring. ✓

c) No, since donkeys and horses don't normally mate on their own. Besides, mules are usually sterile and can't breed to give more mules. ✓ (A mule is the offspring of a male donkey and a female horse. A mule is easier to obtain than a hinny, which is the offspring of a male horse and a female donkey. All male mules and most female mules are infertile.)

45. a) button H ✓

b) i) oval, large, light coloured, two holes, unstriped ✓

ii) square, small, light coloured, two holes, striped ✓

c) by the number of holes; one is striped, the other unstriped ✓

d) i) A and Q ✓

ii) Shading: is it light coloured or dark coloured? ✓

e) J and O; M and S ✓

46. This is a first-class lever. ✓ It has a mechanical advantage as the effort arm of the lever is longer than the load arm. ✓

47. a) Yes. ✓ It has a mechanical advantage as much less effort is required to lift the crate than just lifting using your muscles alone. ✓

b) Through 5 m. ✓ Besides the length of rope the man is pulling, there are five vertical lengths associated with the crate. Each of these needs to be shortened by 1 m, a total of 5 m. ✓

48. for 5 months ✓ (October shows exactly 75%, but the question asked for over 75%)

49. a) kitchen and toilet ✓

b) approximately 40% ✓

c) The bath and shower sector occupies about 20%. So 20% of 300 kL = 60 kL. ✓

d) There are a number of answers possible, such as: take shorter showers; plant drought-tolerant plants in the garden (especially natives that are adapted to Australian conditions); don't use washing machines or dishwaters without a full load; wash single items by hand; and repair leaky taps. ✓

50. P, condensation ✓; Q, precipitation ✓; R, percolation ✓; S, groundwater ✓; T, transpiration ✓; U, evaporation ✓; V, stream flow ✓

51. a) friction between the box and the shelf ✓

b) The polished shelf presents less friction to the box. ✓

c) Roughing up shoes increases the friction, thereby lessening slip. ✓

52. a) Light travels through a vacuum, so the astronaut sees it. ✓ But sound does not travel through a vacuum, so no sound is heard. ✓

b) He might detect the vibration through the ground on impact. ✓

53. a) weight = 72 × 9.8 ✓ = 706 newtons ✓

b) one-sixth of 706 is 118 newtons ✓

c) Its weight would change. ✓ (It depends on the acceleration of gravity.)

54. P = gas ✓; Q = oil ✓ ; R = water ✓

55. A/K/P ✓; B/M/U ✓; C/H/Q ✓; D/I/R ✓; E/J/T ✓; F/N/S ✓; G/L/O ✓

Index

Page numbers in **bold** are definitions of terms.

D

E

F

G

H

I

K

L

M

N

O

P

R

S

T

U

V

W

Z